Oklahoma Notes

Basic Sciences Review for Medical Licensure
Developed at
The University of Oklahoma College of Medicine

Suitable Reviews for:
United States Medical Licensing Examination
(USMLE), Step 1

Oklahoma Notes

Biochemistry

Third Edition

Edited by
Thomas Briggs
Albert M. Chandler

Springer-Verlag
New York Berlin Heidelberg London Paris
Tokyo Hong Kong Barcelona Budapest

Thomas Briggs, Ph.D.
Department of Biochemistry and Molecular Biology
University of Oklahoma
P.O. Box 26901
Oklahoma City, OK 73190
USA

Albert M. Chandler, Ph.D.
Department of Biochemistry and Molecular Biology
University of Oklahoma
P.O. Box 26901
Oklahoma City, OK 73190
USA

Library of Congress Cataloging-in-Publication Data
CIP applied for

Printed on acid-free paper.

Production managed by Laura Carlson; manufacturing supervised by Joe Quatela.
Camera-ready copy prepared by the authors.
Printed and bound by Edwards Brothers, Inc., Ann Arbor, MI.
Printed in the United States of America.

9 8 7 6 5 4 3 2 1

ISBN 0-387-94398-6 Springer-Verlag New York Berlin Heidelberg

Preface to the
Oklahoma Notes

In 1973, the University of Oklahoma College of Medicine instituted a requirement for passage of the Part I National Boards for promotion to the third year. To assist students in preparation for this examination, a two-week review of the basic sciences was added to the curriculum in 1975. Ten review texts were written by the faculty: four in anatomical sciences and one each in the other six basic sciences. Self-instructional quizzes were also developed by each discipline and administered during the review period.

The first year the course was instituted the Total Score performance on National Boards Part I increased 60 points, with the relative standing of the school changing from 56th to 9th in the nation. The performance of the class since then has remained near the national candidate mean. This improvement in our own students' performance has been documented (Hyde et al: Performance on NBME Part I examination in relation to policies regarding use of test. J. Med. Educ. 60: 439–443, 1985).

A questionnaire was administered to one of the classes after they had completed the Boards; 82% rated the review books as the most beneficial part of the course. These texts were subsequently rewritten and made available for use by all students of medicine who were preparing for comprehensive examinations in the Basic Medical Sciences. Since their introduction in 1987, over 300,000 copies have been sold. Obviously these texts have proven to be of value. The main reason is that they present a *concise overview* of each discipline, emphasizing the content and concepts most appropriate to the task at hand, i.e., passage of a comprehensive examination covering the Basic Medical Sciences.

The recent changes in the licensure examination that have been made to create a Step 1/Step 2/Step 3 process have necessitated a complete revision of the Oklahoma Notes. This task was begun in the summer of 1991 and has been on-going over the past three years. The book you are now holding is a product of that revision. Besides bringing each book up to date, the authors have made every effort to make the tests and review questions conform to the new format of the National Board of Medical Examiners. Thus we have added numerous clinical vignettes and extended match questions. A major revision in the review of the Anatomical Sciences has also been introduced. We have distilled the previous editions' content to the details the authors believe to be of greatest importance and have combined the four texts into a single volume. In addition a book about neurosciences has been added to reflect the emphasis this interdisciplinary field is now receiving.

I hope you will find these review books valuable in your preparation for the licensure exams. Good luck!

Richard M. Hyde, Ph.D.
Executive Editor

Preface to the Third Edition

This book is intended to be a *quick* review for those who are studying for their licensure examination in Biochemistry (USMLE, Step 1). As in the first two editions, we have left out a great deal of detail in the interest of making the book digestible in a reasonable time and to keep the price affordable. Those who need a more thorough reference should consult any of several excellent and up-to-date texts that are now available.

Significant improvements in this edition, besides a general updating and correction of the text, are as follows:

- **Test Questions.** In addition to many new sample questions, we have supplied an annotation for each, to explain briefly why the selected answer is correct. Although this took up extra space in the book, we felt the teaching value was worth it.

- **Medical Genetics.** We have added a new broadly based chapter, professionally illustrated, on this rapidly developing field, describing recombinant DNA techniques and their applications in diagnosis of genetic disease and in gene therapy. The chapter also includes a review of basic genetics and the chemistry of mutations.

As before, each part is written by a colleague who is either an expert in the field or an experienced teacher of the topic. The words are the authors' own, but as editors we accept responsibility for what to include or omit. We welcome comments and suggestions for future editions.

Thomas Briggs
Albert M. Chandler

Contents

Preface to the *Oklahoma Notes* v
Preface vii

1. Amino Acids and Proteins   *A. Chadwick Cox*
Amino Acids 1
Protein Structure: General 3
Collagen and Fibrous Proteins 4
Hemoglobin 5
Plasma Proteins 6
Hemostasis and Blood Coagulation 7
Genetic Diseases 12
Review Questions on Amino Acids and Proteins 14

2. Enzymes   *A. Chadwick Cox and Wai-Yee Chan*
Nature of Enzymes 19
Enzyme Kinetics 21
Effects of Kinetic Parameters on Enzyme Activity 24
General Aspects 26
Review Questions on Enzymes 27

3. Carbohydrates   *Albert M. Chandler*
Introduction 31
Structural Aspects of Carbohydrates 31
Digestion 34
Glycolysis 35
Entry of Other Hexoses into the Glycolytic Pathway 38
Metabolism of Pyruvate 39
Regulation of Glycolysis 40
Pentose Phosphate Pathway 41
Gluconeogenesis 42
Glycogen Metabolism 44
Glycogen Storage Diseases 47
Hormonal Regulation of Carbohydrate Metabolism 47
Review Questions on Carbohydrates 49

4. Energetics and Biological Oxidation   *Thomas Briggs*
Concepts in Biological Oxidation 61
Metabolism of Pyruvate to Carbon Dioxide 63
Electron Transfer via the Respiratory Chain 67
Chemi-Osmotic Theory of Oxidative Phosphorylation 71
Review Questions on Energetics and Biological Oxidation 72

5. Amino Acid Metabolism   *Albert M. Chandler*
   Functions of Amino Acids in Man.   80
   Essential and Non-Essential Amino Acids   80
   Nitrogen Balance.   81
   Protein Quality   81
   Protein Digestion.   81
   Amino Acid Absorption.   82
   Amino Acid Degradation   83
   The Urea Cycle (Krebs-Henseleit Cycle)   84
   Excretion of Free Ammonia   86
   Degradation of the Carbon Skeletons.   87
   One-Carbon Fragment Metabolism   88
   Metabolism of Phenylalanine and Tyrosine   91
   General Precursor Functions of Amino Acids.   93
   Review Questions on Amino Acid Metabolism   94

6. Porphyrins   *Thomas Briggs*
   Structure and Chemistry.   104
   Metabolism   105
   Disorders of Porphyrin and Heme Metabolism.   106
   Genetic Diseases   107
   Review Questions on Porphyrins.   109

7. Lipids   *Chi-Sun Wang*
   Classification of Lipids   113
   Functions of Lipids.   113
   Digestion and Absorption of Lipids.   114
   Fatty Acids   114
   Ketone Bodies   121
   Triacylglycerols.   122
   Phospholipids.   123
   Glycosphingolipids (Glycolipids)   125
   Lipoproteins   125
   Genetic Diseases   127
   Review Questions on Lipids.   130

8. Steroids   *Thomas Briggs*
   Cholesterol: Structure and Chemistry   139
   Occurrence and Function.   140
   Biosynthesis of Cholesterol   140
   Metabolism of Cholesterol   142
   Action of Steroid Hormones.   144
   Cholesterol Levels.   145
   Disorders of Cholesterol and Steroid Metabolism.   146
   Cholesterol in Perspective   147
   Review Questions on Steroids   148

9. Membranes   *Thomas Briggs*
   Overview.   154
   Chemistry and Structure.   155
   Functions.   157
   Review Questions on Membranes   165

10. Nutrition   *Thomas Briggs*
    Major Nutrients.................................... 171
    Micronutrients   174
    Review Questions on Nutrition 185

11. Purines and Pyrimidines   *Leon Unger*
    Structure and Nomenclature 193
    Synthesis of Nucleoside Diphosphates and Triphosphates 194
    Purine Metabolism................................. 195
    Pyrimidine Metabolism: Biosynthesis................ 199
    Deoxyribonucleotides Are Formed by Reduction of Ribonucleoside
    Diphosphates by Ribonucleoside Diphosphate Reductase
    (Ribonucleotide Reductase) 199
    Some Anticancer Drugs Act by Blocking Deoxythmidylate Synthesis . 201
    Functions of Nucleotides 202
    Review Questions on Purines and Pyrimidines 203

12. Nucleic Acids: Structure and Synthesis   *Jay Hanas*
    DNA... 207
    RNA... 215
    Review Questions on Nucleic Acids 219

13. Protein Biosynthesis   *Jay Hanas and Albert M. Chandler*
    Introduction 227
    Cellular Machinery................................ 227
    Information Transfer.............................. 232
    Peptide Bond Formation........................... 233
    Postribosomal Modification of Proteins............ 236
    Some Antibiotics Acting as Inhibitors of Protein Synthesis......... 238
    Review Questions on Protein Biosynthesis.......... 239

14. Medical Genetics   *Sara L. Tobin, with figures by
    Ann Boughton*
    Basic Genetics.................................... 247
    Mutations 252
    Recombinant DNA Techniques 254
    Genetic Diagnostic Techniques 270
    Gene Therapy 274
    Well-Known Genetic Diseases 279
    Review Questions on Medical Genetics............. 280

1. AMINO ACIDS AND PROTEINS

A. Chadwick Cox

I. AMINO ACIDS

A. The 20 Amino Acids. These are coded for in DNA:

Alanine	(Ala)	Leucine	(Leu)
Arginine	(Arg)	Lysine	(Lys)
Asparagine	(Asn)	Methionine	(Met)
Aspartic Acid	(Asp)	Phenylalanine	(Phe)
Cysteine	(Cys)	Proline	(Pro)
Glycine	(Gly)	Serine	(Ser)
Glutamine	(Gln)	Threonine	(Thr)
Glutamic Acid	(Glu)	Tyrosine	(Tyr)
Histidine	(His)	Tryptophan	(Trp)
Isoleucine	(Ile)	Valine	(Val)

B. Stereochemistry

The absolute configuration is the L (or S) enantiomer. The carbon adjacent to the carboxyl group is the α-carbon, to which the amino group is attached. Other carbon atoms are also attached to the α-carbon to form an R group. Amino acids differ from one another because they have different α-carbon R groups.

C. Classification: According to properties important to protein structure.

1. Hydrophobic (nonpolar): Ala, Cys, Gly, Ile, Leu, Met, Phe, Trp, Val.

2. Hydrophilic (polar and form H-bonds except for Pro)

   ♦ neutral: Asn, Gln, Pro, Ser, Thr, Tyr

   ♦ acidic (negative charge): Asp, Glu

   ♦ basic (positive charge): Arg, His, Lys

D. Post-translational Modifications

Several amino acids can be modified after being incorporated into proteins. Common post-translational modifications include formation of disulfide crosslinks, glycosylation of Ser or Asn residues, and phosphorylation of Ser, Thr and Tyr. The functions of these modifications will be discussed later.

E. Amino Acids as Ampholytes

All the amino acids at neutrality possess both positive and negative charges and are called ampholytes and zwitterions. Because proteins contain mixtures of the acidic and basic amino acid residues indicated above, they are also ampholytes. Each of the ionizable groups on amino acids and proteins behaves according to the same principles that apply to any weak acid.

$$HA \rightleftharpoons H^+ + A^- \qquad K_a = [H^+][A^-]/[HA]$$

This equation can be written in a form that directly relates the concentrations to the pH — the Henderson-Hasselbalch equation:

$$pH = pK_a + Log([conjugate\ base]/[conjugate\ acid])$$

When the conjugate base and acid are equal in concentration, their ratio is one, the log is zero, so $pH = pK_a$. The pH is buffered best by any ionizable group at or near its pK_a.

For carboxyl groups,

$$pH = pK_a + Log([COO^-]/[COOH]) \qquad pK_a\ usually\ in\ the\ 2\text{-}4\ range$$

For amino groups,

$$pH = pK_a + Log([NH_2]/[NH_3^+]) \qquad pK_a\ usually\ in\ the\ 6\text{-}10\ range$$

In histidine the N is part of an imidazole ring but it acts like an amino group with pK_a value around neutrality. For this reason His can participate in acid and base catalysis and is part of the catalytic mechanism of many enzymes.

As these equations indicate, the **charge ratio** varies with pH. An amino acid that has one amino group and one carboxyl group will be positively charged at low pH values. The net charge will diminish as the pH proceeds through values about the pK_a of the carboxyl group, because the latter are becoming negatively charged. It will equal zero at some pH intermediate between the pK values of the two ionizable groups, then will become negative as the pH proceeds through values about the pK_a of the amino group because these are losing their positive charge. The pH at which the net charge is zero is called the **isoelectric point** (pI) because the molecules in solution will not migrate in an electric field.

Figure 1-1 illustrates the points made above. The figure shows the number of equivalents of NaOH consumed by an amino acid in titrating the solution from pH 0 to pH 14. The amino acid has one basic and two acidic ionizable groups. (The latter, in this example, are too close to resolve clearly).

TITRATION OF AMINO ACID

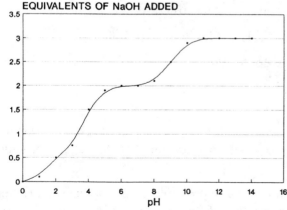

1. How many pK_a's are there? The same number as equivalents. (3)

2. What are their values? The same as the pH at each half equivalence. (About 2, 4, and 9 in this example).

3. Where is the isoelectric point? With two pK_a values in the acid region it must be an acidic amino acid and the pI must be in the acid region halfway between the two pK_a values, that is, at pH 3.

Figure 1-1. Titration of an Amino Acid having Three Ionizable Groups.

II. PROTEIN STRUCTURE: GENERAL

A. The Peptide Bond

The peptide bond has partial double bond character because the double bond of the carbonyl group contributes electrons which allow different resonant forms to occur. Therefore, it is **planar** and predominantly in the *trans* configuration in proteins.

B. Steric Hindrance

The only rotation along the backbone of a polypeptide is the rotation of planes of the peptide bonds with respect to each other about the alpha carbons. However, this rotation is severely restricted by the steric hindrance between the different R groups and the carbonyl oxygens. One type of allowed rotation occurs when the carbonyl oxygens are pointed in the same directions and the R groups radiate out from the resulting spiral (as in the α-helix). The other type occurs when alternate carbonyl oxygens point in opposite directions forming a pleated plane, with alternate R groups pointing up and down from this plane (as in β-pleated sheets).

C. Non-Covalent Bonds

The interactions between different parts of a protein are responsible for protein structure. These interactions are weak because they compete with interactions with water. They are:

Hydrogen bonds (H-bonds) between -C=O and -NH- of different peptide bonds; also involving some other amino acids capable of H-bonding.

Salt bonds (ionic) between acidic and basic amino acids.

Hydrophobic interactions among non-polar R groups.

Van der Waals forces.

D. Structural Levels of Proteins

Primary structure is the amino acid sequence within a peptide chain, including disulfide bonds and other covalent modifications. Under proper conditions proteins spontaneously fold into their native conformations, hence primary structure dictates all levels of structure. *Homology*, the similarity between the sequences of two proteins, indicates probable similar conformations.

Secondary structure consists of recognizable structures adopted by the backbone of the peptide segments. These are composed of the allowable sterically restricted types that also align H-bonds. That is, they are constrained by steric and H-bond requirements. The common secondary structures are:

- ♦ right-handed α-helix
- ♦ β-pleated sheet
- ♦ β-turn

Tertiary structure is the specific three-dimensional conformation of a particular peptide chain. Most proteins are globular; hydrophobic interactions are primarily responsible for producing a compact conformation with a hydrophobic core.

Quaternary structure is the arrangement of subunits that are held together by non-covalent associations. An example is the tetrahedral arrangement of the four subunits of hemoglobin.

E. Domains

Most modern proteins are composites of evolutionarily older, smaller proteins. These smaller building blocks, called domains, often correspond to exons of the gene.

F. Conformational Changes

When ligands such as substrates, effectors, cofactors, etc. bind to a protein, the tertiary structure changes subtly, reversibly affecting the activity of an enzyme. Changes in activity of an enzyme caused by changes in conformation are called **allosteric** effects. Denaturation refers to a much more extensive change, such as unfolding, caused by heat, acid, or some other extreme condition, and is usually irreversible.

G. Zymogens

Many enzymes are secreted as inactive precursors. These are either given the prefix "pro" as in prothrombin or the suffix "ogen" as in trypsinogen. Removal of small peptide fragment(s) by proteolysis activates the protein by causing a marked conformational change.

H. Isozymes

Often enzymes having multiple subunits but with the same catalytic function may differ from one another by slight variations in the primary sequence of one or more of the subunits. Lactic dehydrogenase is an example of an enzyme that occurs in different forms, called **isozymes**.

III. COLLAGEN AND FIBROUS PROTEINS

A. Collagen

Collagen is one of the most abundant of all proteins and is an important constituent of tendons, ligaments, cartilage, skin and the interstitial matrix.

Basic structure. Collagen is composed of three helical polypeptides which are wound tightly around each other to form a long, rigid fibrous molecule very unlike the usual globular proteins. The helical structures found in these polypeptides are not α-helices because proline and hydroxyproline (~25% of the amino acids in collagen) do not fit the steric requirements of α-helices.

Post-translational modifications

♦ Two of the amino acids in collagen can be hydroxylated. These are **lysine** and **proline**. The formation of hydroxylysine and hydroxyproline increases the overall stability of the molecule. The hydroxylations are post-translational events and require the participation of **ascorbic acid** (Vitamin C). Ascorbic acid deficiency leads to the condition known as **scurvy** which is characterized by defective function of connective tissues and skin. (See Chapter 10). The hydroxylysine residues can also be glycosylated.

♦ Lysine can also undergo oxidative deamination of its ε-amino group to form an aldehyde at this position. Two of these aldehyde residues on adjacent peptides can interact in an aldol condensation to form a covalent bond and increase the overall stability of the collagen molecule. Deficiency in the copper cofactor or in lysyl oxidase itself (Ehlers-Danlos syndrome) reduces the crosslinking in collagen and the desmosine crosslink in elastin. Another form of Ehlers-Danlos syndrome is caused by a deficiency in the propeptidase that converts procollagen to tropocollagen. This condition is characterized by stretchable skin and hypermobile joints.

Glycine is every third residue in a collagen peptide chain. Glycine is important because the degree of intertwining of the three chains is so tight that only glycine is small enough to fit at these positions.

Gene and exon structure. The collagen gene contains approximately 90 exons, the majority of which appear to have been constructed from an exon of (X-Y-Gly)$_6$.

B. Laminin and Fibronectin

These are fibrous proteins composed of multiple but dissimilar domains connected by flexible polypeptide segments. They are involved in cell-to-matrix interactions, each domain being specific in its binding to components of the extracellular matrix like collagen or to sites on cells. These proteins are examples of a class of *cell adhesion molecules* (CAM's) which have Arg-Gly-Asp recognition sites for cellular receptors.

IV. HEMOGLOBIN

A. Oxygen-Binding Proteins

Hemoglobin (Hb) is the oxygen-carrying system found in erythrocytes. It transports oxygen from the lungs to all tissues of the body, and aids in buffering blood.

Myoglobin (Mb) is a protein related to hemoglobin and is found in muscle. It can take oxygen from hemoglobin and acts as the intracellular oxygen-transport system.

Myoglobin	Hemoglobin
monomer	tetramer, 2α and 2β
binds one O_2	binds 4 O_2, one per subunit
binding property: hyperbolic	binding property: sigmoidal

B. Binding Curves (Figure 1-2)

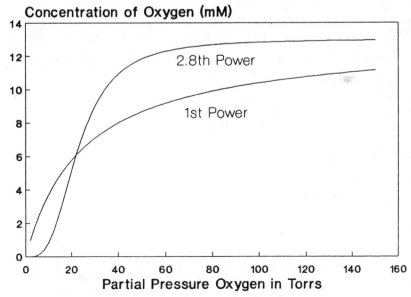

The binding curve for hemoglobin measures the amount of O_2 bound by Hb *vs* the partial pressure of O_2. It is **sigmoidal** or S-shaped, indicating that hemoglobin is an allosteric and cooperative protein. The binding of O_2 to the first subunit causes a conformational change in the molecule which allows the other three O_2 molecules to bind much more readily. In the physiological range of oxygen concentrations, this sigmoidal binding curve is sharper (2.8th power) than the hyperbolic binding curve (1st power) exhibited by the monomeric and non-cooperative myoglobin. This sharpness allows more O_2 to be delivered for the same drop in oxygen tension between the lungs and other tissues.

Figure 1-2. Oxygen-Binding Curves of Hemoglobin and Myoglobin.

C. Regulation of O_2 Binding

The binding of O_2 to hemoglobin is regulated physiologically by three negative allosteric effectors: hydrogen ions, CO_2, and BPG. In the parlance of enzymologists, these effectors convert the conformation of hemoglobin from the "relaxed" to the "tense" state.

Hydrogen ions. When deoxygenated Hb binds O_2 in the lungs, hydrogen ions are released:

$$Hb(H^+)_2 + 4O_2 \rightleftharpoons Hb(O_2)_4 + 2H^+$$

In the tissues, metabolism releases acidic waste products (such as CO_2) causing an increase in hydrogen ion concentration. By mass action in the equation above this causes more O_2 to be released from the Hb. In the lungs CO_2 is excreted, the blood pH increases and Hb binds more O_2. This overall process, called the **Bohr effect**, has the result of increasing the amount of O_2 delivered to the more active tissues. At the same time more protons are removed from tissues and delivered to the lungs.

Carbon dioxide binds preferentially to deoxygenated Hb. It also has an allosteric effect like that produced by H^+. In the tissues both H^+ and carbon dioxide act additively to promote the release of oxygen.

2,3-Bisphosphoglyceric acid (BPG) also has an allosteric effect on Hb. BPG binds between the two β-subunits of deoxygenated Hb (one BPG per Hb) and stabilizes the molecule. BPG causes the steepest part of the oxygen binding curve to occur at an oxygen tension between that in the peripheral tissues and that in the lung. This arrangement permits the greatest amount of oxygen to be delivered to the tissues.

D. Sickle Cell Anemia

A reduction in the amount of hemoglobin in blood is referred to as anemia and of course leads to a reduction in the amount of O_2 carried. Sickle cell anemia is a heritable type resulting from a single amino acid mutation of a Val for a Glu in the β subunit (Hb S). While this has no effect on O_2 binding, Hb S in the deoxygenated state precipitates and forms fibers that distort the erythrocytes, which are then destroyed leading to anemia. Although the homozygous state is detrimental to survival, heterozygotes possess increased resistance to malaria. Hemoglobin S thus has net survival value in some geographic areas.

V. PLASMA PROTEINS

Plasma contains many proteins of a non-enzymatic nature. These include the immunoglobulins, blood coagulation proteins, transport proteins and proteins of unknown function. The functions of some of these proteins are described in other sections.

A. Transport or Carrier Proteins

Functions:

- ♦ Increase the water-solubility of hydrophobic molecules.
- ♦ Decrease the loss of small molecules in the kidney.
- ♦ Transport bound molecules to specific tissues.
- ♦ Assist in detoxification.

Selected examples:

Albumin makes up 50 to 55% of the proteins of plasma, and has very broad and non-specific binding properties. It binds and transports fatty acids released from adipose tissue, but also binds many drugs which compete with fatty acid binding. It does not bind steroids. Albumin is the main contributor to the osmotic pressure of blood. Severe loss of albumin causes edema and disturbances in blood volume and pressure.

Transferrin carries two Fe^{+++} ions and transfers them to cells having receptors for transferrin. **Ferritin** is the intracellular form of iron storage, but small amounts also occur in blood.

Haptoglobin and **hemopexin** bind methemoglobin and hemin, respectively. Binding hemin not only protects the kidneys from its toxic effects but also conserves the iron for re-use in metabolism.

B. Trace Enzymes

In trace amounts, blood contains many enzymes that come from dying cells. An increased level of an enzyme is diagnostic for trauma to those tissues that are rich in the specific enzyme. Some important examples are listed under the enzyme section of this text.

VI. HEMOSTASIS AND BLOOD COAGULATION

A. The Hemostatic Process

Hemostasis, the stopping of blood flow at a wound site, is brought about by the combined effect of platelets, the vessel wall, and plasma coagulation factors. The chronology of hemostasis is that a wound exposes the collagen layer just below the endothelial cells that line the vessel wall. Platelets recognize and bind to the collagen, become activated, recruit other platelets to form a hemostatic plug, secrete many substances that promote coagulation, and support coagulation once it starts.

Cast of characters. Table 1-1 lists the hemostatic and fibrinolytic factors, the helper proteins, and the protease inhibitors that control the proteolytic factors.

Factor	Trivial Name	Helper Protein	Inhibitor in Plasma
I	Fibrinogen		
II	Prothrombin, Thrombin	Thrombomodulin	Antithrombin III
III	Tissue Factor		
IV	Calcium ions		
V	Plasma accelerator globulin		Protein C
VI	(Not Assigned)		
VII	Proconvertin	Tissue Factor	Antithrombin III
VIII	Antihemophilic factor A		Protein C
IX	Antihemophilic factor B	F VIII	Antithrombin III
X	Stuart factor	F V	Antithrombin III
XI	Thromboplastin antecedent		
XII	Contact factor, Hageman		
XIII	Transglutaminase		
	Tissue plasminogen activator	Fibrin clot	Inhibitors of plasminogen activator
	Plasminogen/plasmin		Anti$_2$-plasmin
	Protein C	Protein S	Protein C inhibitor
	Protein S		

Table 1-1. Hemostatic and Fibrinolytic Factors, their Helper Proteins, and Inhibitors.

The plasma coagulation proteins are numbered by Roman numerals and a subscripted "a" indicates they have been converted into their active forms. Prothrombin (Factor II, or F II), thrombin (F II_a, the active form of prothrombin), and fibrinogen (F I), are almost always still referred to by their trivial names. Plasma also contains many protease inhibitors some of which control coagulation.

Wound recognition and formation of platelet plug. When a wound occurs, the thin layer of endothelial cells lining blood vessels is disrupted, exposing collagen in the basement membrane below. Platelets bind to this collagen and become active. To build the plug, activated platelets recruit other platelets by secreting the platelet agonists ADP, serotonin, and thromboxane A_2. The latter two help hemostasis by causing smooth muscle cells of vessel walls to contract. The aggregated platelets occlude the vessels but do not provide a sufficiently stable plug to prevent re-bleeding. Coagulation of the blood provides the added stability.

Initiation of coagulation. In addition to exposing collagen, the wound also exposes cells that have a receptor, called tissue factor, that binds F VII and F VII_a. In Figure 1-3, adventitial cells are shown supplying tissue factor but other cells provide it as well. Exposure of tissue factor starts coagulation.

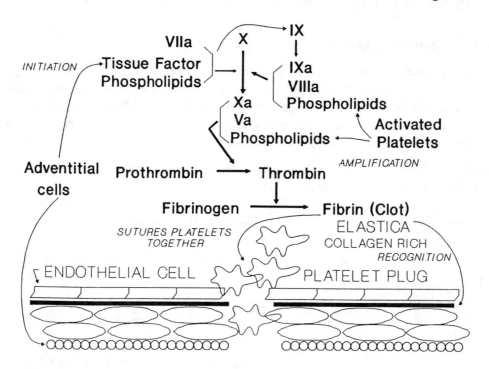

Figure 1-3. The Coagulation Cascade.

The cascade of zymogen activation — amplification: When F VII_a is bound to tissue factor, which is also a helper protein, it proteolytically activates F X and to a lesser extent F IX. F IX_a, with its helper protein F VIII_a, also activates F X. In turn, F X_a, with its helper F V_a activates prothrombin to thrombin. Because this cascade is composed of a series of enzymes activating other enzymes, it takes only a small amount of F VII_a bound to tissue factor to produce large quantities of thrombin. Furthermore, F X activates F VII on tissue factor and thrombin activates F V and F VIII, each to promote its own production.

This scheme also provides a degree of redundancy. That is, the step involving Factors IX and VIII is only supportive in many tissues. Classical hemophilia results from production of a mutant, inactive F VIII. Bleeding in tissues rich in tissue factor during mild trauma is not a major problem for hemophiliacs because of the direct activation of F X. Hemophiliacs suffer mainly chronic bleeding into joints and muscles where cells rich in tissue factor are sparsely distributed.

The activations are all proteolytic cleavages, each requiring the following agents:

- The *activating factor*, in each case a serine protease: $F VII_a$, $F IX_a$, and $F X_a$.
- *Substrate*: F X, F X, and prothrombin, respectively.
- *Helper protein*: tissue factor, $F VIII_a$, and $F V_a$ respectively. A helper protein aids in holding the substrate in place during the activation (it lowers the apparent K_M). Tissue factor is an integral protein of the membranes of certain cells. Factors V_a and VIII_a are not integral proteins but are tightly bound to platelet membranes.
- *Acidic phospholipid surfaces*, needed for sufficient reaction velocities.
- *γ-Carboxy-glutamyl residues*, whose post-translational production is vitamin K-dependent.
- *Calcium ions* which, through the γ-carboxy-glutamyl residues, form bridges between the proteases and phospholipid surfaces, and between the latter and the substrates of the proteases.

Factors II, VII, IX, X, and proteins C and S (to be discussed below) share the requirement for the modified glutamyl groups because binding to phospholipids is an integral part of their function. Warfarin, a competitive inhibitor of vitamin K - dependent processing, and vitamin K deficiency prevent the formation of the γ-carboxy-glutamyl residues and thereby reduce hemostasis.

Coagulation is restricted to the wound site because the activating reactions occur only when damaged tissue binds the activated platelets that then supply acidic phospholipid surfaces for the proteolytic steps catalyzed by $F IX_a$ and $F X_a$. While bound with their helper proteins to the platelets, these proteases are also protected from the inhibitors of plasma proteases, antithrombin III and α-macroglobulin. Furthermore, the activation reactions occur too slowly if the activating factors are not bound to activated platelets.

Clot formation. The clot forms as fibrin is produced between the aggregated platelets. Thrombin is the first protease that is freed from the platelets in a viable, active form. It causes the clot by cleaving fibrinogen, removing fibrinopeptides A and B. Fibrin, the product of the cleavage, polymerizes on its own. As with other fibrous proteins, fibrin polymers are cross-linked by $F XIII_a$, a transglutaminase that joins the side chains of a Gln on one fibrin molecule to a Lys on a different fibrin molecule. Thrombin also activates F XIII. Platelets aggregate by binding to either end of fibrinogen which then forms a bridge between two platelets. Fibrin strands still have these same ends, therefore platelets bind fibrin, help in organizing the fibrin strands and even pull these into a tighter form to maintain a strong clot. These fibrin strands suture the platelets together to provide the needed stability that the platelet plug alone does not possess.

Summary of the hemostatic process. There are two central elements in coagulation. One is the platelet that recognizes the wound, provides the surface for activation steps, secretes many substances that affect coagulation, and organizes and retracts the clot. The other is thrombin which causes fibrinogen to clot, further activates platelets, stimulates its own production by activating Factors V and VIII, activates Factor XIII, and even initiates the reactions that regulate its own production.

B. Regulation of Hemostasis

Obviously, from the description above something is required to prevent the thrombin and unincorporated, activated platelets from spreading the clot throughout the body. This something is a separate collection of reactions primarily under the control of the monolayer of endothelial cells that line the blood vessels.

Termination of coagulation. Any thrombin that escapes from the area of the clot is rapidly bound to the surface of intact endothelial cells whereupon it activates the anticoagulation scheme shown in Figure 1-4 (next page). Thrombin binds to thrombomodulin, its receptor on endothelial cells, which also binds to protein C. Thrombomodulin acts like a helper protein in the activation by thrombin of protein C, which in turn with its helper protein, protein S, inhibits $F V_a$ and $F VIII_a$ on activated platelets. This is the reaction mechanism that ensures that even if an activated platelet escapes the hemostatic plug, the coagulation process will be terminated. Eventually these factors terminate the clotting reactions at the wound site as well.

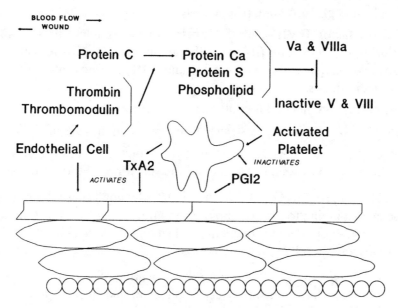

Figure 1-4. Regulation of Coagulation Downstream from Wound.

Inactivation of activated platelets. Endothelial cells also deactivate activated platelets. Thrombin and thromboxane both stimulate endothelial cells to produce prostacyclin I_2 (PGI_2). Prostacyclin I_2 is a potent antagonist that deactivates platelets. It also is a potent vasodilator that overcomes the contraction of the vessel wall induced by the platelet-derived serotonin and thromboxane A_2.

C. Fibrinolysis or Removal of the Clot

The clot is removed as part of the process of wound healing (Figure 1-5, next page). Fibrin clots are removed proteolytically by plasmin. Plasmin is activated on the clots by plasminogen activator, a protein released from endothelial cells in response to the clotting process. The fibrin clot acts as a helper protein for this activation. Both plasminogen, the inactive form of plasmin, and the activator are bound to the clot and the activation is enhanced by the clot. This leaves plasmin on the clot until it finishes the hydrolysis. Antiplasmin, a protein that inhibits plasmin, is present in plasma to inhibit any plasmin that escapes from the clot prematurely. In this way, clot removal is also restricted to the wound site.

D. Thrombosis

Thrombosis is abnormal coagulation that occurs on atherosclerotic plaques and contributes to a variety of vascular problems including myocardial infarctions, strokes and deep-vein thrombosis. Several methods of controlling this disorder have been tried, most based on the material presented above. Another method is based on reducing risk factors. This approach depends primarily on educating the public as to the benefits of diet and exercise in decreasing the severity of atherosclerosis that comes with age.

Regulating platelet function. Attempts to control thrombosis by regulating platelet function have been of limited value. **Aspirin** has received the most attention but it appears to provide limited aid in preventing strokes or recurrent heart attacks. Aspirin is an irreversible inhibitor of *cyclooxygenase*, an enzyme necessary for the synthesis of thromboxane A_2, the "message for help" sent out by activated platelets. One problem with aspirin is that it blocks cyclooxygenase in endothelial cells as well and thereby decreases the production of the natural anticoagulant, prostacyclin I_2.

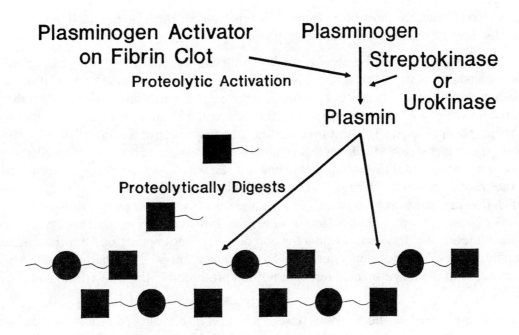

Figure 1-5. Fibrinolysis (clot digestion).

Regulating coagulation. **Warfarin** and other coumarin drugs have been used in the past for prophylaxis against thrombosis. Recall that warfarin prevents the post-translational modification necessary for binding of some coagulation factors to the reactive phospholipid surfaces of platelets. However, the efficacy of coumarin drugs was poor and the discovery of the regulatory role of proteins C and S, both of which contain the vitamin K-dependant modifications, explains the deficiency in effect. But since these drugs can be administered orally, and do have some prophylactic effect, they are useful for certain types of patient care.

Heparin, an anticoagulant, also is used to control thrombosis. (Chelators of calcium ions cannot be used because the calcium ion concentration in blood is very critical for many processes). Heparin acts as a catalyst for the inactivation of thrombin by antithrombin III, the plasma protease inhibitor. Heparin is one of the favored agents for regulating deep vein thrombosis in many types of patients and in preventing recurrent clot formation after clot removal. Heparin must be injected for this treatment and is not suitable as a long-term prophylactic.

Increasing fibrinolysis. Another approach used to treat thrombosis is to activate the fibrinolysis reactions. **Plasminogen activator**, *urokinase*, and *streptokinase*, all activators of plasminogen, have been used in treating thrombosis on an acute basis. Streptokinase, a bacterial product, was the only one inexpensive enough for routine use until all were cloned into bacteria or culture cells. At first, recombinant tissue plasminogen activator (TPA) was preferred because it was thought that there would be reduced problems of bleeding since it requires a clot for activation. But several studies suggest that there are few differences among fibrinolytic agents in removing clots. However, the foreign nature of streptokinase can add complications for patients previously exposed to this agent.

E. Coagulation Deficiencies

Thrombosis is the most common problem associated with the clotting system but there are examples of deficiencies in all the factors (see Table 1-2, page 13). There are also various platelet disorders and defects in vessel wall reactions. The best known example is classical **hemophilia**, which results from deficient activity of F VIII. These conditions are most often treated by replacement therapy, in which major complications can arise from blood-borne diseases and immune responses.

Clinical Tests. Coagulation can be assayed readily in the clinical laboratory using plasma samples prepared from anticoagulated blood. These tests are useful in screening for possible coagulation deficiencies, the nature of which can be established later by more specific tests. **Prothrombin time** is a test that measures the time required for the plasma to clot after the reaction is initiated by addition of thromboplastin, a mixture of tissue factor and phospholipids. This test bypasses the step that requires a complex of F IX and its helper F VIII. Because tissue factor was added and this is not a plasma protein, this pathway is called the **extrinsic pathway**. In contrast, the **partial thromboplastin time** is a test which measures the time required for a clot to form after the reaction is initiated by addition of phospholipid and a contact activator. The contact activator activates F XII which activates F XI, which in turn activates F IX. The rest of this pathway, called the **intrinsic pathway** because no non-plasma factors were added, is the same as the process discussed above. The physiological relevance of this pathway of activation of F XII is not understood but the diagnostic virtue is. Thus hemophilia A (F VIII deficiency) or B (F IX deficiency) prolong the partial thromboplastin time (the longer pathway with the longer-named test) but not the prothrombin time. Warfarin and the coumarins prolong both times.

Platelet function is tested by the **bleeding time**. Cessation of bleeding from a standardized cut in the skin depends on formation of a plug by platelets. Aspirin, which blocks production of thromboxane A_2, prolongs bleeding time, whereas hemophilias and warfarin do not.

VII. GENETIC DISEASES

A. Inherited Structural Variants of Hemoglobin

A single base change (point mutation) may occur in the gene for the α, β, γ, or δ-globin chain. This may cause a normal (non-deleterious) amino acid substitution, e.g., Hb C $\beta^{6\ Glu \to Lys}$; Hb E $\beta^{26\ Glu \to Lys}$, or it may result in production of an abnormal protein. It may also be a nonsense mutation, causing chain termination. Deletion (or insertion) of a single base causes a frameshift mutation. Mutant proteins occur in which amino acids are deleted or inserted, resulting from deletion or insertion of entire codons.

The defective molecule in **sickle cell anemia** is Hb S, which has a single amino acid substitution in the β chain — 6 Glu → 6 Val (see page 6).

B. Quantitative Disorders of Globin Synthesis: The Thalassemias

These diseases are characterized by abnormalities in the amounts of the different globin chains synthesized. The chains themselves are normal in structure.

α-Thalassemias: deficiency of α chain synthesis.

- homozygous for α thalassemia: all four α globin genes — Hb Barts, death in utero
- heterozygous for α thalassemia: three globin genes affected — Hb H disease
- α_1-thalassemia: two α globin genes affected
- α_2-thalassemia: one α globin gene affected.

β-Thalassemias: decreased (β^+) or absent (β^0) synthesis of β-chain results in increased levels of Hb F and Hb H2.

- homozygous condition: thalassemia major — Cooley's anemia, severe microcytic anemia.
- heterozygous condition: thalassemia minor.

C. Collagen metabolism

Ehlers-Danlos syndrome, types I - VII. As a group, these diseases are characterized by deficiencies at various stages in the formation of collagen, from synthesis of protein to cross-linking of chains.

Osteogenesis imperfecta, types I - III: defective synthesis or secretion of types I and/or III collagen.

Cutis laxa. An X-linked recessive defect causing deficient lysyl oxidase with abnormal copper metabolism.

D. Blood coagulation: Deficiency of Clotting Factors

Disorder	Factor	Genetics
Afibrinogenemia	Fibrinogen	Autosomal recessive
Dysfibrinogenemia	Fibrinogen	Autosomal recessive
Parahemophilia	V	Autosomal recessive
Factor VII deficiency	VII	Autosomal recessive
Classic hemophilia	VIII	X-linked recessive
Hemophilia B	IX	X-linked recessive
Factor X deficiency	X	Autosomal recessive
Factor XI deficiency	XI	Autosomal recessive
von Willebrand	von Willebrand	Mostly autosomal dominant

Table 1-2. Heritable Diseases of Blood Coagulation.

VIII. REVIEW QUESTIONS ON AMINO ACIDS AND PROTEINS

> **DIRECTIONS:**   For each of the following multiple-choice questions (1 - 16), choose the ONE BEST answer.

Questions 1 - 3: In the figure below, curve 3 represents the normal oxygen binding curve for hemoglobin. Which curve represents the binding curve after the indicated parameter in questions 1 - 3 is changed but all other conditions are held constant?

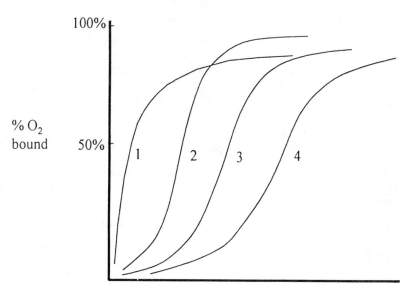

1. The pH is raised.

A. 1
B. 2
C. 3
D. 4
E. remains in the same position but is reduced in height.

Ans B: An increase in pH means a reduction in H^+ concentration which in turn increases oxygen binding to hemoglobin, and the curve shifts to the left.

2. The concentration of BPG is raised.

A. 1
B. 2
C. 3
D. 4
E. remains in the same position but is reduced in height.

Ans D: An increase in the concentration BPG causes more oxygen to be released and the curve shifts to the right.

3. The number of red blood cells is reduced.

A. 1
B. 2
C. 3
D. 4
E. remains in the same position but is reduced in height.

Ans C: A reduction in RBC number does not affect the *percentage* of oxygen bound to hemoglobin but does reduce the total amount bound.

4. Factor XIII$_a$

A. activates tissue factor
B. digests clots
C. helps Factor IX$_a$ activate Factor X
D. recruits other platelets
E. stabilizes the polymerized fibrin.

Ans E: Polymerized fibrous proteins are stabilized by covalent crosslinkages and fibrin is no exception. Factor XIII$_a$ introduces the transpeptide bonds.

5. All the following statements about fibronectin are true EXCEPT:
Fibronectin:

A. binds to cells
B. contains an RGD (Arg-Gly-Asp) recognition site
C. is a heterodomainal protein
D. is a CAM (cell adhesion molecule)
E. is self-polymerizing.

Ans E: Fibrous proteins, such as tropocollagen, elastin, and fibrinogen, often self-polymerize. Fibronectin and laminin do not, but bind to various extracellular components to organize their matrices.

6. Thrombomodulin supports

A. activation of Factors V and VIII by thrombin
B. cleavage of fibrinogen by thrombin
C. activation of protein C by thrombin
D. inactivation by heparin by thrombin
E. activation of prothrombin by Factor X.

Ans C: As part of the regulation of coagulation, endothelial cells contribute thrombomodulin as a helper protein for the activation of protein C by thrombin.

7. A baby is brought to you on the complaint of cyanosis (indicating hypoxia, lack of oxygen) with severe shortness of breath upon exertion. The red blood cell count is normal. The electrophoretic pattern of the hemoglobin indicates the net charge is increased by 4 negative charges, suggesting a substitution of Asp or Glu for a Lys or Arg. Too much O_2 remains on hemoglobin at tissue levels of pO_2. Which of the following tests would you order?

A. blood pH
B. blood pCO_2
C. blood BPG
D. hemoglobin BPG binding assay
E. hemoglobin heme content.

Ans D: The cause of the improper release of O_2 resides in the hemoglobin molecule itself. The alteration of positive charges could reduce binding of BPG which then would affect the release of O_2.

8. A child is brought to you on the complaint of serious bruising from what is described as a game of tag. APTT and PT were moderately elevated but bleeding time was not. Values for both Factors VII and IX were reduced. Without further information, a prudent approach, besides continued observation, would be to administer:

A. heparin
B. tissue plasminogen activator
C. a unit of whole blood
D. vitamin K
E. warfarin.

Ans D: Multiple factors are involved and the problem appears to be recently developing. Warfarin would *produce* this response; an injection of vitamin K would be harmless and could overcome some of the effects. A unit of blood is uncalled for; the other options would make things worse.

9. A child is brought to you on the complaint of lack of growth. The child is shorter than average for his age but also presents very distensible skin and hypermobile joints. A skin biopsy shows adequate levels of lysyl oxidase. Which of the following would be further useful information about the biopsy?

A. lysyl hydroxylase activity
B. prolyl hydroxylase activity
C. collagenase activity
D. amount of procollagen in cells
E. amount of procollagen in collagen.

Ans E: The symptoms are compatible with the Ehlers-Danlos syndrome, one type of which is caused by inadequate removal of the peptides from procollagen.

10. A unique aspect of collagen is the repeating nature of its primary structure. These repetitions can be represented as:

A. $(X\text{-}Y\text{-}Pro)_{30}$
B. $(X\text{-}Y\text{-}His)_{30}$
C. $(X\text{-}Y\text{-}His)_{300}$
D. $(X\text{-}Y\text{-}Gly)_{300}$
E. $(X\text{-}Y\text{-}Lys)_{300}$

Ans D: Gly is repeated at the third position in tropocollagen as required by the tight winding; no other side chain is small enough to fit at these positions.

11. All of the following statements of the functions of hemoglobin are true EXCEPT:

Hemoglobin:

A. decreases osmotic load on red blood cells by self-association.
B. enhances oxygen delivery specifically to active tissues.
C. helps hold red blood cells in biconcave form (except in sickle cell anemia).
D. increases the oxygen carrying capacity of blood.
E. prevents oxidation of ferrous atom to ferric state.

Ans C: It is cytoskeletal elements (not Hb) that normally hold RBC in the biconcave shape. But to pass through capillaries RBC must be deformable, which is prevented by the exaggerated polymerization of sickle-cell hemoglobin.

12. Cellular control is exerted on the reaction scheme of hemostasis. Which of the following statements about cellular function is INCORRECT?

A. Adventitial cells initiate coagulation.
B. Endothelial cells deactivate platelets.
C. Endothelial cells terminate coagulation and initiate clot digestion.
D. Platelets recognize the wound site and support coagulation.
E. Platelets deactivate endothelial cells.

Ans E: Platelets activate endothelial cells downstream from the wound site by producing TXA_2. Endothelial cells then produce PGI_2 which inhibits inappropriate clot formation.

13. The effectors of hemoglobin function, H^+, CO_2, and BPG:

A. are positive, allosteric regulators
B. increase the amount of oxygen bound to hemoglobin
C. preferentially bind to oxygenated hemoglobin
D. all compete for the same site of binding
E. exert a greater effect in peripheral tissues compared to lungs at normal oxygen tension.

Ans E: The pO_2 in lung is usually so high that these agents have little effect on the amount bound. They displace the curve so as to primarily affect unloading of O_2 in peripheral tissues, where H^+ and CO_2 are also at greater concentrations and more effective.

14. Binding of the first molecule of oxygen to sickle-cell hemoglobin produces all of the following effects EXCEPT:

A. a conformational change in the quaternary structure
B. a cooperative effect comparable to that of "normal" hemoglobin
C. a decrease in the probability of sickling
D. an enhancement in carbamate formation
E. release of protons.

Ans D: Sickle-cell hemoglobin has normal oxygen-binding properties. One of these is that oxygen-binding blocks the reaction of CO_2 with N-terminal amino groups to form a carbamate.

15. Which of the following proteins initiates clot digestion by forming plasmin at a specific clot site?

A. Factor XIII_a
B. streptokinase
C. thrombin
D. tissue plasminogen activator
E. urokinase.

Ans D: Plasminogen activator secreted by endothelial cells binds to the clot along with plasminogen. This is the only activator that uses the clot as a helper protein.

16. What is the origin of the acidic phospholipids that are required for activation of Factor X by Factor IX_a?

A. adventitial cell membranes
B. endothelial cell membranes
C. platelet membranes
D. smooth muscle cell membranes
E. all of the above.

Ans C: Acidic phospholipids are provided by the cells whose membranes contain these lipids. Thus tissue cells control the initiation and platelets control the amplification steps of coagulation.

DIRECTIONS: For each set of questions, choose the ONE BEST answer from the lettered list above it. An answer may be used one or more times, or not at all.

Questions 17 - 20:

A. Primary structure
B. Secondary structure
C. Tertiary structure
D. Quaternary structure
E. Domain

17 Directed by steric hindrance and H-bond alignment.

Ans B: Secondary structure is almost entirely the result of minimizing the steric hindrance associated with rotation of the planes of the peptide bonds about the alpha carbons and aligning the H-bonds for maximum interactions.

18. Determines all levels of structure.

Ans A: Proteins contain all the information to correctly fold up, given proper conditions, so the primary structure is sufficient to produce all other levels of structure.

19. Restricted to complex proteins.

Ans D: Only complex proteins have true quaternary structure because they have more than one subunit (polypeptide).

20. Total conformation of a polypeptide.

Ans C: The conformation of a polypeptide is tertiary structure.

Questions 21 - 25:

    A. conformational change
    B. denaturation
    C. N-terminus
    D. posttranslational modification
    E. side chain

21. Contains alpha amino group (not in amide form).

Ans C: The N-terminus is the end of the polypeptide that has the free amino group and the C-terminus is the other end with the free carboxyl group.

22. Conversions between tertiary structures promoted by associations with other molecules.

Ans A: Conformational changes are detected when proteins associate with other molecules and are important to protein function.

23. Enzymic addition to a completed polypeptide.

Ans D: Most posttranslational modifications are catalyzed by enzymes directed by recognition sites within the sequence of the protein.

24. Loss of tertiary structure, leading to inactivity.

Ans B: Unfolding or other radical conformational changes that convert the enzyme to inactive forms is referred to as denaturation and is often irreversible.

25. R group.

Ans E: The side-group that differs among the amino acids is referred to as the R group or side chain.

2. ENZYMES
A. Chadwick Cox and Wai-Yee Chan

I. NATURE OF ENZYMES

A. Introduction

Enzymes are **biological catalysts**, produced by living tissues, that increase the **rates** of reactions. They do not affect the nature of an equilibrium; they merely speed up the rate at which it is achieved. By enzymically increasing the rates of certain reactions, organisms can select the reactions required for their life from a large collection of undesirable but competing, spontaneous reactions.

B. Definition of Terms

Substrate: substance acted upon by an enzyme.

Activity: amount of substrate converted to product by the enzyme per unit time (e.g. micromoles/minute).

Specific activity: activity per quantity of protein (e.g. micromoles/minute/mg protein).

Catalytic constant: proportionality constant between the reaction velocity and the concentration of enzyme catalyzing the reaction. Unit: activity/mole enzyme.

Turnover number: catalytic constant/number of active sites/mole enzyme.

International Unit (IU): quantity of enzyme needed to transform 1.0 micromole of substrate to product per minute at $30°C$ and optimal pH.

C. Nomenclature

Some enzymes have trivial names, e.g. *pepsin*, *trypsin*, etc. Others are named by adding the suffix -ase to the name of the substrate, e.g. *arginase*, which catalyzes the conversion of arginine to ornithine and urea. All enzymes have systematic names, of which there are 6 major classes:

1. *Oxidoreductase*: oxidation-reduction reactions, e.g., *alcohol: NAD oxidoreductase* for the enzyme catalyzing the reaction:

$$RCH_2OH + NAD^+ \rightleftharpoons RCHO + NADH + H^+.$$

2. *Transferase*: transfer of functional groups including amino, acyl, phosphate, glycosyl, etc. Example: *ATP: Creatine phosphotransferase* for the enzyme catalyzing the reaction:

$$ATP + Creatine \rightleftharpoons Phosphocreatine + ADP.$$

3. *Hydrolase*: Cleavage of bond between carbon and some other atoms by the addition of water. Example: *Peptidase* for the enzyme catalyzing:

$$R_1CONHR_2 + H_2O \rightleftharpoons R_1COOH + R_2NH_2.$$

4. *Lyase*: Addition or removal of the elements of water, ammonia or CO_2 to or from a double bond. Example: *Phenylalanine ammonia lyase* for the enzyme catalyzing:

$$phenylalanine \rightleftharpoons cinnamic\ acid + ammonia.$$

5. *Isomerase*: Racemization of optical or geometric isomers. Two types:

epimerase or *racemase* for optical isomers (asymmetric carbon), as in:

$$D\text{-lactic acid} \rightleftharpoons L\text{-lactic acid} \ (racemase)$$

mutase for geometric isomers or intramolecular group transfer, as in:

$$2\text{-phosphoglycerate} \rightleftharpoons 3\text{-phosphoglycerate}$$

6. *Ligase*: Formation of C-O, C-S, C-N, or C-C with the hydrolysis of ATP. Example: *Pyruvate carboxylase* for the enzyme catalyzing the reaction:

$$pyruvate + ATP + CO_2 \rightleftharpoons oxaloacetate + ADP + P_i$$

D. Basic Enzyme Structure

Enzymes may be composed of a **single** polypeptide chain, or several identical or different **subunits**. Different enzymatic activities may be contributed by independent **domains** of a single polypeptide.

A compound (organic or inorganic) other than an amino acid side chain may be required for activity and is not modified at the end of the reaction When tightly bound to an enzyme, such as the heme of cytochrome, it is known as a **prosthetic group**. When less tightly bound, or removable by dialysis, e.g. metal ions, it is known as a **cofactor**. The catalytically active enzyme complex consisting of a protein **apoenzyme** and a non-protein cofactor is called a **holoenzyme**.

Coenzymes are organic molecules fulfilling the role of substrate, being modified at the end of the reaction, but readily regenerated by another linked reaction. Examples: biotin, NAD, ATP, TPP, FAD, pyridoxal phosphate, coenzyme A, etc.

When an enzyme is first made as an activatable precursor, it is called a **zymogen**, or **proenzyme**. The active mature enzyme is generated by specific cleavage of a peptide bond. Examples:

$$chymotrypsinogen \rightarrow chymotrypsin; \quad pepsinogen \rightarrow pepsin$$

E. Characteristics of Enzymatic Reactions

Enormous catalytic power.

Optimum pH (Figure 2-1): the pH at which the enzyme activity is at its maximum. This depends on the acid-base behavior of the enzyme and substrate. It can vary widely; the majority of enzymes have optima between 4 and 8. The kinetic parameters K_M and V_{max} vary independently at different pH's. At extreme pH or temperature, enzymes are **denatured**.

Optimum temperature (Figure 2-1): the temperature at which the enzyme activity is at its maximum. This varies according to conditions such as salt content, pH, etc. But enzymes from different tissues of the same organism do not necessarily have the same temperature profile. The rate of most enzymatic reactions about doubles for each $10°C$ rise in temperature.

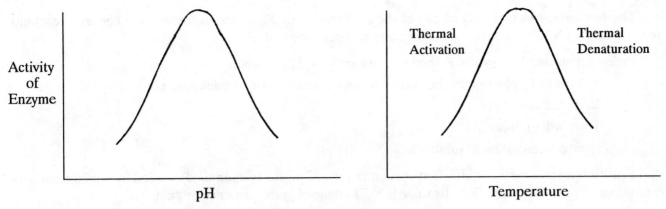

Figure 2-1. Enzyme Activity as a Function of pH and Temperature.

Saturability. A maximum velocity is not exceeded even in the presence of excess substrate.

Reaction velocity is directly proportional to enzyme concentration, provided there is enough substrate. The enzymatic reaction continues until substrate is exhausted.

Substrate specificity can be high or even absolute (one particular compound). Specificity can apply broadly to a class of compounds sharing a type of linkage, steric structure (cis-trans), or optical activity (D,L).

Regulation of activity is by **feedback inhibition**, altered **availability of substrate**, or altered **kinetic parameters**.

II. ENZYME KINETICS

A. Basic Principles

Energetics of catalysis (Figure 2-2). The rate of the reaction depends on the number of activated molecules in the **transition state** (activated complex A - - - B). The **free energy of activation** (E_A, or $\Delta G^{\ddagger}$ or $\Delta F^{\ddagger}$) is the amount of energy which must be put into the system to reach the activated transition state. A catalyst forms a transition complex with a **lower E_A(cat)**. Since an enzyme lowers the E_A for both the forward and back reactions, the velocity for both reactions is faster and equilibrium is achieved sooner.

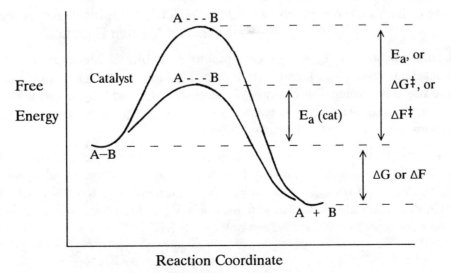

Figure 2-2. Energetics of Enzyme Catalysis.

The **free energy change** (ΔG or ΔF) of the reaction is the difference in free energy between reactants and products. It is NOT changed in the presence of a catalyst.

Order of reaction. If S $\rightleftharpoons$ P, then the forward v $= k[S]^R$, where

v = velocity of reaction (Forward v = net v when there is no back reaction.)

k = rate constant

R = order of reaction

[S] = concentration of substrate.

Equilibrium constant — equilibrium concentrations of products multiplied, divided by equilibrium concentrations of reactants multiplied. Its value is NOT changed in the presence of a catalyst:

$$K_{eq} = \frac{[P_1] \cdot [P_2] \cdot \ldots}{[A_1] \cdot [A_2] \cdot \ldots}$$

B. Kinetics

Michaelis-Menten Equation

Given the reaction:

$$E + S \underset{k_2}{\overset{k_1}{\rightleftharpoons}} ES \overset{k_3}{\longrightarrow} E + P$$

then $v = \dfrac{V_{max}(S)}{K_M + (S)}$

v (Figure 2-3) is the **initial velocity** of the reaction at time essentially 0, when there is no P. Therefore the back reaction ES ← E + P can be disregarded.

V_{max} is the **maximal velocity** achieved when the enzyme is saturated with substrate. It is proportional to the concentration of enzyme, and measures the catalytic efficiency of the enzyme: the bigger the V_{max} the more efficient the enzyme.

where:

E = enzyme

S = substrate

ES = enzyme-substrate complex

P = product

k_1, k_2, k_3 = rate constants

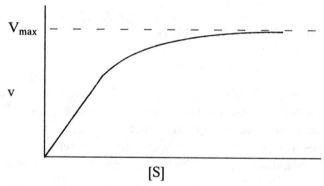

Figure 2-3. Graphical Representation of the Michaelis-Menten Equation.

K_M, the **Michaelis constant**, or $(k_2 + k_3)/k_1$, is equal to the substrate concentration at which the reaction rate is half its maximal value (V_{max}), and is in units of moles/liter. A high K_M indicates weak binding between enzyme and substrate; when dissociation of ES complex to E and P is the rate-limiting step (i.e., k_1, $k_2 \gg k_3$), K_M becomes the dissociation constant of ES. K_M indicates the concentration range where changing concentrations of substrate have decreasing effect on v.

Lineweaver-Burk plot (Figure 2-4). When the reciprocal of velocity is plotted against the reciprocal of substrate concentration, the graph is in the form of y = b + xm, and the slope m and y-intercept b are easily obtained. They are K_M/V_{max} and $1/V_{max}$, respectively. The x-intercept is $-1/K_M$. Note that the intercept on the 1/v axis must be in the same units as the axis and must be $1/V_{max}$. Likewise, the intercept on the 1/[S] axis must be minus, in the same units as the axis, and can only be $-1/K_M$.

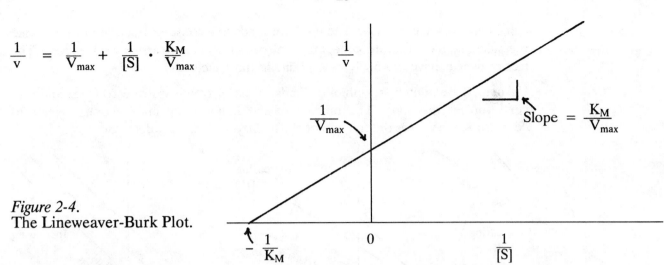

$$\frac{1}{v} = \frac{1}{V_{max}} + \frac{1}{[S]} \cdot \frac{K_M}{V_{max}}$$

Figure 2-4.
The Lineweaver-Burk Plot.

C. Enzyme Inhibition

There are two major types: irreversible and reversible. Drugs are designed to inhibit specific enzymes in specific metabolic pathways.

1. *Irreversible inhibition* involves destruction or covalent modification of one or more functional groups of the enzyme. Examples:

   –Diisopropylfluorophosphate and other fluorophosphates bind irreversibly with the -OH of the serine residue of acetylcholine esterase.

   –Para-chloromercuribenzoate reacts with the -SH of cysteine.

   –Alkylating agents modify cysteine and other side chains.

   –Cyanide and sulfide bind to the iron atom of cytochrome oxidase.

   –Fluorouracil irreversibly inhibits thymidine synthetase.

   –Aspirin acetylates an amino group of the cyclooxygenase component of prostaglandin synthase.

2. *Reversible inhibition* (Figure 2-5) is characterized by a rapid equilibrium of the inhibitor and enzyme, and obeys Michaelis-Menten kinetics. K_I, the dissociation constant of the complex E–I, is a measure of the effective concentration of the inhibitor, whether a drug or poison. There are three major types of reversible inhibition:

   ♦ **Competitive** inhibition: resembling the substrate, the inhibitor competes with it for binding to the active site of the enzyme. A competitive inhibitor at a fixed concentration can be overcome by increased substrate concentration. Hence V_{max} is not affected, only K_M. (See figure 2-5).

Examples:	Inhibitor	Enzyme Inhibited
	malonate	succinate dehydrogenase
	sulfanilamide	dihydropteroate synthetase
	methotrexate	dihydrofolate reductase
	allopurinol	xanthine oxidase

   ♦ **Noncompetitive** inhibition: inhibitor does not resemble substrate, binds to the enzyme at a locus other than the substrate binding site. Examples: heavy metal ions — silver, mercury, lead, etc. Metalloenzymes are inhibited by metal-chelating agents that bind metal cofactors, e.g. EDTA. The inhibitor binds to a constant fraction of the enzyme regardless of the substrate concentration so that V_{max} appears to be lowered. The free enzyme behaves normally so K_M remains the same.

♦ **Uncompetitive** inhibition: inhibitor binds to the enzyme-substrate complex and prevents further reaction. The binding of the inhibitor by mass action appears to decrease K_M as well as the fraction of active enzyme, i.e., V_{max} decreases also.

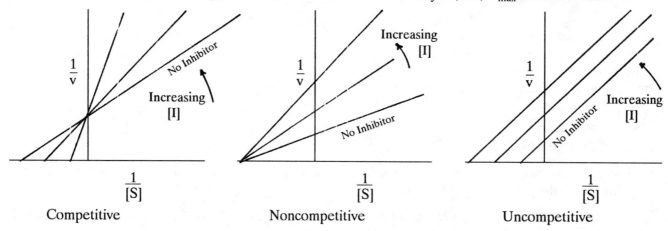

Figure 2-5. Lineweaver-Burk Plots Showing the Effect of Inhibitor in Three Types of Enzyme Inhibition.

D. Active Site

The **active site** of an enzyme is the region that binds the substrate and contributes the amino acid residues that directly participate in the making and breaking of bonds. It is a three-dimensional entity having nonpolar clefts or crevices. These contain two types of amino acid residues: the **contact** (or catalytic) and the **auxiliary** amino acids. A substrate can induce a conformational change in the active site. This is known as the induced-fit model.

E. Catalytic Efficiency

A number of factors contribute to the efficiency of enzymes —

–Catalysis by distortion: a conformational change occurs in the substrate.

–A conformational change is induced in the enzyme.

–Covalent catalysis: a highly reactive covalent intermediate is formed.

–Acid and base catalysis.

–Proximity and orientation effect.

III. EFFECTS OF KINETIC PARAMETERS ON ENZYME ACTIVITY

A. Allosteric Enzymes

General properties. Enzyme activity is modulated through the noncovalent binding of a specific metabolite (allosteric effector, modulator or modifier) to the protein at a site **other than** the catalytic site (*allo* = other).

All known allosteric enzymes are oligomeric, i.e., they have 2 or more polypeptide **subunits**, often four. Binding of the modulator to the allosteric site affects the binding of substrate to the catalytic site by changing the **quaternary structure** of the allosteric enzyme. The effect can be either **positive**, i.e., to increase the binding of substrate, or **negative**, i.e., to decrease the binding of substrate.

Regulation frequently occurs at the first, or **committed step** of a metabolic pathway, or at a **branch point**, with the final product of the pathway as a negative effector. This is end-product or feedback inhibition.

Examples:	Enzyme	Allosteric Effector
	homoserine dehydrogenase	threonine (–)
	homoserine succinylase	methionine (–)
	threonine deaminase	isoleucine (–)
	aspartate transcarbamoylase	cytidine triphosphate (–)
		adenosine triphosphate (+)
	phosphofructokinase	fructose-6-phosphate (+)
	pyruvate carboxylase	acetyl-CoA (+)

Kinetics (Figure 2-6). Allosteric enzymes do not follow Michaelis-Menten kinetics. A **sigmoidal** rather than hyperbolic curve is obtained when reaction rate is plotted against substrate concentration. This kinetic behavior is analogous to the binding of oxygen to hemoglobin. (Oxygen binding to myoglobin follows Michaelis-Menten kinetics instead.) The sigmoidal shape of the curve is caused by positive cooperativity, i.e., binding of the first substrate molecule enhances the binding of subsequent molecules at other sites.

An activator has the effect of shifting the curve to the left; an inhibitor shifts it to the right. Thus the sigmoidal response of allosteric enzymes facilitates a more vigorous control of enzyme activity compared to enzymes displaying Michaelis-Menten kinetics. The greater control results from the steeper activity curves of cooperative enzymes that produce bigger changes in activity with the same shift along the concentration axis.

Mechanism of the Regulatory Activity. There are two major models proposed for cooperative binding of allosteric enzymes:

♦ Symmetry-concerted model. Enzymes are generally composed of 2 or more identical subunits arranged in a symmetrical manner. If there is a change in conformation of one subunit, all other subunits must change their conformation in a concerted way to preserve symmetry.

♦ Sequential model. The binding of substrate changes the shape of the subunit to which it is bound. However, the conformation of the other subunits in the enzyme molecule are not appreciably altered. The con-

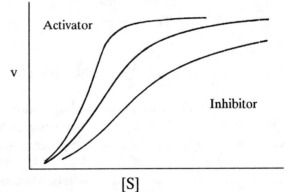

Figure 2-6. Allosteric Enzyme Kinetics.

formational change caused by the binding of substrate in one subunit can increase or decrease the substrate binding affinity of the other subunits in the same enzyme molecule.

B. Covalently Modified Regulatory Enzymes

Some regulatory enzymes may be modified by phosphorylation. Covalent binding of modifier (phosphate) to the enzyme changes its activity. Example: phosphorylation of active glycogen synthetase turns the enzyme into the inactive form, while phosphorylation of inactive glycogen phosphorylase turns the enzyme into the active form.

IV. GENERAL ASPECTS

A. Isozymes

Isozymes are enzymes that catalyze the **same reactions** and have similar molecular weights but **differ in subunit composition** and physical chemical properties. They may also differ in K_M, V_{max}, optimal temperature and pH, or substrate specificity. They often contain **multiple polypeptide subunits** of 2 or more types.

An example is *L-Lactate Dehydrogenase* (LDH). It is a tetramer of 2 types of subunit, M (muscle) and H (heart). Five isozymes occur: H_4, H_3M, H_2M_2, HM_3, and M_4. Various mixtures of isozymes are found in different tissues. H_4 occurs predominantly in cardiac tissue, and M_4 in skeletal muscle and liver.

B. Medical Aspects of Enzymology

As diagnostic tools. Enzymes or isozymes normally found intracellularly in various organs can be used as indicators of organ damage when they are found in blood. These enzymes, although always present in blood at low levels, are elevated far above normal in pathological conditions. Some examples are listed below (Table 2-1).

Condition	Enzymes with Elevated Levels in Blood
Myocardial Infarction	glutamic-oxaloacetic transaminase (SGOT) lactic dehydrogenase — H_4 and H_3M isozymes creatine phosphokinase
Bone disease	alkaline phosphatase
Obstructive liver disease	sorbitol dehydrogenase lactic dehydrogenase M_4 and M_3H isozymes
Prostatic cancer	acid phosphatase
Acute pancreatitis	amylase
Muscular dystrophy	aldolase glutamic pyruvate transaminase (SGPT)

Table 2-1. Some Enzymes Useful in Diagnosis.

As Laboratory Reagents. In **simple assays**, enzymes may be used for accurate determination of small quantities of blood constituents. In **coupled assays**, combinations of enzymes are often used to measure concentrations of specific substrates, coenzymes, or products. For example, an enzyme reaction may be coupled to the conversion of NADH to NAD^+, the removal or production of which can be followed easily by measuring the absorbance of the solution spectrophotometrically.

If the assay is for substrate (or product), then enzyme, coenzymes, etc. must be in excess; if the assay is for enzyme, then substrate, etc. must be in excess.

V. REVIEW QUESTIONS ON ENZYMES

DIRECTIONS: For each of the following multiple-choice questions (1 - 7), choose the ONE BEST answer.

The following plot accompanies ***Questions 1 - 3:***

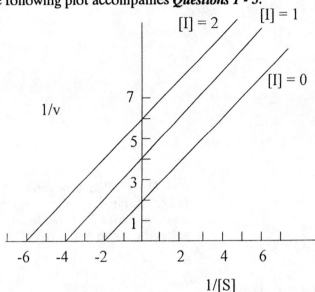

1. What type of inhibitor produces the pattern in the plot above?

A. competitive inhibitor
B. denaturant
C. negative effector
D. noncompetitive inhibitor
E. uncompetitive inhibitor.

Ans E: Uncompetitive because both K_M and V_{max} are affected.

2. What is the value of K_M of the uninhibited enzyme obtained from the plot?

A. 0.25
B. 0.5
C. 1
D. 2
E. 4.

Ans B: The value of the intercept on the 1/[S] axis is –2 and the negative reciprocal is 0.5.

3. What is the value of V_{max} of the uninhibited enzyme obtained from the plot?

A. 0.25
B. 0.5
C. 1
D. 2
E. 4.

Ans B: The value of the intercept on the 1/v axis is 2 and the reciprocal is 0.5.

Questions 4 - 6: If curve 3 represents a cooperative enzyme under normal conditions, which curve would represent the enzyme after changing the condition in each question but with all else held constant?

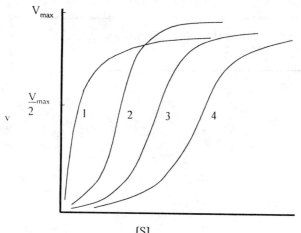

4. Increased concentration of CTP in aspartate transcarbamoylase reaction mixture.

A. 1
B. 2
C. 3
D. 4
E. remains at the same position but is reduced in height

Ans D: CTP is a negative, feedback inhibitor of the committed step in pyrimidine synthesis, and shifts the curve to the right.

5. Increased concentration of acetyl-CoA in pyruvate carboxylase reaction mixture.

A. 1
B. 2
C. 3
D. 4
E. remains at the same position but is reduced in height

Ans B: Acetyl-CoA is a positive effector of the reaction, and so shifts the curve to the left.

6. Complete loss of cooperativity.

A. 1
B. 2
C. 3
D. 4
E. remains at the same position but is reduced in height

Ans A: On loss of cooperativity the enzyme would behave as a Michaelis-Menten enzyme, with a hyperbolic curve.

7. All of the following processes are ways by which temperature may affect enzymic reactions EXCEPT:

A. promotes formation of transition state
B. reduces the rate as the temperature is lowered from physiological temperature down to the freezing point
C. increases the rate as the temperature is increased above physiological temperature up to the boiling point
D. raises the energy level of substrates
E. raises the energy level of products.

Ans C: The rate increases with temperature until the enzyme denatures, almost always well before the boiling point of water.

DIRECTIONS: For each set of questions, choose the ONE BEST answer from the lettered list above it. An answer may be used one or more times, or not at all.

Questions 8 - 11:

    A. defined by the energy levels and concentrations of substrates and products prior to reaction
    B. dependent on the kinetic energies of substrates
    C. mimicked by substrate at active site
    D. obtained earlier with enzyme present
    E. reduced by enzymic activity

8. Transition state.

Ans C: The substrate, when bound at the active site looks more like the transition state and reacts more readily, a part of what an enzyme does to lower the energy of activation.

9. Equilibrium.

Ans D: Catalysis occurs in both forward and back directions so equilibrium is reached more quickly with the same concentrations of reactants and products as with the non-catalyzed reaction.

10. Energy of activation.

Ans E: Enzymes lower the energy of activation so the reaction occurs more readily.

11. Free energy change for reaction.

Ans A: The free energy of the reaction is set by the energy levels of all the reactants and products and is independent of whether the reaction is catalyzed or not.

Questions 12 - 15:

    A. k_2
    B. k_3
    C. $K_{equilibrium}$
    D. K_I
    E. K_M

12. Unrelated to properties of an enzyme.

Ans C: Equilibrium concentrations are independent of catalysis.

13. Related to doses of some drugs.

Ans D: K_I indicates the concentration of inhibitor required to half-saturate the enzyme. For a drug to be effective its concentration in the presence of the enzyme must be at least in this concentration range.

14. Related to allosteric effector concentrations.

Ans E: The apparent K_M is decreased by positive effectors and increased by negative effectors; these statements are equivalent to stating that the v versus [S] curves are shifted to the left and to the right, respectively.

15. Related to a reduction in the energy of activation.

Ans B: The rate constant for the reaction that actually converts S to P is k_3, the catalytic rate constant. This constant is related to how well the enzyme reduces the energy of activation that leads to the catalysis.

Questions 16 - 19:

    A. equilibrium
    B. initial velocity
    C. saturation
    D. $[S] = K_M$
    E. $[S]$ much less than K_M

16. Forward velocity can not be increased.

Ans C: Increasing the concentration of substrate does not increase the forward rate when the enzyme is already saturated, but would if added to the reaction under initial or equilibrium conditions if the enzyme were not saturated.

17. Forward velocity is half maximum velocity.

Ans D: The forward velocity is half the maximum velocity when $[S] = K_M$ because the enzyme is half saturated at that concentration.

18. When $[S]/[P]$ is at a maximum.

Ans B: The initial velocity must be measured when little or no product is present and represents the reaction at "zero" time. Initial velocities are necessary so that the back reaction can be ignored in deriving the Michaelis-Menten equation which represents only the forward velocity of the reaction.

19. When $[S]/[P]$ is constant.

Ans A: The reaction will continue until enough substrate is converted to product so that the back reaction just equals the forward reaction, i.e., equilibrium. Since no more product will be produced, this is the point at which the ratio will be constant.

3. CARBOHYDRATES

Albert M. Chandler

I. INTRODUCTION

Carbohydrate metabolism is the core of intermediary metabolism, providing a large part of the energy requirements for the organism, short-term storage of energy in the form of glycogen, and carbon skeletons for biosynthesis of other compounds. Amino acids and components of the citric (tricarboxylic) acid cycle feed into the pathways of carbohydrate metabolism. The following material is a concentrated essence of carbohydrate metabolism organized to show the pathways and their regulation.

II. STRUCTURAL ASPECTS OF CARBOHYDRATES

A. Major Families: Aldoses and Ketoses (Table 3-1).

	Aldoses	Ketoses
3C	Glyceraldehyde	Dihydroxyacetone
4C	Erythrose	Erythrulose
	Threose	
5C	Ribose	Ribulose
	Arabinose	Xylulose
	Xylose	
	Lyxose	
6C	Allose	Psicose
	Altrose	Fructose
	Glucose	Sorbose
	Mannose	Tagatose
	Gulose	
	Idose	
	Galactose	
	Talose	

Table 3-1. Aldoses and Ketoses through Six Carbons.

Aldoses. The parent compound is **glyceraldehyde**. C-2 is an asymmetric carbon, therefore, the OH group can be on the left or the right, forming two families of related but distinct D and L sugars. There is a total of 30 aldoses, 15 of the D-form and 15 of the L-form. Those aldoses most commonly seen in metabolism or in biological structures are **all of the D-form** and include **glyceraldehyde**, **erythrose**, **ribose**, **glucose**, **mannose** and **galactose**.

Ketoses. the parent compound is **dihydroxyacetone** (DHA). Since the 2-C of DHA is not asymmetric, DHA is not optically active. Only when the asymmetric C-3 is added to make **erythrulose** does the compound become optically active. Thus there are fewer ketoses than there are aldoses. **D-Ribulose** and **D-xylulose** are found in the pentose phosphate pathway. **D-Fructose** is also an important ketose.

With the exception of DHA, all of these sugars are optically active, that is, in solution they turn the plane of polarized light either clockwise or counterclockwise (+ or −). There is **NOT** necessarily a correlation between the absolute configuration (D or L) and the direction of optical activity (+ or −).

B. Cyclization of Sugars (Figure 3-1)

An alcohol can react with an aldehyde to form a **hemiacetal**. When the aldehyde group of a hexose forms an internal hemiacetal with its C-5 hydroxyl, the cyclic compound formed is call a **pyranose**.

Ketones can also react with alcohols to form a **hemiketal**. In a pentose the C-2 ketone can form an internal hemiketal with the C-5 hydroxyl to form a **furanose**. In a hexose, it can also react with the C-6 hydroxyl to form a **pyranose**.

When sugars are depicted as their Haworth projections, they can be looked at "edge on". The formation of a hemiacetal or hemiketal creates a new asymmetric center (anomeric carbon) at C-1 or C-2. The OH group produced can be either above or below the ring. If it is projected **below** the ring it is in the α-**position**; **above** the ring, the β-**position**.

In solution, the linear and ring forms are in equilibrium. Which form predominates depends upon the sugar involved. Free fructose prefers to form pyranose rings and this is its predominant form; its derivatives produce furanose rings.

The α and β forms of these sugars are also in equilibrium. Glucose in solution, for example, is in an equilibrium mixture of about 30% α and 70% β, with the open form less than 1%. The form that predominates is sugar specific. There are enzymes that accelerate the changes between α and β forms (**mutarotases**).

Conformation of the rings. The ring forms of the pyranoses and the furanoses are not planar. They form "boat" or "chair" forms in the case of pyranoses and "envelope" forms for furanoses. This is important for recognition purposes on the carbohydrate side chains of glycoproteins and glycolipids.

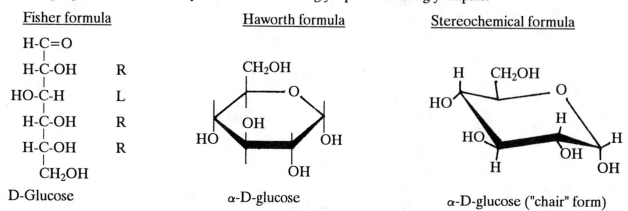

Figure 3-1. Conformation of Glucose, Depicted in Different Ways.

C. Sugar Derivatives

Glycosides are formed by splitting out a hydroxyl group between an alcohol and the -OH of a hemi-acetal to form an **acetal**. This can be done either chemically or enzymatically. The bond formed is called a glycosidic bond and in this case is an O-glycosidic bond. N-glycosidic bonds can also be formed. The structures shown (Figure 3-2) are methylglycosides, α below the ring, β above.

This is a common way for sugars to link together. Cellobiose, for example, is a disaccharide composed of two glucoses linked (1→4) in a β linkage. It is a glycoside. If we string many glucoses together in this exact way we have cellulose.

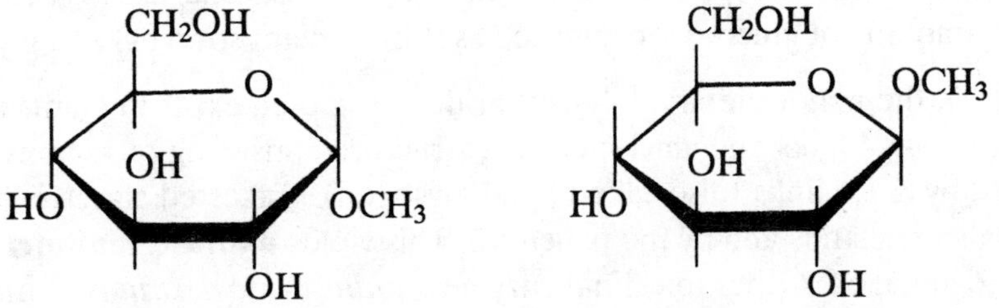

Figure 3-2.
Sugar Glycosides.   Methyl-α-D-glucoside   Methyl-β-D-glucoside

Adenosine contains an example of an **N-glycosidic** bond. All nucleosides are made up of a base linked to either ribose or deoxyribose in β-N-glycosidic linkages (See Chapter 11).

The most common disaccharides of interest are shown in Figure 3-3. Sucrose (table sugar) [glucose-fructose (α-1,2)]; lactose (milk sugar) [galactose-glucose (β-1,4)]; and maltose [2 glucose (α-1,4)].

Note that lactose can have both α and β anomers, depending upon the configuration of the glucose, while sucrose cannot have anomers because it lacks an anomeric carbon atom.

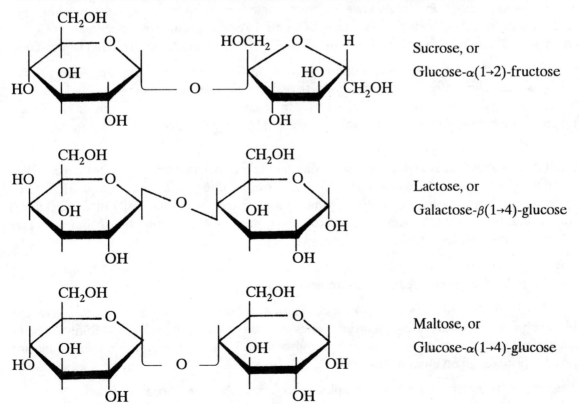

Sucrose, or
Glucose-α(1→2)-fructose

Lactose, or
Galactose-β(1→4)-glucose

Maltose, or
Glucose-α(1→4)-glucose

Figure 3-3. The Most Common Disaccharides.

Sugars are referred to as reducing or non-reducing sugars. This arises from the fact that aldehydes or ketones in alkaline solutions are capable of reducing Cu^{++} to Cu_2O, forming a reddish precipitate. If the C-1 carbon of an aldose is tied up in a glycosidic link it cannot act as a reductant. With regard to glycogen chains, therefore, one can refer to their reducing or non-reducing ends.

III. DIGESTION

Carbohydrates are present in the diet as monosaccharides, disaccharides, and polysaccharides. The greatest amounts of glucose are ingested as starch (plant sources) and glycogen (animal sources).

Starch, the major dietary polysaccharide in the diet, exists in plants in two forms. Amylose, a long linear polymer of α-1,4 links and amylopectin, a structure consisting of regions of α-1,4 linked residues periodically branched by α-1,6 links (also called plant glycogen). Ingested starch is broken down by *α-amylases* secreted by the salivary glands and by the pancreas. This yields **maltose**, **maltotriose** and **oligosaccharides** which contain α-1,6 linkages. Other intestinal enzymes, *maltase* and *dextrinase*, complete the hydrolysis to glucose.

Disaccharides are cleaved in the small intestine by disaccharidase enzymes of varying specificities to monosaccharides. Examples include *maltase*, which cleaves maltose to **glucose**, *lactase*, which cleaves lactose to **glucose** and **galactose** and *sucrase* which cleaves sucrose to **glucose** and **fructose**. The resulting monosaccharides are then absorbed through the intestine into the bloodstream. There are several genetic diseases resulting from a deficiency in one of the disaccharidases, the most common one being **lactase deficiency**.

Lactase Deficiency. While infants of all races possess sufficient lactase in the intestine, adults of many races lack sufficient lactase, thereby giving rise to **lactose intolerance**. Intolerant individuals who consume more than a few ounces of milk will experience diarrhea and intestinal gas caused by microbial fermentation of lactose in the gut. Lactase persists in adults whose ancestors came from parts of the world with a long history of consuming milk, such as Northern Europe, parts of Central Asia, and parts of Africa. Most Blacks and Orientals are lactose-intolerant, as are significant proportions of Semitic and Mediterranean Caucasians.

Glycogen. Monosaccharides, especially glucose, are a readily available source of energy. Free glucose molecules cannot be stored efficiently in cells; therefore, they are condensed together as highly branched, insoluble structures; glycogen in animals, starch in plants. In glycogen the glucose molecules are linked together by α-1,4 and α-1,6 bonds. Glucose is freed from glycogen by the combined action of phosphorylase and debranching enzymes.

Cellulose. Cellulose is the most abundant polysaccharide in the world. *Animals do not synthesize cellulases (enzymes capable of hydrolyzing β-1,4 links) and, therefore, cannot utilize cellulose directly.* High fiber diets are mainly cellulose. Termites harbor protozoa which produce cellulases and easily digest cellulose. Animals with four stomachs (ruminants) support bacteria which are capable of digesting cellulose. The ruminants, therefore, can use cellulose as a source of energy.

Glycoproteins. Glycoproteins are proteins which have oligosaccharide side chains covalently attached to the polypeptide backbone. (See the chapter on Protein Synthesis).

Glycolipids. Glycolipids are lipid molecules to which oligosaccharides are covalently linked. They are usually found in membranes with the carbohydrate moieties projecting out into the surrounding environment. Many cell-surface antigens or recognition sites are glycolipids. The oligosaccharide side chains are similar in structure to those found on glycoproteins.

Glycosaminoglycans. These compounds are complex polysaccharides found primarily in connective tissues where they form covalent links to proteins. The protein-carbohydrate compounds are called **proteogly-**

cans and are about 95% polysaccharide and 5% protein. Glycosaminoglycans are composed of multiple repeating units of disaccharides. One of the sugars is an amino sugar, the other is often a uronic acid. Some examples are shown in Table 3-2.

Several diseases exist that are the result of the deficiency of one or more enzymes responsible for the degradation of glycosaminoglycans.

Compound	Sugars
Chondroitin-6-sulfate	Glucuronic acid, GalNAc
Keratan sulfate	Galactose, GlcNAc
Heparin	Iduronic acid, Glucuronic acid, GlcN-SO$_4$
Dermatan sulfate	Iduronic acid, GalNAc
Hyaluronic acid	Glucuronic acid, GlcNAc

Table 3-2. Some Glycosaminoglycans.

The role of these compounds is linked to their chemistry and structure. The repeated, closely spaced negative charges at physiological pH tend to cause the molecules to form stiff rods which form viscous solutions. They also bind divalent cations readily and tend to be highly hydrated and can act as lubricating agents.

IV. GLYCOLYSIS

In the text that follows, many abbreviations are used, summarized in Table 3-3 below.

Abbreviation	Name of Intermediate	Abbreviation	Name of Intermediate
1,3-BPG	1,3-Bisphosphoglyceric acid	G-3-P	Glyceraldehyde 3-phosphate
2,3-BPG	2,3-Bisphosphoglyceric acid	G-6-P	Glucose 6-phosphate
2-PG	2-Phosphoglyceric acid	Gal-1-P	Galactose 1-phosphate
3-PG	3-Phosphoglyceric acid	OAA	Oxaloacetate
DHAP	Dihydroxyacetone phosphate	PEP	Phosphoenolpyruvate
E-4-P	Erythrose 4-phosphate	R-5-P	Ribose 5-phosphate
F-1,6-bisP	Fructose 1,6-bisphosphate	S-7-P	Sedoheptulose 7-phosphate
F-1-P	Fructose 1-phosphate	UDPG	Uridyl diphosphoglucose
F-2,6-bisP	Fructose 2,6-bisphosphate	UDPGal	Uridyl diphosphogalactose
F-6-P	Fructose 6-phosphate	X-5-P	Xylulose 5-phosphate

Table 3-3. Abbreviations of Some Common Intermediates

Monosaccharides released by intestinal degradation of starches and other complex carbohydrates are rapidly absorbed in the small intestine. There are special transport systems for them. Glucose "carriers" in the plasma membranes of intestinal cells bind both glucose and sodium ion. Both are transported across the membrane; Na$^+$ down its concentration gradient, glucose against. ATP is utilized to pump Na$^+$ out again exchanging it for K$^+$. This is an example of **active transport**. The glucose and fructose carriers are rapidly saturated (see chapter on Membranes). The monosaccharides enter the bloodstream and are rapidly carried throughout the organism where they come into contact with all cells of the body. By far the most important sugar in terms of abundance is **glucose**, which may require insulin for uptake by some tissues (Table 3-4).

Readily permeable	Require Insulin
liver, kidney	muscle (all types)
brain	adipose tissue
lens, retina	leukocytes
erythrocytes	mammary gland

Table 3-4 Entry of Glucose into Cells

When the glucose molecule enters the cell it is acted upon by the **Glycolytic Pathway (Embden-Meyerhof Pathway)**. This pathway converts glucose to pyruvate (lactate) giving a net yield of 2 moles of ATP for every mole of glucose processed. The process is called **glycolysis**.

A. Overview of Glycolysis

1. All intermediates after glucose are phosphorylated and remain trapped in the cell. The membrane is quite permeable to the end-product, lactate, which can easily go in and out.

2. Glucose is broken into two trioses which are <u>interconvertible</u>.

3. ATP is involved 4 times; NAD^+ once.

4. The key rate-limiting enzyme and the major point of control is *phosphofructokinase (PFK)*. Control points also exist at *hexokinase* and *pyruvate kinase*.

B. Reactions of Glycolysis (Figure 3-4)

Hexokinase (1 [numbers refer to enzymatic steps in figure 3-4]), an enzyme with a high affinity for glucose (low K_M), catalyzes the phosphorylation of the hydroxyl at C-6. In liver and adipose tissue only, a second enzyme exists called *glucokinase*. It has a low affinity (high K_M) and comes into play when the blood levels of glucose are high and the capacity of hexokinase is saturated; that is, when conditions are favorable for storing excess energy as glycogen and fat. Glucokinase is inducible and the induction is insulin-dependent. The product of either kinase, **glucose-6-phosphate (G-6-P)**, can enter into several different pathways. (For abbreviations, see Table 3-3).

Phosphoglucose isomerase (2) converts G-6-P (an aldose) to **fructose-6-phosphate** **(F-6-P)** (a ketose). This is a freely reversible reaction.

Phosphofructokinase, or *PFK* (3) catalyzes: F-6-P + ATP $\longrightarrow$ **F-1,6 bisphosphate**

This is a complex, multisubunit enzyme subject to control by allosteric regulation and phosphorylation. It is also inducible. In the older literature the product was referred to as a diphosphate but the more accurate term is bisphosphate.

Aldolase (4) catalyzes a freely reversible reaction, splitting F-1,6-bisP into two 3-carbon units, **dihydroxyacetone phosphate (DHAP)** and **glyceraldehyde-3-phosphate (G-3-P)**. These trioses are closely related to one another, one being a ketose and the other an aldose. They are readily interconverted by the enzyme *triose phosphate isomerase* (5). The strategy here is to convert all of the DHAP over to G-3-P which is the triose that is metabolized further.

Glyceraldehyde 3-phosphate Dehydrogenase (6) catalyzes an oxidation-reduction step. The first ATP generated by glycolysis ultimately comes from this reaction. G-3-P interacts with NAD^+ and inorganic phosphate to form **1,3-bisphosphoglyceric acid (1,3-BPG)** and **NADH + H^+**. Note that 1,3-BPG is a mixed

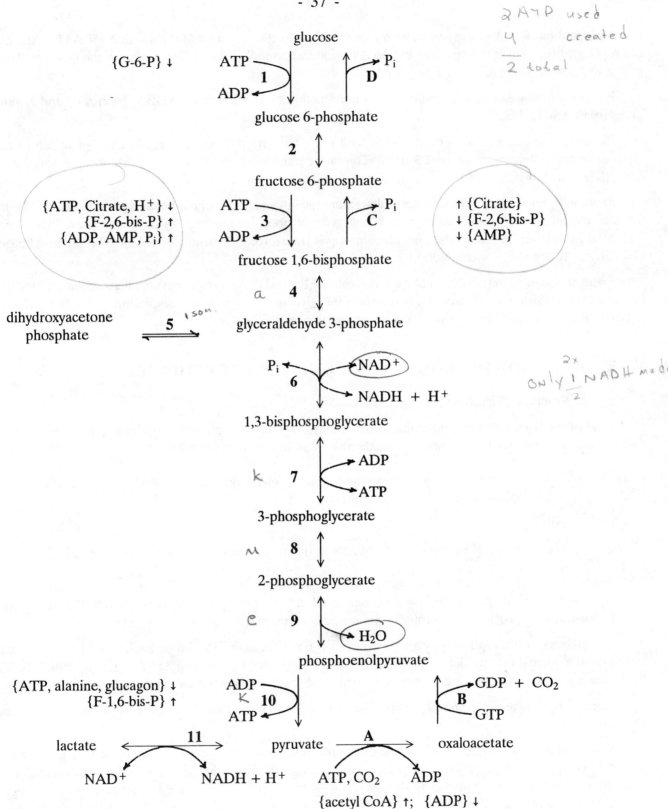

Figure 3-4. Glycolysis - Gluconeogenesis. Letters and numbers in boldface identify enzymes; see correspondingly numbered enzymes in Section IV B of text. A - D are enzymes of gluconeogenesis; see Section IX. Items in curved brackets refer to regulatory effects: ↑ = stimulatory; ↓ = inhibitory.

anhydride and has a high-energy bond with a $\Delta G°'$ of hydrolysis even greater than that for ATP. Arsenite can replace phosphate in this reaction but no ATP will be formed. The NADH can enter the electron transport system to form an additional 3 ATP under aerobic conditions.

Phosphoglycerate Kinase (7) then catalyzes the reaction of 1,3-BPG with ADP to form **ATP** and **3-phosphoglyceric acid (3-PG)**.

Phosphoglyceromutase (8) converts 3-PG to **2-PG**. This interconversion requires the presence of catalytic amounts of the intermediate, **2,3-BPG**. This latter compound also plays an important regulatory role in oxygen transport by hemoglobin. (See chapter on Amino Acids and Proteins).

Enolase (9) converts 2-PG into a second high-energy intermediate, **phosphoenolpyruvate (PEP)**. Dehydration instead of oxidation creates a high energy enol phosphate group. Resonance states have been increased in number and when the phosphate group is transferred, the enol converts to the favored ketone state driving the reaction to completion.

Pyruvate Kinase (10) transfers a phosphate from PEP to ADP to form **pyruvate** plus a second **ATP**. This last reaction is essentially irreversible and drives the glycolytic pathway to completion. The formation of pyruvate is considered to be the end point of the glycolytic pathway.

V. ENTRY OF OTHER HEXOSES INTO THE GLYCOLYTIC PATHWAY

A. Fructose (Figure 3-5)

Most of the fructose in the diet comes from the hydrolysis of sucrose obtained from cane or beet sugar. Sweeteners and some fruits also contain considerable quantities of free fructose.

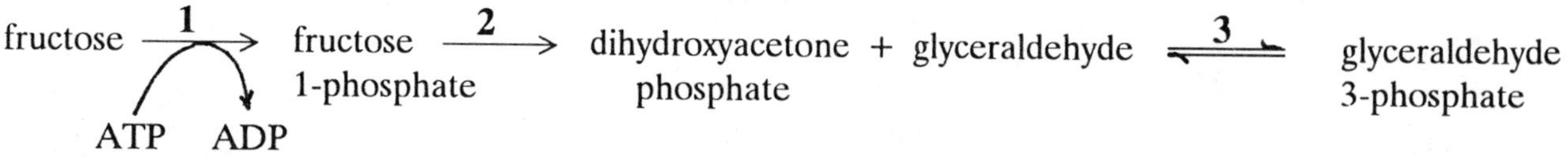

fructose $\xrightarrow{\quad 1 \quad}$ fructose 1-phosphate $\xrightarrow{\quad 2 \quad}$ dihydroxyacetone + glyceraldehyde phosphate $\underset{\quad}{\overset{3}{\rightleftharpoons}}$ glyceraldehyde 3-phosphate

(ATP → ADP)

Figure 3-5. Metabolism of Fructose. (Numbers in boldface identify enzymes described below).

Fructokinase (1). Fructose can be phosphorylated by hexokinase but the K_M is extremely high so this is probably not a significant reaction. Fructokinase, found in liver, kidney and intestine uses ATP to form **fructose-1-phosphate (F-1-P)**. This is probably the major entry point for fructose.

Fructokinase will not phosphorylate glucose. It is not affected by insulin or fasting and fructose is metabolized at a normal rate by diabetics. Because fructose bypasses PFK it is metabolized more rapidly in the liver than is glucose. A genetic deficiency of fructokinase leads to **essential fructosuria**, a benign disorder.

Fructose-1-phosphate Aldolase (2) (also called aldolase B, or aldolase 2) cleaves F-1-P into DHAP and **glyceraldehyde**.

A deficiency of aldolase B results in **hereditary fructose intolerance**, a disease characterized by fructose-induced hypoglycemia and liver damage. In the presence of fructose, the resulting high level of fructose 1-phosphate inhibits liver phosphorylase, stopping glucose production from glycogen (see glycogen metabolism). It also sequesters all of the cell's phosphate and virtually stops ATP synthesis in the liver, which apparently prevents the maintenance of normal ionic gradients and leads to osmotic damage to hepatocytes.

Triose Kinase (3) catalyzes the phosphorylation of glyceraldehyde by ATP. The G-3-P formed can either be converted to pyruvate or to glucose.

B. Galactose (Figure 3-6)

Galactose, obtained primarily from lactose in milk, is converted to **galactose-1-phosphate (Gal-1-P)** by *Galactokinase* (1, figure 3-6). ATP is the phosphate donor.

Galactose-1-phosphate Uridyltransferase (2) catalyzes the reaction between Gal-1-P and **uridyl diphosphoglucose (UDPG)** to form **UDPGal** and **G-1-P**. This is a reversible exchange reaction: the UDP group is transferred from glucose to galactose, and the glucose 1-phosphate can be converted (*phosphoglucomutase*) to the glycolytic intermediate, glucose 6-phosphate.

UDPGal-4-epimerase (3) catalyzes the epimerization of UDPGal to UDPG, regenerating the UDP-glucose consumed by reaction (2), making a cycle. The reaction is reversible and requires NAD^+ as cofactor.

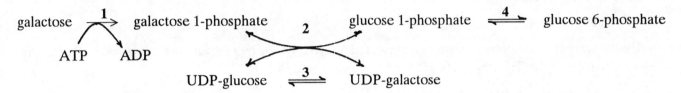

Figure 3-6. Metabolism of Galactose. (Boldface numbers identify enzymes described in text).

The UDPG formed from UDPGal can also enter the glycogen cycle and the glucose can be transferred to the non-reducing end of a glycogen molecule by glycogen synthase. When the glycogen is degraded the glucose is then converted to G-1-P by phosphorylase and then to G-6-P by phosphoglucomutase (4) and then on to pyruvate by the glycolytic pathway.

Galactosemia is an autosomal recessive trait in which Gal-1-P accumulates in the cell and is converted to galactitol which collects in the lens and causes cataracts. Other symptoms include failure to thrive, vomiting and diarrhea after drinking milk, liver damage, and mental retardation. Severe **galactosemia** results from a deficiency of the *uridyltransferase* which allows the accumulation of large amounts of galactose 1-phosphate. It can also be caused, in a relatively mild form, by a deficiency of *galactokinase*.

C. Glycerol

Glycerol is released from adipose tissue and liver primarily from the hydrolysis of triglycerides.

Glycerol kinase catalyzes the following reaction, but its activity is low in muscle and adipose tissue:

$$glycerol + ATP \longrightarrow glycerol\text{-}3\text{-}P + ADP$$

Glycerol phosphate dehydrogenase catalyzes the reaction:

$$glycerol\text{-}3\text{-}P + NAD^+ \longrightarrow dihydroxyacetone\ phosphate + NADH + H^+$$

The dihydroxyacetone phosphate can then enter the glycolytic pathway to be converted to pyruvate or can be converted to glucose via the gluconeogenic pathway.

VI. METABOLISM OF PYRUVATE

The pyruvate formed by glycolysis can go in three major directions.

A. Alcohol

In yeast and other microorganisms grown under anaerobic conditions the pyruvate is converted to alcohol. *Pyruvate decarboxylase* converts the pyruvate to **acetaldehyde** and CO_2. Then *alcohol dehydrogenase* reduces the acetaldehyde to **ethanol** using NADH as a cofactor. The NAD$^+$ formed from this reaction can then go back to the G-3-P dehydrogenase step to keep glycolysis going.

B. Lactate

Erythrocytes lack mitochondria, therefore their metabolism is purely anaerobic. Under prolonged, intense activity muscles are unable to obtain sufficient oxygen to convert all of the pyruvate formed to CO_2 and water via the tricarboxylic acid cycle and shift over to anaerobic metabolism. Under anaerobic conditions both muscle and erythrocytes convert the pyruvate to **lactate**.

Formation of lactate from pyruvate is catalyzed by *lactate dehydrogenase* which uses NADH as a cofactor. The function of this conversion is to regenerate NAD$^+$ which can then go back to oxidize G-3-P to form two more ATP molecules.

C. Acetyl CoA

Under aerobic conditions pyruvate is converted to **acetyl CoA** by *pyruvate dehydrogenase*. The acetyl CoA can then enter the Tricarboxylic Acid Cycle. (See Chapter 4).

VII. REGULATION OF GLYCOLYSIS

A. Phosphofructokinase

An important control point in glycolysis, this enzyme is allosterically inhibited by high concentrations of ATP and of citrate. Hydrogen ion is also inhibitory. When the enzyme is saturated, F-6-P accumulates; excess F-6-P is shunted though a second pathway to form **fructose-2,6-bisphosphate (F-2,6-bisP)**. F-2,6-bisP is a potent **allosteric activator** of PFK. The regulation of synthesis and degradation of this compound is itself very complex but is primarily governed by the levels of F-6-P and ATP. Because F-2,6-bisP acts at micromolar concentrations, it is now thought to be the most important compound regulating PFK and, therefore, glycolysis.

B. Hexokinase: High levels of G-6-P inhibit hexokinase allosterically.

C. Pyruvate kinase exists in three different forms (isozymes) in different tissues:

Liver (L-form): regulated by F-1,6-bisP (+) and ATP (–) and alanine (–). This form can also be phosphorylated (inactivated) and dephosphorylated (activated). Phosphorylation (inhibition) occurs when the cell is in the gluconeogenic mode.

Muscle (M-form): this enzyme is not phosphorylated because muscle does not carry out gluconeogenesis.

Other tissues (A-form): regulation by phosphorylation and dephosphorylation varies depending on the type of tissue.

D. 2,3-Bisphosphoglyceric acid (2,3-BPG)

In red blood cells the glycolytic step from 1,3-BPG to 3-PG can be bypassed. An additional enzyme, *bisphosphoglycerate mutase*, converts 1,3-BPG to **2,3-BPG**. 2,3-BPG can, in turn, be converted to 3-PG by *2,3-BPG phosphatase*. One ATP is lost when glycolysis goes by this route. The energy normally trapped as ATP is dissipated as heat. The 2,3-BPG can then combine with hemoglobin to assist in oxygenating the peripheral tissues. (See chapter on Amino Acids and Proteins).

VIII. PENTOSE PHOSPHATE PATHWAY

A second major route for G-6-P is through the **Pentose Phosphate Pathway** (Pentose Shunt, Hexose Monophosphate Shunt). The major functions for this pathway are to produce **NADPH** for reductive biosyntheses and **pentoses** used primarily for nucleic acid synthesis. The pathway also provides a mixture of 3, 4, 5, 6, and 7-C sugars which may be used as precursors for other compounds. (These are often abbreviated; see Table 3-4).

A. Oxidative Phase (G-6-P to Ribulose-5-P).

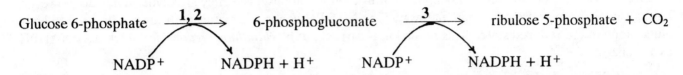

Three enzymes are involved:

G-6-P dehydrogenase (1) catalyzes the irreversible dehydrogenation of G-6-P to **6-P-gluconolactone**, forming an internal ester between C-1 and C-5. *This is the committed, rate-limiting step in the pathway.* The enzyme is highly specific for **NADP$^+$** and the level of this compound controls the rate of the reaction. High concentrations of NADPH will compete with NADP$^+$ binding and slow the reaction down.

Lactonase (2) hydrolyzes 6-P-Gluconolactone to **6-P-gluconic acid**.

6-P-gluconic acid dehydrogenase (3) also uses NADP$^+$ as an electron acceptor and converts 6-P-gluconate to **ribulose-5-P** and **CO$_2$**. The release of CO$_2$ drives the reaction to completion. For every G-6-P committed to this pathway, 2 NADPH, 1 CO$_2$ and 1 ribulose-5-P are obtained.

B. Non-oxidative Phase

Under most conditions the amount of pentoses formed is in excess of requirements; therefore, this excess is put back into glycolytic intermediates. This is accomplished by the non-oxidative phase of the pentose phosphate pathway, which converts 3 pentoses to 2 hexoses and 1 triose.

The non-oxidative branch consists of five reactions catalyzed by four enzymes, summarized as follows:

$$3 \text{ Ribulose 5-P} \xrightleftharpoons{1} 3 \text{ ribose 5-P} \xrightleftharpoons{2, 3, 4, 3a} 2 \text{ F-6-P } + 1 \text{ glyceraldehyde 3-P}$$

Phosphopentose isomerase (1) converts ribulose-5-P to **ribose-5-P (R-5-P)**. R-5-P is a key intermediate for the formation of nucleosides, nucleotides and nucleic acids.

Phosphopentose epimerase (2) can convert ribulose-5-P to **xylulose-5-P**.

Transketolase (3) transfers the top two carbons from **xylulose-5-P (X-5-P)** to R-5-P to form G-3-P and **sedoheptulose-7-P (S-7-P)**. The X-5-P comes from the epimerization of ribulose-5-P. Transketolase requires **thiamin pyrophosphate** as cofactor.

Transaldolase (4) transfers the top three carbons from S-7-P to G-3-P to form **erythrose-4-P (E-4-P)** + F-6-P.

Transketolase (3a) also transfers the top two carbons from X-5-P to E-4-P to form G-3-P again and another F-6-P.

NOTE: Since these enzymes catalyze freely reversible reactions it is possible to synthesize R-5-P without going through the oxidative phase of the pathway.

C. G-6-P Dehydrogenase Deficiency and Hemolytic Anemia

Red blood cells contain high concentrations of glutathione. In its reduced form (free -SH) glutathione acts as a protective scavenging agent against hydrogen peroxide and other strong oxidizing agents. If these agents are not neutralized they will oxidize hemoglobin iron to the ferric state and also oxidize membrane lipids. The hemoglobin will not carry oxygen and the membranes will become very fragile. Thus reduced glutathione helps to keep hemoglobin in the Fe^{++} state and helps to maintain red blood cell membrane integrity. It can be used up, however, and when fully converted to G-S-S-G it no longer functions as a protective agent.

Red blood cells do not have mitochondria, thus their metabolism is mainly glycolytic. They do, however, have a very active pentose phosphate pathway. The NADPH generated by this pathway can be used to keep glutathione in the reduced state. This reduction is carried out by *glutathione reductase*, which uses NADPH as a coenzyme.

A significant number of people have a sex-linked genetic defect that results in a deficiency in G-6-P dehydrogenase. It particularly affects blacks and like the sickle cell trait, apparently persists because the deficiency affords protection against the malarial organism. The erythrocytes of people with this deficiency have difficulty generating enough NADPH to keep the glutathione reduced. The red blood cells of these individuals are readily hemolyzed, particularly after exposure to certain drugs. These include antimalarials, aspirin and sulfonamides. Fava beans also contain a compound which is highly toxic to these cells. The hemolysis can be so great as to be fatal.

D. Adipose Tissue and Phagocytic Cells

The activity of the pentose phosphate pathway is very high in adipose tissue. A major purpose of this tissue is to synthesize fatty acids and store them as triglycerides. This synthesis requires the tissue to have a very active source of reducing power. For every molecule of fatty acid made from acetyl CoA about 14 - 16 NADPH are needed, about half of which comes from the pentose phosphate pathway. This pathway is also very active in cells undergoing **phagocytosis**.

IX. GLUCONEOGENESIS

Gluconeogenesis is the synthesis of glucose from non-carbohydrate precursors. These precursors are primarily lactate, glycerol and certain amino acids.

A. Function and Importance

This is an extremely important pathway for the maintenance and functioning of the **central nervous system**. The brain uses glucose as its primary fuel. Calculations have shown that the total daily body requirement for glucose is over 150 g/24 hours and the brain uses about 75% of this. The total body reserve in the form of glucose and glycogen is about 200 g or a little over one day's requirement. If one fasts for longer than a day or uses up these reserves faster through intense exercise, glucose must be supplied by the conversion of other compounds to glucose. (In prolonged fasting the supply of glucose is partially spared when the brain adapts to using ketone bodies).

The circulating blood glucose must be kept within fairly narrow limits. If the concentration drops too low (hypoglycemia) brain dysfunction occurs. If severe enough it can lead to coma and death. Glucose is also required for adipose tissue to produce glycerol for triglyceride formation and for the metabolism of erythrocytes. The concentrations of citric acid cycle intermediates can be maintained only if some glucose is being metabolized (amphibolic pathway). Thus there is a basal requirement for glucose even when most of the calories are supplied from triglycerides. Skeletal muscle operating under anaerobic conditions primarily uses glucose as a fuel and during intense exercise releases large amounts of lactate. Erythrocytes also release considerable quantities of lactate. Adipose tissue releases glycerol continuously. The lactate and glycerol must be recycled to glucose. Finally, under severe stress or starvation, glucose levels are maintained by breaking down skeletal muscle proteins so that some of the amino acids released can be converted to glucose. This recycling and conversion is carried out by the gluconeogenic pathway.

B. Energy Barriers to the Reversal of Glycolysis

Pyruvate is a common key compound for both glycolysis and gluconeogenesis. However, gluconeogenesis is **NOT** the reversal of glycolysis. In the glycolytic pathway, the conversion of PEP to pyruvate greatly favors the formation of pyruvate. There is a very large energy barrier inhibiting the reversal of this reaction. The formation of G-6-P and its conversion to F-1,6-bisP are also irreversible. Specific gluconeogenic enzymes make "end-runs" around these energy barriers.

Pyruvate carboxylase [reaction (a), below, and enzyme A, Figure 3-4]

$$\text{Pyruvate} + CO_2 + ATP \longrightarrow OAA + ADP + P_i + 2\,H^+ \qquad \text{(a)}$$
$$OAA + GTP \rightleftharpoons PEP + GDP + CO_2 \qquad \text{(b)}$$

$$\text{Sum: } \text{Pyruvate} + ATP + GTP \longrightarrow PEP + ADP + GDP + P_i + 2\,H^+$$

All carboxylations require **biotin** as a cofactor and pyruvate carboxylase is a biotin-containing enzyme, binding CO_2 (HCO_3^-) in an "active" form. This activation step requires the utilization of 1 ATP. Pyruvate carboxylase transfers the activated CO_2 to pyruvate to form **oxaloacetate (OAA)**.

Acetyl CoA is an **obligatory allosteric effector** of pyruvate carboxylase. A high level of acetyl CoA is a signal that more OAA must be synthesized. Acetyl CoA links both glucose metabolism and fatty acid metabolism to the TCA cycle. If the concentrations of both acetyl CoA and ATP are high, the OAA will be directed toward glucose formation (gluconeogenesis). If acetyl CoA concentration is high but ATP is low, the OAA will be shunted to the TCA cycle to form CO_2 and water (ATP formation).

Pyruvate Carboxykinase [reaction (b) and enzyme B, Figure 3-4]

The OAA is concurrently decarboxylated and phosphorylated by *PEP carboxykinase*. GTP donates the phosphate group. The removal of CO_2 drives the reaction to completion. Thus the sum of reactions (a) and (b) is the conversion of pyruvate to the high-energy compound, PEP.

Fructose-1,6,bisphosphatase [enzyme C, Figure 3-4] catalyzes the second major step toward glucose. The phosphate on the 1 position of F-1,6-bisP is cleaved forming F-6-P.

Glucose-6-phosphatase [enzyme D, Figure 3-4] catalyzes the hydrolysis of the phosphate to form free glucose. Glucose-6-phosphatase is membrane-bound ensuring that the glucose released leaves the cell.

*NOTE: Glucose-6-phosphatase does NOT exist in brain, adipose tissue or muscle, therefore, these tissues are not gluconeogenic. The major gluconeogenic tissues are **liver**, **kidney** and **intestinal epithelium**.*

C. Compartmentalization of the Reactions

All enzymes of glycolysis are cytosolic. All enzymes of the TCA cycle and of oxidative phosphorylation reside in mitochondria. Enzymes of the gluconeogenic pathway are in both: pyruvate carboxylase is mitochondrial and PEP carboxykinase is cytosolic. But the mitochondrial membrane is not permeable to OAA, an intermediate of the TCA cycle. However, it is permeable to **malate**, a derivative of OAA, and to pyruvate. Thus pyruvate in the cytosol enters mitochondria where it encounters pyruvate carboxylase and is converted to OAA. The OAA can be reduced to malate by *malate dehydrogenase* (NADH involved). The malate leaves the mitochondrion by means of an antiport system in exchange for one of several substrates (P_i, citrate, α-KG or other dicarboxylic acids) and is reoxidized back to OAA by a cytosolic form of malate dehydrogenase (NAD^+ involved). A second transport mechanism involves the transamination of oxaloacetate to aspartate which is exchanged by another antiport system for glutamate. The aspartate is transaminated back to OAA in the cytosol. The OAA is then acted upon by PEP carboxykinase to form PEP.

D. Energy Requirements and Regulation of Gluconeogenesis

Gluconeogenesis is an energy-expensive process. It takes six high energy bonds (4 ATP + 2 GTP) to make one molecule of glucose from 2 pyruvate molecules.

The pathways of glycolysis and gluconeogenesis are regulated reciprocally. It would make no sense to catabolize and synthesize glucose simultaneously, that is, run a "futile cycle". The key enzymes of glycolysis and gluconeogenesis, and their allosteric regulators are shown in Table 3-5.

Glycolysis	Gluconeogenesis
PFK	*F-1,6-bisPase*
ATP($-$)	ATP($+$)
AMP($+$)	AMP($-$)
citrate($-$)	citrate($+$)
F-2,6-bisP($+$)	F-2,6-bisP($-$)
Pyruvate kinase	*Pyruvate carboxylase*
F-1,6-bisP($+$)	Acetyl CoA($+$)
ATP($-$)	ADP($-$)

(handwritten margin note: slows or speeds up)

Table 3-5. Regulators of Glycolysis and Gluconeogenesis

E. Lactate Dehydrogenase and the Cori Cycle

Lactate dehydrogenase (LDH) catalyzes the interconversion of pyruvate to lactate and *vice versa*. LDH exists in several isozyme forms; each tissue has a distinct isozyme distribution. Analysis of which of the isozyme forms predominates in serum can indicate which tissue or organ has suffered significant damage.

When peripheral tissues undergo anaerobic metabolism they release lactate into the bloodstream. This is carried to the liver where it is converted back to glucose by gluconeogenesis. The glucose is released into the bloodstream and is again carried to the peripheral tissues. This cycle is referred to as the **Cori Cycle**.

F. The Alanine Cycle

Under conditions of stress, glucocorticoids cause the induction of gluconeogenic enzymes in the liver. They also cause the breakdown of muscle proteins, releasing amino acids into the bloodstream. These are carried to the liver where the glycogenic amino acids (characterized by alanine) are converted to glucose via the gluconeogenic pathway. The glucose is released into the bloodstream where it can be utilized by the peripheral tissues. This is known as the **Alanine Cycle**.

X. GLYCOGEN METABOLISM

A. Glycogen: Structure and Location

Glycogen consists of relatively long chains of glucose residues linked together by α-1,4 bonds. These chains are branched at about every 10 residues via α-1,6 bonds. Glucose residues are added to or removed from the non-reducing ends of the chains.

Glycogen is stored in cells, mainly liver and muscle, as granules. The enzymes responsible for glycogen metabolism are also bound to these granules. Several inherited deficiencies involving these enzymes are responsible for the several different glycogen storage diseases (see Table 3-6).

B. Glycogenolysis (Figure 3-7)

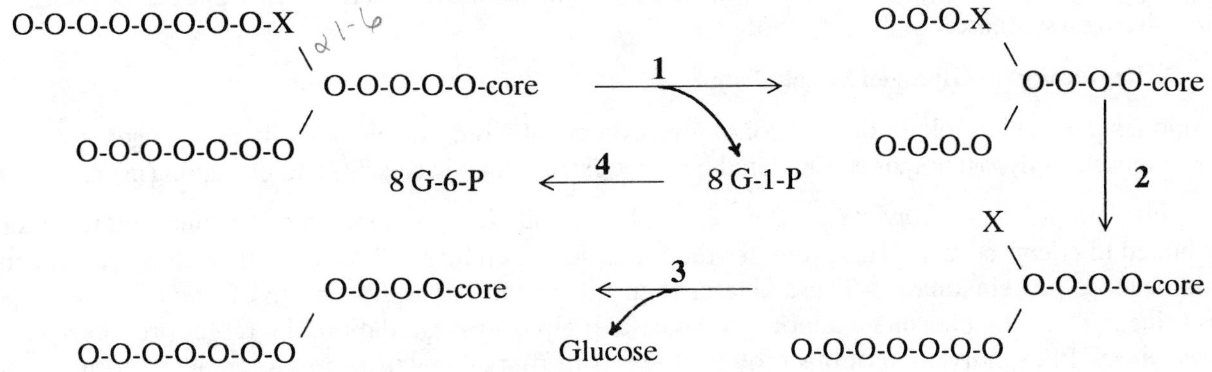

Figure 3-7. Glycogenolysis. Numbered steps correspond to numbered enzymes in text.

Glycogen Phosphorylase (1), by the process of **phosphorolysis**, catalyzes removal of glucose molecules one by one from the outermost ends of the chains (non-reducing ends), releasing each glucose residue as a molecule of G-1-P. Theoretically, this is a reversible reaction but in the cell the ratio of P_i / G-1-P is high enough to drive the reaction only toward glycogen breakdown. **Pyridoxal phosphate (Vit B$_6$)** is a cofactor for this enzyme.

Glucan Transferase (2). Phosphorolysis of the glucose residues continues down the chain until the enzyme gets within 4 residues of an α-1,6 link. At this point a second enzyme, *glucan transferase*, removes three of the four residues, transferring them to the non-reducing end of another chain.

Debranching enzyme (α-*1,6 glucosidase*, 3) removes the remaining residue. Phosphorylase can then continue down the chain until another branch point is reached. The transferase and debranching activities appear to be on the same polypeptide in separate domains.

Phosphoglucomutase (PGM, 4) converts the released G-1-P to G-6-P. The G-1-P to G-6-P conversion requires the participation of an intermediate compound, **G-1,6-bisP**. The residue released by the debranching enzyme is free glucose, not G-1-P, and must be phosphorylated by hexokinase before being metabolized further.

The energy conservation from the breakdown of glycogen is 90 % efficient, since only about 1 residue in 10 is released in the form of free glucose which must be re-phosphorylated before it can be metabolized. Since the metabolism of 1 mole of G-1-P yields 37 moles of ATP but requires the expenditure of only 1 mole of ATP, the recovery of ATP-energy is over 97 %.

C. Glycogen Synthesis

$$G\text{-}1\text{-}P + UTP \xrightarrow{\ \ 1\ \ } UDPG + PP_i$$

$$Glycogen_{(n)} + UDPG \xrightarrow{\ \ 2\ \ } glycogen_{(n+1)} + UDP$$

UDPG pyrophosphorylase [(1) above] catalyzes production of UDPG, the activated form of glucose. The release of PP$_i$, which is readily hydrolyzed in the cell, drives this reaction toward UDPG formation.

Glycogen Synthase [(2) above] carries out the transfer of a glucose residue to the non-reducing end of a glycogen chain. A chain of at least four glucose residues, bound to protein, is required as a primer.

Branching Enzyme creates α-1,6 linkages. Fragments of chain of about seven residues long are broken off at an α-1,4 link and re-attached elsewhere through an α-1,6 link. The branching process increases the solubility of glycogen and also increases the number of terminal residues which can be attacked by phosphorylase and glycogen synthase.

D. Regulation of Glycogen Metabolism

Hormones play a vital role in the control of glycogen metabolism. Insulin stimulates glycogen synthesis in liver and muscle. Glycogenolysis is stimulated by epinephrine (mainly muscle) and glucagon (mainly liver).

Phosphorylation-Dephosphorylation. Both epinephrine and glucagon bind to cell membrane receptors that are linked to *adenyl cyclase*. This promotes the formation of **cyclic AMP** which in turn allosterically activates one or more protein kinases. These kinases then catalyze the phosphorylation of one or more target proteins using ATP as the phosphate donor. In the case of glycogen degradation, the target protein is *phosphorylase kinase*. Phosphorylation of this protein converts it from an inactive form to an active form. Activated phosphorylase kinase in turn phosphorylates inactive *phosphorylase b* converting it to *phosphorylase a*. This is the **active** form of phosphorylase.

The cyclic AMP-dependent protein kinase that activates phosphorylase kinase is also known as *phosphorylase kinase kinase*. This enzyme has broad specificity and can also phosphorylate *glycogen synthase*. In the case of glycogen synthase, the non-phosphorylated form is active (a) and the phosphorylated form is inactive (b). Thus this one enzyme can reciprocally control both the synthesis and breakdown of glycogen.

Allosteric Regulators (Muscle). Phosphorylase b can also be activated allosterically by an increased concentration of AMP. The initial contraction of muscle breaks down ATP to ADP. *Myokinase* takes 2 ADP and forms 1 ATP and 1 AMP. If the formation of ADP occurs rapidly enough, AMP concentrations will rise and will activate phosphorylase b. On the other hand, high concentrations of ATP and G-6-P allosterically inhibit phosphorylase b.

Phosphorylase kinase is a multi-subunit enzyme. One of these subunits is **calmodulin**. Calmodulin has a great affinity for Ca^{++} and participates in the regulation of many different enzymes. Ca^{++} is released when muscle contracts and the increased Ca^{++} concentration binds to the calmodulin subunit partially activating phosphorylase kinase b allosterically.

Phosphatases. All of the phosphorylations described above can be reversed by specific *phosphatases*. *Protein phosphatase 1* converts phosphorylase a back to phosphorylase b. It also converts glycogen synthase b back to a.

Phosphatase 1 is blocked by a protein called **inhibitor 1**. Inhibitor 1 binds when it is phosphorylated but does not bind when dephosphorylated. The phosphorylation of inhibitor 1 is catalyzed by a specific cAMP-dependent protein kinase. Insulin decreases the amount of phosphorylated inhibitor 1.

Regulation of Liver Glycogen. Liver cells monitor the concentration of blood glucose. Liver phosphorylase a acts as a **blood glucose sensor**. When glucose binds to phosphorylase a it causes an allosteric change that exposes the usually cryptic or hidden phosphate group. *Phosphatase 1*, which is tightly bound to the phosphorylase a, then cleaves the phosphate off, inactivating the phosphorylase a by converting it to the b-form. This conversion releases the phosphatase 1, allowing it to now attack the inactive, phosphorylated glycogen synthase and converting it to its active form.

XI. GLYCOGEN STORAGE DISEASES

A number of diseases of glycogen metabolism have been identified. Depending upon which enzyme is deficient, these diseases can manifest either storage of excessive levels of glycogen or the synthesis of glycogen of abnormal structure, or both. Their characteristics are summarized in Table 3-6.

Type	Name	Defective enzyme	Organ affected	Glycogen levels and structure *
I	von Gierke's Disease	Glucose 6-Pase	Liver & Kidney	I, N
II	Pompe's Disease	α-1,4-glucosidase	All organs	much I, N
III	Cori's Disease	debranching enzyme	Muscle and Liver	I, A: short outer branches
IV	Andersen's	branching enzyme	Liver and Spleen	I, A: v. long outer branches
V	McArdle's Disease	phosphorylase	Muscle	I, N
VI	Hers' Disease	phosphorylase	Liver	I, N
VII		phosphofructokinase	Muscle	I, N
VIII		phosphorylase kinase	Liver	I, N

Table 3-6. Glycogen Storage Diseases: Nomenclature and Effects on Glycogen
*I = increased; N = normal levels or structure; A = abnormal structure.

For example, the dynamics of two types of glycogen storage diseases that are sometimes encountered are as follows:

- ◆ von Gierke's Disease: Due to the failure of glycogen breakdown, the liver is greatly enlarged and loaded with glycogen. This is accompanied by hypoglycemia because lack of glucose-6-Pase prevents glucose from leaving the liver.

- ◆ McArdle's Disease: Lack of glycogen phosphorylase prevents muscle from using glycogen as an energy source, leading to increased levels of glycogen in muscle cells, cramps, and intolerance to exercise. Lactic acid is not produced during exercise, but instead the pH actually rises due to the breakdown of phosphocreatine.

XII. HORMONAL REGULATION OF CARBOHYDRATE METABOLISM

A. Insulin

Insulin is secreted by the β-cells of the pancreas. It is released by high levels of blood glucose and by stimulation of the parasympathetic nervous system. It stimulates the dephosphorylation of key enzymes. Insulin affects different pathways as follows: (+ means increase,– means decrease) glycogen synthesis (+), gluconeogenesis (–), glycolysis (+), fatty acid synthesis (+), protein synthesis (+) and protein degradation (–).

B. Glucagon

Glucagon is secreted by the α-cells of the pancreas. Its secretion is stimulated by **low** levels of glucose. The liver is the main target organ. Glucagon affects different pathways as follows: glycogen synthesis (–), glycogen degradation (+), fatty acid synthesis (–), gluconeogenesis (+) and glycolysis (–). Glucagon works through cAMP-dependent protein kinases causing the release of glucose from the liver and mobilization of fatty acids from adipose tissue.

C. Epinephrine and Norepinephrine

These hormones are secreted from the adrenal glands and the sympathetic nerve endings in response to low blood glucose levels. They act primarily on muscle tissues and prevent the uptake of glucose into muscle. The muscle then uses fatty acids for fuel. This spares glucose for use by the brain.

D. Steroids

Produced by the adrenal cortex, glucocorticoids induce expression of genes coding for enzymes of gluconeogenesis. The result is a rise in the level of blood glucose, and, if glucocorticoids are excessive, there may be muscle wasting due to continued conversion of amino acids to glucose.

XIII. REVIEW QUESTIONS ON CARBOHYDRATES

DIRECTIONS: For each of the following multiple-choice questions (1 - 54), choose the ONE BEST answer.

---

1. The immediate products of oxidation of one mole of glucose 6-phosphate through the oxidative portion of the pentose phosphate pathway are:

A. 2 moles of reduced NAD, one mole of ribulose 5-phosphate and one mole of CO_2
B. 2 moles of oxidized NADP, one mole of ribulose 5-phosphate and one mole of CO_2
C. 2 moles of reduced NADP, one mole of xylulose 5-phosphate and one mole of CO_2
D. 2 moles of reduced NADP, one mole of ribulose 5-phosphate and one mole of CO_2
E. one mole of fructose 6-phosphate and five moles of CO_2.

Ans D: The products of the oxidative branch include 2 NADPH plus the removal of C-1 of the G-6-P as CO_2, yielding the pentose, ribulose-5-P.

2. Which one of the following statements regarding glycogen metabolism is INCORRECT?

A. Glycogen consists of α-1,4-glycosidic bonds and α-1,6-glycosidic bonds.
B. Glycogen phosphorylase catalyzes the hydrolytic cleavage of glycogen into glucose-1-phosphate.
C. A debranching enzyme is needed for the complete breakdown of glycogen.
D. Phosphoglucomutase converts glucose-1-phosphate into glucose-6-phosphate.
E. Glycogen is synthesized and degraded by different pathways.

Ans B: Glycogen phosphorylase catalyzes a *phosphorolytic* cleavage of glycogen, not a hydrolytic cleavage.

3. The Cori Cycle:

glucose → 2 lactate + 2 ATP (muscle), and
2 lactate + 6 ATP → glucose (liver)

is important because:

A. there is a net destruction of ATP, restoring the energy balance between muscle and liver.
B. it results in the net generation of glucose in liver and ATP in muscles without build up of high lactate levels in the bloodstream.
C. it enables muscle mass to be used for energy in conditions of extreme starvation.
D. it serves to prevent lactate levels from dropping too low in the blood, which would impair brain function.
E. it enables G-6-P to be transported across the liver cell plasma membrane.

Ans B: The Cori cycle allows anaerobic glycolysis to proceed in muscle without causing toxic metabolic acidosis from accumulation of lactate in the bloodstream.

4. Which one of the following statements regarding gluconeogenesis is INCORRECT?

A. Glucose can be synthesized from non-carbohydrate precursors.
B. Gluconeogenesis is not a reversal of glycolysis.
C. Gluconeogenesis and glycolysis are reciprocally regulated.
D. Lactate and alanine formed by contracting muscle are converted into glucose in muscle.
E. Six high energy phosphate bonds are spent in synthesizing one molecule of glucose from pyruvate.

Ans D: Muscle lacks glucose-6-phosphate activity and, therefore, cannot carry out gluconeogenesis.

5. A patient presenting with a suspected metabolic disorder shows (1) abnormally high amounts of glycogen with normal structure in liver, and (2) no increase in blood glucose levels following oral administration of fructose. From these two findings, which one of the following enzymes is likely to be deficient?

A. phosphoglucomutase
B. UDP-glycogen transglucosylase
C. fructokinase
D. glucose-6-phosphatase
E. glucokinase.

Ans D: Failure of the liver to produce glucose from fructose while piling up glycogen indicates that glucose-6-phosphatase activity is inhibited or missing.

6. Glucagon has which one of the following effects on glycogen metabolism in the liver?

A. The net synthesis of glycogen is increased.
B. Glycogen phosphorylase is activated, whereas glycogen synthase is inactivated.
C. Both glycogen phosphorylase and glycogen synthase are activated.
D. Glycogen phosphorylase is inactivated, whereas glycogen synthase is activated.
E. Phosphoglucomutase is phosphorylated.

Ans B: Glucagon stimulates production of cAMP which activates protein kinases. These phosphorylate phosphorylase, activating it, and glycogen synthase, inactivating it. The two enzymes are thus reciprocally controlled.

7. There are three irreversible steps in glycolysis. These are:

A. hexokinase, phosphoglycerate kinase, and pyruvate kinase
B. hexokinase, phosphofructokinase, and pyruvate kinase
C. phosphofructokinase, aldolase, and phosphoglyceromutase
D. phosphoglucose isomerase, glyceraldehyde 3-phosphate dehydrogenase, and enolase
E. triose phosphate isomerase, phosphoglycerate kinase, and enolase.

Ans B: These enzymes are the major control points of glycolysis.

8. The first step in the gluconeogenic pathway (starting with pyruvate) results in the formation of:

A. phosphoenolpyruvate
B. malate
C. aspartate
D. oxaloacetate
E. lactate.

Ans D: Pyruvate carboxylase converts pyruvate to oxaloacetate as the first step in gluconeogenesis.

9. Which one of the following statements concerning the pentose phosphate pathway is true?

A. NADH is a major product of the oxidative branch.
B. The entire pathway is "off" when NADPH levels are high
C. It is not present in brain tissue.
D. When glucose 6-phosphate dehydrogenase is inhibited by one of its products, ribose 5-phosphate biosynthesis still occurs.
E. It provides an alternative pathway for the synthesis of glucose from glycerol.

Ans D: Transaldolase and transketolase catalyze freely reversible reactions, which means that ribose-5-P can be synthesized from fructose-6-P and glyceraldehyde-3-P.

10. If a pure crystal of β-D-glucose is dissolved in water and left in solution for a long period of time, it is likely that the solution will contain:

A. the open-chain form of D-glucose, α-D-glucose, and β-D-glucose
B. the open-chain form of D-glucose and β-D-glucose
C. α-D-glucose and β-D-glucose only
D. α-D-glucose only
E. β-D-glucose only.

Ans A: All three forms of glucose will be in an equilibrium mixture of about 70%-β-D-glucose, 30% α-D-glucose and less than 1% of the linear form.

11. The enzyme aldolase catalyzes the:

A. formation of fructose-6-phosphate from glucose-6-phosphate
B. oxidation of the aldehyde group of glucose
C. oxidation of the aldehyde group of glyceraldehyde-3-phosphate
D. conversion of glyceraldehyde-3-phosphate to dihydroxyacetone-phosphate
E. formation of dihydroxyacetone phosphate and glyceraldehyde-3-phosphate from fructose-1,6-bisphosphate.

Ans E: Aldolase A (aldolase 1) splits the symmetrical fructose-1,6-bisP into dihydroxyacetone phosphate and glyceraldehyde 3-phosphate which are readily interconvertible.

12. Galactosemia is caused by

A. a deficiency of galactose-1-phosphate uridyltransferase
B. a deficiency of UDP-galactose 4-epimerase
C. the high content of lactose in artificial feeding formulae for babies
D. absorption of non-hydrolyzed lactose through the intestinal mucosa
E. excessive conversion of glucose-1-phosphate into galactose-1-phosphate.

Ans A: Failure to form UDP-galactose because of deficient transferase activity allows the accumulation of galactose-1-P resulting in the symptoms of galactosemia.

13. Biotin is required as a coenzyme in which one of the following reactions?

A. α-ketoglutarate + NAD$^+$ + CoA $\rightarrow$ succinyl CoA + CO$_2$ + NADH
B. pyruvate + CO$_2$ + ATP $\rightarrow$ oxaloacetate + ADP + P$_i$
C. pyruvate + NAD$^+$ + CoA $\rightarrow$ acetyl CoA + CO$_2$ + NADH
D. 6-phosphogluconate $\rightarrow$ ribulose-5-phosphate + CO$_2$
E. α-ketoglutarate + alanine $\rightleftharpoons$ glutamate + pyruvate.

Ans B: All carboxylases require biotin as a coenzyme. Decarboxylases do not require a coenzyme.

14. In glucose 6-phosphate dehydrogenase deficiency, increased red cell lysis is ultimately due to:

A. problems with ATP production in mitochondria
B. a deficiency in ability to carry out glycolysis
C. increased leakage of K ion into the cells
D. an intrinsic deficiency of membrane structure
E. inability of the cell to maintain normal levels of NADPH.

Ans E: NADPH is required for glutathione reductase activity to convert GSSG back to 2 GSH.

15. During starvation, as gluconeogenesis increases to maintain the levels of blood glucose, which one of the following will be enhanced?

A. liver pyruvate kinase activity
B. the secretion of insulin by the pancreas
C. muscle phosphoglucomutase activity
D. the metabolism of acetyl CoA to pyruvate
E. the metabolism of glutamate to glucose-6-phosphate.

Ans E: Glutamate, derived from the breakdown of glycogenic amino acids can be converted to glucose-6-P via the gluconeogenic pathway.

16. In hereditary fructose intolerance, the primary biochemical defect is:

A. a deficiency in activity of an aldolase isozyme
B. increased allosteric sensitivity of phosphofructokinase to AMP
C. inhibition of glycogen synthetase
D. an inability to absorb fructose
E. a deficiency in the activity of fructokinase.

Ans A: Aldolase B (aldolase 2) catalyzes cleavage of F-1-P to dihydroxyacetone P + glyceralde-hyde-3-P. Deficiency causes accumulation of F-1-P which is toxic to liver in high concentrations.

17. The major rate-limiting step of glycolysis is the:

A. conversion of glucose to glucose 6-phosphate
B. conversion of glucose 6-phosphate to fructose 6-phosphate
C. conversion of fructose 6-phosphate to fructose 1,6-bisphosphate
D. aldolase reaction
E. epimerase reaction.

Ans C: The key rate-limiting step in glycolysis is the phosphofruc-tokinase reaction which converts fructose-6-P to F-1,6 bis P.

18. Amylose is:

A. a branched homopolysaccharide
B. a linear homopolysaccharide
C. a linear heteropolysaccharide
D. a salivary enzyme
E. a pancreatic enzyme.

Ans B: Amylose is a linear poly-mer of glucose molecules linked α 1-4.

19. In the metabolism of glycerol to glycogen, the first intermediate of glycolysis encountered is:

A. glyceraldehyde-3-phosphate
B. dihydroxyacetone phosphate
C. 3-phosphoglyceric acid
D. ribulose-5-phosphate
E. 1,3-bisphosphoglyceric acid.

Ans B: Glycerol is phosphorylated to glycerol-3-Phosphate which is then converted to dihydroxyace-tone phosphate by glycerol-3-phosphate dehydrogenase.

20. Glucose, labeled with ^{14}C in different carbon atoms is added to a tissue that is rich in the enzymes of the hexose monophosphate shunt. Which one will give the most rapid initial production of $^{14}CO_2$?

A. glucose-1-^{14}C
B. glucose-2-^{14}C
C. glucose-3,4-^{14}C
D. glucose-5-^{14}C
E. glucose-6-^{14}C.

Ans A: Glucose-6-P, when flowing through the oxidative branch of the pentose phosphate pathway loses the C-1 as CO_2. The other carbons go to ribulose-5-P.

21. Which one of the following is a glycolytic enzyme of liver and is activated by protein phosphate phosphatase in response to decreasing glucagon levels?

A. glycogen synthetase
B. glycogen phosphorylase
C. pyruvate kinase
D. triose phosphate isomerase
E. lactate dehydrogenase.

Ans C: Pyruvate kinase of liver is phosphorylated by a glucagon-activated kinase during gluconeogenic conditions. Falling glucagon (rising insulin) levels allow the P_i to be removed by a phosphatase, restoring glycolysis.

22. In a patient suffering from Type III glycogen storage disease, an abnormal glycogen exhibiting short outer branches is observed. Which enzyme is most likely to be defective?

A. glycogen synthase
B. amylo-1,6-glucosidase (debranching enzyme)
C. glycogen phosphorylase
D. phosphoglucomutase
E. UDP-glucose pyrophosphorylase.

Ans B: Debranching enzyme has catalytic activities for glucan transferase and α-1,6 glucosidase. Defective enzyme leads to incomplete removal of the branch points.

23. Von Gierke's disease is characterized by massive enlargement of the liver, severe hypoglycemia, ketosis, hyperuricemia and hyperlipemia. It is caused by defective:

A. amylo-α-1,6 glucosidase
B. branching enzyme (α-1,4 $\rightarrow$ α-1,6)
C. glucose 6-phosphatase
D. α-1,4 glucosidase
E. phosphorylase.

Ans C: Von Gierke's disease is due to lack of hepatic glucose 6-phosphatase activity resulting in a non-functional gluconeogenic pathway.

24. The absence of which one of the following reactions is responsible for the inability of man to use fatty acids in the *de novo* net synthesis of glucose?

A. oxaloacetate $\rightarrow$ pyruvate
B. oxaloacetate + acetyl CoA $\rightarrow$ citrate
C. acetyl CoA $\rightarrow$ pyruvate
D. pyruvate $\rightarrow$ phosphoenolpyruvate
E. phosphoenolpyruvate $\rightarrow$ oxaloacetate.

Ans C: The pyruvate dehydrogenase complex is not reversible so Acetyl CoA cannot go back to pyruvate directly.

25. Which one of the following substrates can NOT contribute to net gluconeogenesis in mammalian liver?

A. alanine
B. stearate
C. α-ketoglutarate
D. glutamate
E. pyruvate.

Ans B: Stearate is catabolized to acetyl CoA, a substrate that cannot participate in gluconeogenesis to increase the NET production of glucose.

26. The major site of carbohydrate digestion is:

A. mouth
B. stomach
C. small intestine
D. large intestine
E. pancreas.

Ans C: Although saliva contains an amylase, the major site of carbohydrate digestion is the small intestine.

27. Activated sugar residues utilized for the biosynthesis of complex glycoproteins include all EXCEPT:

A. GDP-mannose
B. GDP-fucose
C. UDP-glucuronic acid
D. CMP-N-acetylneuraminic acid
E. UDP-galactose.

Ans C: Glucuronic acid can be found as a component of glycosaminoglycans but not glycoproteins.

28. The tissue with the lowest activity for the oxidation of glucose-6-phosphate by the pentose phosphate pathway is:

A. liver
B. lactating mammary gland
C. striated muscle
D. adrenal cortex
E. adipose tissue.

Ans C: All the tissues mentioned, except muscle, carry out reactions requiring large amounts of NADPH.

29. Glyceraldehyde-3-phosphate dehydrogenase produces which one of the following as a product?

A. dihydroxyacetone phosphate
B. 1,3-bisphosphoglyceric acid
C. glycerol
D. phosphoenolpyruvic acid
E. NADPH.

Ans B: Glyceraldehyde 3-phosphate dehydrogenase catalyzes:
Glyceraldehyde 3-phosphate $+ NAD^+ + P_i \rightarrow NADH + H^+ + $ 1,3-bisphosphoglyceric acid.

30. Which one of the following is a ketose?

A. D-glucose
B. D-ribose
C. D-galactose
D. D-fructose
E. N-acetylglucosamine.

Ans D: D-fructose is a ketose, the other sugars are aldoses or derivatives of aldoses.

- 55 -

31. Hyaluronic acid is a:

A. glycoprotein
B. high molecular weight, positively charged polysaccharide
C. polymer which contains sulfate
D. repeating disaccharide of glucuronic acid and N-acetyl glucosamine
E. lipoprotein.

Ans D: Hyaluronic acid is a non-sulfated glycosaminoglycan.

32. The increase of glycogenolysis in muscle produced by epinephrine may be attributed to:

A. decreased Ca^{++}
B. activation of aldolase
C. reduction in total NAD^+ plus NADH
D. conversion of phosphorylase b to phosphorylase a
E. conversion of glycogen synthase b to glycogen synthase a.

Ans D: Epinephrine triggers the protein kinase-mediated cascade which eventually phosphorylates phosphorylase b converting it to the active form, phosphorylase a.

33. Glucagon:

A. has actions similar to those of insulin
B. is secreted by the beta cells of the pancreatic islets
C. increases triglyceride concentrations in many tissues
D. targets liver primarily
E. targets muscle primarily.

Ans D: With respect to glycogenolysis the liver is affected primarily by glucagon, muscle by epinephrine.

34. The activity of glycogen phosphorylase in muscle is affected primarily by:

A. glucagon
B. melanocyte stimulating hormone
C. adrenocorticotropic hormone
D. epinephrine
E. somatostatin.

Ans D: See question 33.

35. The catabolism of 1 mole of glucose in the glycolytic pathway under anaerobic conditions results in the formation (or NET gain) of:

A. 1 mole of UDP-galactose
B. 2 moles of ATP + 2 moles of lactic acid
C. 1 mole of glucose-1,6-diphosphate
D. 6 moles of CO_2 + 6 moles of H_2O
E. 1 mole of ethanol + 1 mole of lactic acid.

Ans B: Glucose is split into 2 trioses, each when oxidized can yield 1 ATP and 1 pyruvate. Under anaerobic conditions the pyruvate is reduced to lactate.

36. Which one of the following is a gluconeogenic enzyme?

A. glucose-6-phosphate dehydrogenase
B. phosphoglucomutase
C. phosphofructokinase
D. pyruvate carboxylase
E. lactate dehydrogenase.

Ans D: Pyruvate carboxylase is a key gluconeogenic enzyme in mitochondria that converts pyruvate to oxaloacetate. Acetyl CoA is a positive effector.

37. The most significant role of the oxidative portion of the hexose monophosphate shunt is the generation of:

A. hexose monophosphate from free hexose
B. NADPH
C. NAD⁺
D. ATP from ADP and P_i
E. UDP-gluconic acid from UDP-glucose.

Ans B: About one-half of the requirement for NADPH comes from the pentose phosphate pathway. Especially important for cells conducting fatty acid biosynthesis.

38. A nine-carbon acid found as the non-reducing termini of oligosaccharide side chains of many glycoproteins is:

A. hyaluronic acid
B. sialic acid
C. iduronic acid
D. glucuronic acid
E. gluconic acid.

Ans B: Sialic acid is a nine carbon acid sugar also referred to as N-acetyl neuraminic acid.

39. Which one of the following carbohydrates contains a monosaccharide unit other than glucose?

A. glycogen
B. cellulose
C. maltose
D. lactose
E. starch

Ans D: Lactose is composed of glucose and galactose. The others listed are composed of glucose only.

40. In mammals glucose 6-phosphate is converted to all of the following compounds EXCEPT:

A. glucose
B. fructose 1-phosphate
C. 6-phosphogluconolactone
D. fructose 6-phosphate
E. glucose 1-phosphate.

Ans B: Fructose-1-P is formed only by hepatic fructokinase from fructose and ATP.

41. Glycolysis in the red blood cell produces:

A. citric acid
B. NADH
C. GTP
D. CO_2
E. glucose-1-phosphate.

Ans B: Glycolysis produces one NADH at the glyceraldehyde-3-P dehydrogenase step. The other compounds are not part of the glycolytic pathway.

42. Sucrose is:

A. a disaccharide containing glucose and fructose
B. a reducing disaccharide of plant origin
C. a disaccharide containing glucose and galactose
D. a fructose polymer
E. a product of digestion of cellulose.

Ans A: Sucrose is a non-reducing disaccharide composed of glucose and fructose found primarily in sugar cane and beets.

43. Which one of the following sugars is found exclusively in glycosaminoglycans?

A. N-acetylglucosamine
B. N-acetylgalactosamine
C. sialic acid
D. L-iduronic acid
E. L-arabinose.

Ans D: L-iduronic acid is found in humans only in glycosaminoglycans.

44. Which one of the following enzymes is found mainly in liver and kidney cells?

A. glucose 6-phosphatase
B. hexokinase
C. phosphoglucoisomerase
D. pyruvate kinase
E. glycogen synthase.

Ans A: Gluconeogenesis occurs mainly in liver and kidney. Other cells lack this enzyme.

45. Sialic acid is:

A. found only in mammalian tissues
B. the major carbohydrate found in heparin
C. a normal constituent of glycoproteins
D. an ϵ-carboxyl amino acid
E. a cofactor for neuraminidase.

Ans C: Sialic acid (neuraminic acid) is a common moiety often found at the termini of the oligosaccharide side chains of complex glycoproteins.

46. All of the following are high energy compounds EXCEPT:

A. phosphoenolpyruvate
B. acetyl-CoA
C. glucose 6-phosphate
D. acetyl phosphate
E. creatine phosphate.

Ans C: All the compounds shown except glucose-6-P have standard free energies of hydrolysis greater than or equal to that of ATP.

47. Which one of the following enzymes is inhibited by an accumulation of NADH?

A. aldolase
B. enolase
C. pyruvate kinase
D. glyceraldehyde 3-phosphate dehydrogenase
E. pyruvate decarboxylase.

Ans D: NADH is a product of the glycolytic enzyme glyceraldehyde 3-phosphate dehydrogenase. Rising levels of NADH can cause product inhibition of the enzyme.

48. Which one of the following glycogen storage diseases does NOT involve a defect in the glycogen degradation pathway?

A. Type I (von Gierke's Disease)
B. Type IV (Andersen's Disease)
C. Type VI (Hers' Disease)
D. Type V (McArdle's Disease)
E. Type III (Cori's Disease).

Ans A: Type I (von Gierke's) disease involves glucose-6-phosphatase, a member of the gluconeogenic pathway.

49. Epinephrine (in muscle) and glucagon (in liver):

A. activate adenylate cyclase
B. inactivate phosphorylase and activate glycogen synthetase
C. stimulate triglyceride synthesis
D. stimulate glycogen synthesis
E. act synergistically with insulin.

Ans A: Both epinephrine and glucagon bind to receptors that activate adenylate cyclase.

50. An allosteric inhibitor of phosphofructokinase is:

A. α-ketoglutarate
B. oxaloacetate
C. citrate
D. isocitrate
E. succinate.

Ans C: High citrate and ATP concentrations are signals of a high energy charge and allosterically inhibit the glycolytic pathway at the level of PFK.

51. The number of residues bound by glycosidic linkages to a glucose residue that forms a branch point in glycogen is:

A. 1
B. 2
C. 3
D. 4
E. 5

Ans C: A glucose moiety at a branch point is attached to a glucose at C-4, to another glucose at C-1 and to a third at C-6.

52. What is the NET yield of ATP when glucose 1-phosphate is oxidized by anaerobic glycolysis to lactate?

A. 0
B. 1
C. 2
D. 3
E. 4

Ans D: G-1-P released from glycogen does not need to be phosphorylated by hexokinase. Therefore, the *net* ATP yield is 3.

53. All of the following statements about phosphofructokinase (PFK) are true EXCEPT:

A. it is a major control enzyme in glycolysis
B. ATP is a substrate for PFK
C. fructose 2,6-bisphosphate is a negative effector of PFK
D. ATP is a negative effector of PFK
E. it catalyzes a metabolically irreversible reaction; i.e., its equilibrium point lies far in one direction.

Ans C: F-2,6-bisP is a *positive* effector of PFK.

54. Transketolase requires which one of the following coenzymes?

A. pyridoxal phosphate
B. lipoamide
C. thiamin pyrophosphate
D. cobalamin
E. tetrahydrofolic acid.

Ans C: Transketolase belongs to the group of enzymes, including pyruvate dehydrogenase and α-ketoglutarate dehydrogenase, that transfer carbon units containing a keto group. All require thiamin pyrophosphate as a cofactor.

DIRECTIONS: For each set of questions, choose the ONE BEST answer from the lettered list above it. An answer may be used one or more times, or not at all.

Questions 55 - 58:

    A. glucose-6-P dehydrogenase deficiency
    B. glucose-6-phosphatase deficiency
    C. galactokinase deficiency
    D. UDP-galactose epimerase deficiency
    E. hexokinase deficiency

55. Increased concentration of galactose-1-P.

Ans D: Failure to convert UDP-galactose to UDP-glucose due to epimerase deficiency would ultimately cause accumulation of galactose-1-P.

56. Increased lipid peroxides in erythrocytes.

Ans A: Glucose-6-P dehydrogenase deficiency in erythrocytes causes reduced production of NADPH which is a coenzyme for glutathione reductase.

57. Increased concentration of liver glycogen.

Ans B: Glucose-6-phosphatase deficiency causes increased concentration of G-6-P which is then shunted toward glycogen synthesis.

58. Essentially benign excretion of a glucose epimer

Ans C: Failure to phosphorylate galactose causes galactosuria, a relatively benign condition.

Questions 59 - 61:

    A. glucose-6-phosphate
    B. UDP-galactose
    C. lactate
    D. acetyl CoA
    E. 1,3-bisphosphoglycerate

59. Regulates glycogen synthase.

Ans A: G-6-P can act as a positive effector of glycogen synthase b.

60. Regulates pyruvate carboxylase.

Ans D: Acetyl CoA is an obligatory positive effector of pyruvate carboxylase.

61. Cannot be converted to glucose.

Ans D: Because the pyruvate dehydrogenase reaction is irreversible, acetyl CoA cannot be used for the *net* synthesis of glucose via gluconeogenesis.

Questions 62 - 64:

    A. ATP is a substrate
    B. ATP is an inhibitor
    C. AMP is an inhibitor
    D. ATP is both a substrate and an inhibitor

62. Phosphofructokinase.

Ans D: PFK uses ATP as one of its substrates and ATP also acts as an allosteric inhibitor under conditions of high energy charge.

63. glucokinase.

Ans A: Glucokinase, like hexokinase, uses glucose and ATP as substrates.

64. Fructose-1,6-bisphosphatase.

Ans C: F-1,6-bisP-ase, a gluconeogenic enzyme, is inhibited by AMP and stimulated by ATP, both acting as allosteric effectors.

Questions 65 - 69

    A. Glucose-1-phosphate
    B. Glucose-6-phosphate
    C. Glucose-1,6-bisphosphate
    D. UDP-glucose

65. Activated form of glucose utilized by glycogen synthetase.

Ans D: Glycogen synthase uses UDPG as one of its substrates.

66. Generated during breakdown of glycogen by glycogen phosphorylase.

Ans A: Glycogen phosphorylase catalyzes a *phosphorolysis* of glycogen, yielding G-1-P as a product.

67. Coenzyme for phosphoglucomutase.

Ans C: Phosphoglucomutase catalyzes the reaction G-6-P $\rightleftharpoons$ G-1-P. G-1,6-bisP is a catalytic intermediate used as a coenzyme.

68. Generated by the hexokinase reaction.

Ans D: Hexokinase and glucokinase yield G-6-P as one of the products.

69. Substrate for phosphohexose isomerase.

Ans B: Phosphohexose isomerase catalyzes G-6-P $\rightleftharpoons$ F-6-P.

4. ENERGETICS AND BIOLOGICAL OXIDATION

Thomas Briggs

I. CONCEPTS IN BIOLOGICAL OXIDATION

A. Oxidation and Reduction

Biological oxidation provides most of the energy for aerobic metabolism. This energy is released when electrons are transferred from fuel molecules to oxygen. Biological oxidation effects this transfer in a controlled way, and conserves much of the energy in the form of phosphoanhydride bonds of ATP, a useful molecule which can then supply the energy to drive a multitude of energy-consuming processes. The process of phosphorylation of ADP to ATP, driven by the transfer of electrons to oxygen, is called **oxidative phosphorylation**.

Oxidation can be defined in three ways:

       1. addition of oxygen;

       2. removal of hydrogen;

       3. **removal of electrons**: this is the most general definition.

Reduction is the converse of the above.

Biological oxidation and reduction are always linked. For every oxidation, there *must* be a reduction; for every electron donor, there *must* be an electron acceptor.

B. Thermodynamics

For a reaction $A + B \rightleftharpoons C + D$ at equilibrium, the **equilibrium constant**, K_{eq}, is given by the expression:

$$K_{eq} = \frac{[C]\,[D]}{[A]\,[B]}$$

Any reaction proceeds with a change in **free energy**, the useful energy produced, the energy available to do useful work. The observed change is denoted by ΔG, (sometimes incorrectly called ΔF). ΔG depends on the **nature** of the reaction and the **concentrations** of reactants and products, and has predictive power: a **negative** ΔG means the reaction is exergonic (energy is released) and will go to the **right** as written, or down an energy hill. The reaction is said to be "spontaneous." A **positive** ΔG means the reaction is endergonic (energy is absorbed) and will go to the **left**. *At equilibrium, $\Delta G = 0$.*

Note that ΔG *tells nothing about rates of reactions*. A "spontaneous" process may not proceed by itself at a measurable speed, due to the presence of an activation energy barrier. One function of enzymes is to lower this barrier and speed up the rate. The final position of equilibrium is related to ΔG and **not** affected by enzymes. Thus an enzyme will not affect the ΔG of a reaction.

If a reaction starts at **standard conditions** of pH 7 and 1 M concentration of each reactant and product, and goes to equilibrium, then the change in free energy is the **standard** free energy change (denoted by superscript zero and prime) and is related to K_{eq} (R is the gas constant; T, the absolute temp.):

$$\Delta G^{\circ\prime} = -RT \ln K_{eq}$$

A reaction with an equilibrium constant >1 will have a **negative** $\Delta G^{\circ\prime}$, and will *tend* to go to the right. But this tendency may be reversed by **concentration** effects or input of **energy** from somewhere else. Thus a reaction can be driven against an unfavorable equilibrium by mass action, or if it is coupled, through a common intermediate, with an energy-releasing (exergonic) reaction.

When reactions are coupled, or occur in series, ΔG's of component reactions are simply added. It is the overall free-energy change, the algebraic sum of individual ΔG's, that determines whether the process as a whole will occur spontaneously. If a written equation is reversed, the component ΔG's change sign.

The degree of **disorder**, or randomness, is called **entropy** (S). ΔS is the entropy change associated with a particular transformation. Entropy is higher in a less-ordered system, such as a denatured protein. Entropy of living systems is kept low (highly ordered) at the expense of an increase in entropy of the surroundings. A spontaneous reaction **can** proceed with a decrease in entropy, if the **total** entropy change, including that of surroundings, is positive.

In an oxidation-reduction reaction, reducing power, E, also called **redox potential**, is the capacity to donate electrons. It can be measured, in volts, by comparing a component of the reaction with a standard hydrogen electrode.

$\Delta E_o{}'$ is the difference in **standard** redox potential (standard conditions) between electron donor and acceptor. The component with the more negative $E_o{}'$ is the stronger electron donor or reducing agent, and will tend to reduce (be oxidized by) the second component. A large K_{eq} (i.e., reaction tends to proceed far to the right) is associated with a large **positive** $\Delta E_o{}'$. Thus $\Delta E_o{}'$ is mathematically related to $\Delta G^{\circ\prime}$ but the two are opposite in sign.

Points to remember:

1. A process that goes "spontaneously" has a **negative** ΔG.

2. An oxidation-reduction reaction that goes "spontaneously" has a **positive** ΔE.

3. A substance with the more **negative** redox potential will reduce another with a less negative or more positive redox potential: *electrons tend to flow from* **E negative to E positive**.

4. The **molecular oxygen** system, toward which electrons flow in biological oxidation, has the **most positive** redox potential.

5. The directional tendency that a reaction might normally have *may* be reversed by:

   ♦ a sufficient difference in **concentration** between reactants and products (mass action)

   ♦ input of sufficient **energy**.

C. High-Energy Compounds

When ATP, an anhydride of phosphoric acid, undergoes hydrolysis and the terminal phosphate is transferred to H_2O, the equilibrium lies far to the right and an unusually large amount of free energy is released:

$$ATP + H_2O \rightleftharpoons ADP + P_i \qquad \Delta G^{\circ\prime} = -7.3 \text{ Kcal/mole}$$

This is in part because the products of the reaction are stabilized by a greater number of resonance forms. The reaction can be reversed by coupling it to another reaction that supplies an equal or greater amount of energy. ATP is said to be a "high-energy" compound.

Substances such as ATP serve an "energy-currency" function by accepting energy from electron transfer and the metabolism of certain substrates, and in turn supplying it to drive various energy-consuming functions: synthesis, muscle contraction, the maintenance of ionic gradients, etc. Table 4-1 shows representative high-energy compounds. Note that ordinary esters of phosphoric acid, such as glucose-6-phosphate, are *not* high-energy compounds. (The SI system, in which the Joule is the unit of energy rather than the calorie, is coming increasingly into use; the table shows values in both types of units: 1 Kcal = 4.184 KJ. In this book we will generally use calories, with which most people are more familiar.)

Compound	Type	$\Delta G^{\circ\prime}$, Kcal/mole	KJ/mole
Phosphoenolpyruvate	Anhydride of enol and phosphoric acid	−14.8	−61.9
1,3-Diphosphoglycerate	Anhydride of carboxylic and phosphoric acids	−11.8	−49.4
Phosphocreatine	Guanidino phosphate	−10.3	−43.1
Acetyl phosphate	Anhydride of carboxylic and phosphoric acids	−10.1	−42.2
Acetyl CoA	Thioester (not a phosphorylated cpd)	−7.5	−31.4
ATP	Anhydride of phosphoric acid	−7.3	−30.5
Glucose-6-phosphate	Ester of phosphoric acid	−3.3	−13.8
Glycerol-1-phosphate	Ester of phosphoric acid	−2.2	−9.2

Table 4-1. Standard Free Energy of Hydrolysis of Some Phosphorylated Compounds.

II. METABOLISM OF PYRUVATE TO CARBON DIOXIDE

A. The Pyruvate Dehydrogenase Complex

In aerobic metabolism, the major fate of the pyruvate produced by glycolysis is transport into mitochondria, then conversion to acetyl CoA and CO_2 by the *pyruvate dehydrogenase complex*, which carries out the over-all reaction:

$$\text{Pyruvate} + \text{CoASH} + \text{NAD}^+ \longrightarrow \text{Acetyl CoA} + CO_2 + \text{NADH} + H^+$$

The reaction is, for practical purposes, irreversible. The complex (from *E. coli*) contains multiple copies of several types of subunits: 24 of *pyruvate decarboxylase* (E1), 24 of *dihydrolipoyl transacetylase* (E2), and 12 of *dihydrolipoyl dehydrogenase* (E3). Five cofactors participate in the reaction. Three are bound to components of the complex: thiamin pyrophosphate (TPP), lipoic acid (Lip), and flavin adenine dinucleotide

(FAD); two come and go from solution: nicotinamide adenine dinucleotide (NAD$^+$) and coenzyme A (CoA). The different subunits act in a coordinated fashion as follows (Figure 4-1):

1. Through the carbonyl carbon, pyruvate binds to TPP and loses CO_2, leaving hydroxyethyl TPP on E1.

2. Hydroxyethyl is oxidized to an acetyl group as it is transferred to lipoamide on E2. The resultant thioester is energy-rich.

3. The acetyl group is transferred to CoA and leaves as acetyl CoA. Lipoamide is left with both thiols in the reduced state.

4. Lipoamide on E2 is reoxidized by FAD on E3.

5. Reduced $FADH_2$ on E3 is reoxidized by NAD$^+$, leaving FAD on E3. The entire complex is back in its original state; two H's from the oxidation of pyruvate are carried away as NADH + H$^+$.

Regulation. The pyruvate dehydrogenase complex is subject to end-product inhibition, i.e., NADH and acetyl CoA inhibit the enzyme. This is even more effective in the presence of long-chain fatty acids. In addition, the complex from eukaryotic sources contains a kinase which can phosphorylate a serine residue on E1 in the presence of ATP; the phosphorylated enzyme has decreased activity. A phosphatase, also present in the complex, restores activity. Thus the supply of ATP, signifying the energy-state of the cell, leads to regulation of this important enzyme in a major energy-producing pathway.

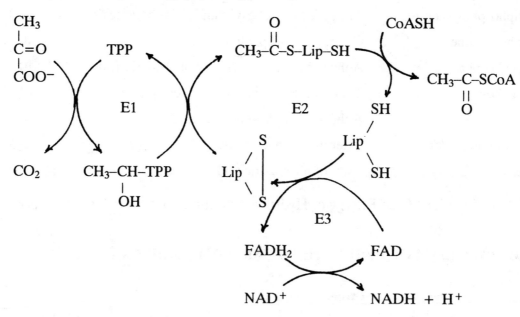

Figure 4-1. Reactions Catalyzed by the *Pyruvate Dehydrogenase Complex.*

Enzyme 1 (E1): *Pyruvate decarboxylase*

Enzyme 2 (E2): *Dihydrolipoyl transacetylase*

Enzyme 3 (E3): *Dihydrolipoyl dehydrogenase*

Cofactors: Thiamin pyrophosphate (TPP)

Lipoamide (Lip)　　　Coenzyme A (CoA)

NAD$^+$　　　　　　　FAD

B. The Citric Acid Cycle (Krebs Tricarboxylic Acid Cycle)

The cycle is only active under aerobic conditions. It metabolizes the remaining carbons of glucose, as well as all carbons of fatty acids, to CO_2. Hydrogens are collected on carriers (NADH, $FADH_2$) for subsequent oxidation by the electron transport system. The cycle also serves to link many metabolic pathways not only in catabolism, but also in an anabolic mode, since it can provide intermediates for gluconeogenesis, synthesis of amino acids, etc.

It functions as a closed loop, in a sense catalytically, in that each turn regenerates the starting material (oxaloacetate). Intermediates are not used up nor do they accumulate. Products of the cycle are two CO_2, one high-energy phosphate, and four *reducing equivalents* (3 NADH and 1 $FADH_2$). Although some individual steps are reversible, the cycle contains enough steps with large negative changes in free energy that as a whole it is not reversible. Nearly all the enzymes are soluble, occurring in the mitochondrial matrix.

The cycle operates as follows (Figure 4-2):

1. *Condensation.* The two-carbon moiety from acetyl CoA condenses with oxaloacetate producing the six-carbon tricarboxylic acid, citrate (*citrate synthase*).

2. *Isomerization.* Citrate is dehydrated to *cis*-aconitate, then rehydrated to isocitrate (*aconitase*, which contains an iron-sulfur center which helps bind citrate).

3. *Oxidative decarboxylation.* Isocitrate is converted to α-ketoglutarate with loss of CO_2. This carbon and that of the CO_2 in the subsequent step are derived from the oxaloacetate in step 1, not from the incoming acetyl group. Two H's are released as NADH + H^+ (*isocitrate dehydrogenase*).

4. A second *oxidative decarboxylation* occurs as α-ketoglutarate is converted to a four-carbon acid (succinate, as the CoA derivative), again with loss of CO_2 and production of a reducing equivalent (NADH + H^+) (*α-ketoglutarate dehydrogenase*). The enzyme is a complex whose constitution and mechanism of action are very similar to pyruvate dehydrogenase and its mode of action.

5. *Phosphorylation.* The high-energy of succinyl CoA is conserved during the generation of GTP from GDP and P_i (*succinyl CoA synthetase*, formerly known as *succinyl thiokinase*).

6. *Oxidation.* Two H's are removed (as $FADH_2$) from succinate to form the unsaturated acid fumarate (*succinate dehydrogenase*, which contains three iron-sulfur centers). It often happens that FAD is the electron acceptor when two H's are removed from two adjacent carbon atoms. This enzyme is the only one of the eight to be membrane-bound. It is *inhibited by malonate*.

7. *Hydration.* Water adds to fumarate to form the hydroxy acid, malate (*fumarase*).

8. *Oxidation.* Oxaloacetate is regenerated as two final H's are carried off from malate as NADH + H^+ (*malate dehydrogenase*).

Regulation. The citric acid cycle is sensitive to the supply of substrates at several points. *Citrate synthase* requires both acetyl CoA and oxaloacetate; if the supply of the former decreases (due to slowed glycolysis or decreased fatty acid oxidation), or if oxaloacetate is diverted for gluconeogenesis, this step and the cycle as a whole will slow down. All four oxidation-reduction reactions need oxidized coenzymes (NAD^+, FAD) in order to function. Thus the cycle is also responsive to the redox state of the cell, and will not operate unless oxidized coenzymes are regenerated by the electron transport system. In addition, there is control by effectors at several points:

- *citrate synthase* is allosterically inhibited by ATP and also by succinyl CoA; activated by ADP;
- *isocitrate dehydrogenase* is also allosterically activated by a high ratio of ATP/ADP;
- the *α-ketoglutarate dehydrogenase* complex is controlled by end-product inhibition very much as is pyruvate dehydrogenase: it is slowed by NADH and, in this case, succinyl CoA.

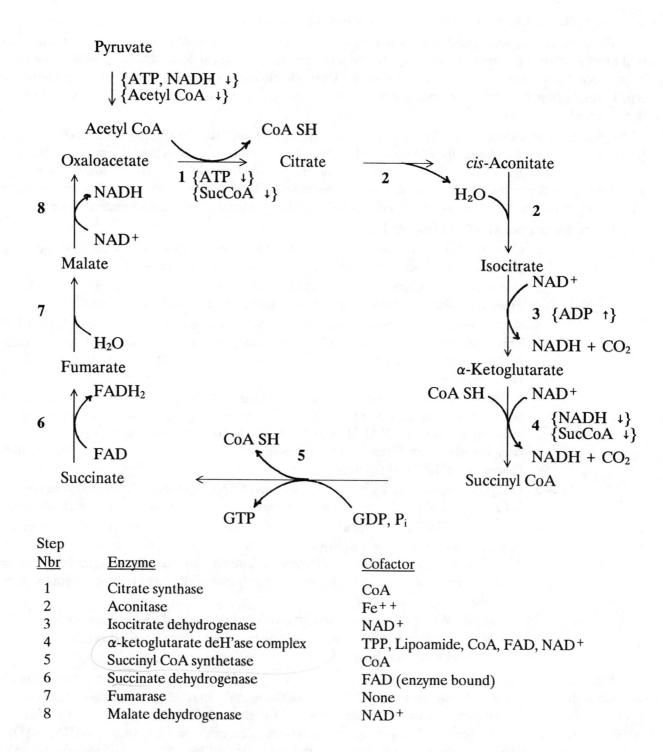

Figure 4-2. The Citric Acid Cycle.

Numbers in boldface refer to enzymes.
Items in curved brackets refer to regulatory effects: ↑ = stimulatory; ↓ = inhibitory.

Control of the pyruvate dehydrogenase complex and of the citric acid cycle illustrates the application of the concept of "energy charge." In analogy with a storage battery, a cell is said to have a high charge if the amount of ATP, as a fraction of the total amount of adenine nucleotides, is high. In the regulatory functions discussed above, a high "energy charge" inhibits those processes that result in production of energy; a state of low "charge" stimulates energy-producing processes.

III. ELECTRON TRANSFER VIA THE RESPIRATORY CHAIN

The Mitochondrion: about 1 x 3 μm (Figure 4-3). A liver cell has about 1000. The inner membrane, of which cristae are extensions, is *very selective in its permeability*: it allows gases and water to pass through, and small hydrophobic molecules, but is impermeable to ions and most polar molecules except through particular gates or transporters. Many enzymes, including those of electron transport and succinate dehydrogenase, are located on or in the inner membrane. The matrix has the other enzymes of the citric acid cycle, pyruvate dehydrogenase, and many others.

Figure 4-3.
The Mitochondrion.

In catabolism, the usual electron (and H) acceptor is NAD^+. Flavoproteins are also used: protein-bound FAD, or FMN. Reduced coenzymes ($FMNH_2$, $FADH_2$, $NADH + H^+$) *must be reoxidized*. In aerobic metabolism, this is by the **electron transport system** (also called the **respiratory chain**).

The *Respiratory Chain*, or *Electron Transport System* (ETS): a chain of enzymes, arranged in several complexes located on the **inner mitochondrial membrane**, specialized to carry out electron transfer from reduced coenzymes to oxygen. A complex may have many subunits. At each electron transfer, a drop in free energy (ΔG is negative) occurs. At complexes I, III, and IV, sufficient free energy is released ultimately to drive the phosphorylation of ADP to ATP by Complex V (Figure 4-4 and Table 4-2).

Several enzymes of the ETS are **cytochromes** — proteins that contain heme (iron protoporphyrin) or a heme derivative. In the cytochrome system, electrons are transferred because the valence of the iron can change from $Fe^{++} \rightleftharpoons Fe^{+++}$, for example:

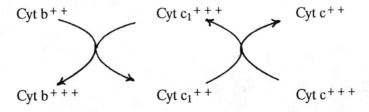

An electron is transferred from Cyt b to Cyt c_1; b is oxidized and c_1 is reduced.

Now the electron passes from reduced Cyt c_1 to Cyt c. c_1 is reoxidized; c is reduced.

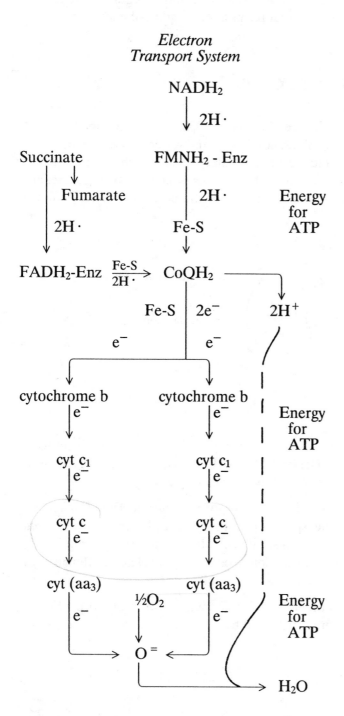

Electron Transport System

NADH$_2$

$\downarrow$ 2H·

Succinate   FMNH$_2$ - Enz

   $\downarrow$ Fumarate $\downarrow$ 2H·

   2H·   Fe-S

FADH$_2$-Enz $\xrightarrow{\frac{Fe-S}{2H\cdot}}$ CoQH$_2$

Fe-S | 2e$^-$ 2H$^+$

e$^-$ e$^-$

cytochrome b   cytochrome b

e$^-$ e$^-$

cyt c$_1$ cyt c$_1$

e$^-$ e$^-$

cyt c   cyt c

e$^-$ e$^-$

cyt (aa$_3$)   cyt (aa$_3$)

e$^-$ ½O$_2$ e$^-$

O$^=$

H$_2$O

Energy for ATP

Energy for ATP

Energy for ATP

Comments

NADH dehydrogenase, **Complex I**, conducts electrons from NADH to CoQ. Inhibited by **rotenone**.

Energy is conserved here for generation of ATP while a pair of electrons (with 2H's) is transferred. This is the **first of 4 complexes** that conduct electrons from reduced coenzymes to oxygen in coupled oxidative phosphorylation.

Succinate dehydrogenase is **Complex II**. Note that electrons from succinate (as FADH$_2$) enter ETS at a lower energy level, bypassing the first energy-producing complex. Thus electrons from succinate can produce only 2 ATP's.

Coenzyme Q is a quinone (ubiquinone). Small and hydrophobic, it can move around in the membrane, and conducts electrons from Complexes I & II to Complex III.

Electron transfer from CoQ to Cyt c is mediated by **b-c$_1$ Complex**, or **Complex III**. Protons and electrons now separate, but it still takes a **pair** of electrons to generate enough energy for 1 ATP. **Antimycin A** inhibits here.

Cytochrome c is a relatively small, mobile protein which conducts electrons from Complex III to IV.

Cytochrome (aa$_3$) is **Complex IV**, also called *cytochrome oxidase*. Contains Cu as well as Fe, and transfers electrons directly to oxygen. Inhibited by **cyanide** (CN$^-$).

Energy for ATP: e$^-$ transfer at I, III, & IV expels H$^+$ from mitochondrion. Re-entry drives phosphorylation by **chemi-osmotic coupling** (page 71).

Figure 4-4. Mitochondrial Electron Transport. See also Table 4-2 for summary.

Iron-Sulfur Proteins, subunits of some carriers of the ETS, have "iron-sulfur centers," non-heme iron linked to cysteine and/or inorganic sulfur atoms, which function in one-electron transfers. Iron-sulfur centers are also found elsewhere, as in aconitase where they help in binding substrate.

Coupled Oxidative Phosphorylation: the process of ATP generation is tightly **coupled** to the process of electron transfer. If one stops, the other stops, like gears that mesh. This is the normal state in respiring mitochondria.

Complex	Alternative Name	Function with Respect to:	
		Electrons	Protons
I	NADH Dehydrogenase	Accepts from NADH; passes to CoQ	Extrudes from mitochondrion to make pH gradient
II	Succinate Dehydrogenase	Accepts from succinate passes to CoQ	
	Co Q (Ubiquinone)	Accepts from I & II passes to III	
III	Cytochrome bc_1 Complex	Accepts from CoQ passes to Cytochrome c	Extrudes from mitochondrion to make pH gradient
	Cytochrome c	Accepts from III passes to IV	
IV	Cytochrome oxidase	Accepts from Cyt c passes to oxygen	Extrudes from mitochondrion to make pH gradient
V	F_oF_1 ATP Synthase		Allows back into mitochondrion to drive phosphorylation

Table 4-2. Summary of Carriers in the Electron Transport System

Shuttles: Because the mitochondrial membrane is impermeable to NADH, extramitochondrial NADH cannot enter directly. Reducing equivalents are passed to a carrier molecule and enter by one of two **shuttle** mechanisms, summarized in Table 4-3. The details are complex, and will not be elaborated here.

	Principal Molecules Involved	Electron acceptor	Located in:	ATP Yield	Comments
Glycerol phosphate shuttle	Glycerol-3-P, Dihydroxyacetone-P	CoQ via FAD	Skeletal muscle, brain	2	No transmembrane passage
Malate-aspartate shuttle	Malate, oxaloacetate, glutamate, α-KG, aspartate	Intra-mito. NAD^+	Liver, Kidney, heart mito.	3	Complicated: uses 2 transporters & 2 transaminations

Table 4-3. Summary of Shuttle Systems for Transporting Extramitochondrial Reducing Equivalents to the Electron Transport System

Transformation	Enzyme	Reducing Equivalent Produced	~P Produced	
			Substrate Level	Electron Transport Level
Glucose $\longrightarrow$ $\longrightarrow$ 1,3-Bisphospho-glycerate	Glyceraldehyde-3-phosphate dehydrogenase	NADH + H$^+$ (x 2)		6 ATP
1,3-Bisphospho-glycerate $\longrightarrow$ 3-phosphoglycerate	Phosphoglycerate kinase		2 ATP	
Phosphoenolpyruvate $\longrightarrow$ pyruvate	Pyruvate kinase		2 ATP	
Pyruvate $\longrightarrow$ acetyl CoA	Pyruvate dehydrogenase	NADH + H$^+$ (x 2)		6 ATP
Isocitrate $\longrightarrow$ α-ketoglutarate	Isocitrate dehydrogenase	NADH + H$^+$ (x 2)		6 ATP
α-Ketoglutarate $\longrightarrow$ succinyl CoA	α-Ketoglutarate dehydrogenase complex	NADH + H$^+$ (x 2)		6 ATP
Succinyl CoA $\longrightarrow$ succinate	Succinyl CoA synthase		2 ATP (via GTP)	
Succinate $\longrightarrow$ fumarate	Succinate dehydrogenase	FADH$_2$ (x 2)		4 ATP
Malate $\longrightarrow$ oxaloacetate	Malate dehydrogenase	NADH + H$^+$ (x 2)		6 ATP
	Gross Total		6 ATP	34 ATP

Subtract 2 ATP's that are consumed at start of glycolysis   −2

Assumptions: (1) tightly-coupled system;

(2) no loss of energy in transferring NADH from outside of mitochondrion (glycolysis) to inside. Actually, yield may be reduced, depending on which "shuttle" mechanism is used (Table 4-3).

♦ **Net Total: 38 ATP** from each glucose

Table 4-4.   Yield of Reduced Coenzymes and ATP from the Complete Oxidation of Glucose

Regulation depends on "energy charge," of which the **ATP/ADP ratio** is in part an indicator. If ATP predominates (high energy charge), the cell doesn't need energy, and electron transfer slows. If the cell needs energy, ADP predominates and is available for coupled phosphorylation, which now speeds up, allowing increased electron transfer, re-oxidation of NADH and FADH$_2$, and speeding up of Krebs (TCA) cycle. A supply of the **reduced coenzymes** and of **P$_i$** are also necessary. Availability of **O$_2$**, the terminal electron acceptor, also limits, since lack of an acceptor would stop everything (Figure 4-4).

Uncoupling: some chemicals, e.g. 2,4-dinitrophenol (DNP), can act as uncouplers, allowing electron transfer to oxygen to proceed without phosphorylation. Energy is wasted as heat; no ATP is formed. Like depressing the clutch on a car, the engine runs but no useful work is done.

P/O Ratio: the number of ATP's formed per atom of O consumed in metabolism of substrate. NADH has P/O of 3; FADH$_2$, only 2 because Complex I is bypassed (Figure 4-4). An uncoupled system has P/O of zero for all substrates.

Yield of ATP: The complete aerobic metabolism of glucose to CO$_2$ and H$_2$O produces 38 ATP. The origin of the reducing equivalents and resulting high-energy phosphates is summarized in Table 4-4, page 70.

IV. CHEMI-OSMOTIC THEORY OF OXIDATIVE PHOSPHORYLATION

The components of the electron transport system are asymmetrically placed in the inner mitochondrial membrane, such that electron transfer results in a **directional extrusion of H$^+$** (Figure 4-5) from inside (matrix) to the intermembrane space. The outer membrane is porous and highly permeable to protons,

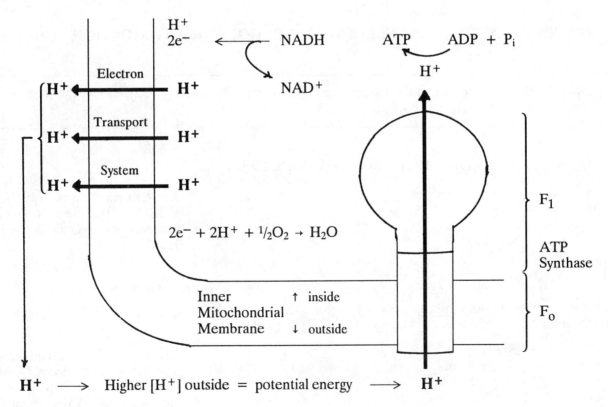

Figure 4-5. Oxidative Phosphorylation via a Gradient of Protons

making the intermembrane space equivalent to the cytoplasm. Since the inner membrane is impermeable to protons (except through the F_o - F_1 complex) this leads to a **pH difference** (more acid outside) and an **electrochemical gradient** across the inner membrane, a condition of potential energy. To release the potential energy, protons are allowed back in through the F_o - F_1 complex in such a way (details unclear) as to drive the phosphorylation of ADP to ATP. This complex (**Complex V**) is therefore known as **ATP Synthase**.

A coupled system is self-regulated by high energy charge: unavailability of ADP prevents further phosphorylation which, in turn, prevents entry of H^+ through the ATPase. As the electrochemical gradient builds up to a maximal level, further extrusion of protons is inhibited, slowing electron transfer. Uncoupling agents act by conducting protons across the membrane so as to bypass the ATPase and "short-circuit" the gradient, allowing electron transfer without phosphorylation.

In this chapter we have considered the oxidation of pyruvate and the production of reducing equivalents from the pyruvate dehydrogenase complex and from the operation of the Krebs Tricarboxylic Acid Cycle. Other oxidative processes, such as those that metabolize fatty acids and amino acids, also release NADH and $FADH_2$; these will be discussed elsewhere.

V. REVIEW QUESTIONS ON ENERGETICS AND BIOLOGICAL OXIDATION

DIRECTIONS: For each of the following multiple-choice questions (1 - 22), choose the ONE BEST answer.

1. The P/O ratio for NADH + H^+ produced in the Krebs Cycle is:

A. 3
B. 2
C. 1
D. 0
E. –1

Ans A: Oxidation of one (NADH + H^+) by the electron transport system uses one O atom and yields 3 ATP.

2. How many high-energy phosphates are generated through the complete metabolism of one acetyl (CoA) unit to CO_2 and H_2O?

A. 1
B. 3
C. 6
D. 11
E. 12

Ans E: 9 ATP from oxidation of 3 (NADH + H^+), 2 from 1 $FADH_2$, and 1 at the substrate level from the succinyl CoA synthetase reaction.

3. In a biological situation, the "criterion of spontaneity" that allows one to predict whether a reaction will proceed, is:

A. $\Delta G^{\circ\prime}$
B. $\Delta E_o^{\prime}$
C. standard free energy change
D. energy of activation
E. ΔG

Ans E: Standard changes can be overcome by concentration effects; activation energy determines kinetics; the only criterion that is *always* predictive is plain ΔG.

4. All of the following are "high-energy" compounds EXCEPT:

A. CTP
B. ATP
C. CH_3CO-SCoA
D. glucose-6-phosphate
E. phosphoenol pyruvate.

Ans D: The standard free energies of hydrolysis of the phosphate esters of ordinary –OH groups are not unusually high.

5. The most general definition of oxidation is:

A. addition of oxygen
B. removal of oxygen
C. removal of electrons
D. addition of hydrogen
E. removal of hydrogen.

Ans C, which also covers A and E. B and D are reduction.

6. An example of an oxidation - reduction reaction is:

A. malate $+ NAD^+ \rightarrow$ oxaloacetate $+ NADH + H^+$
B. succinyl-CoA $+ GDP + P_i \rightarrow$ succinate $+ GTP$
C. acetyl CoA $+$ oxaloacetate $\rightarrow$ citrate $+ CoA$
D. fumarate $+ H_2O \rightarrow$ malate
E. ATP $+ H_2O \rightarrow ADP + P_i$

Ans A because NAD^+ accepts electrons from malate. The other reactions are, respectively: phosphorylation, condensation, hydration, hydrolysis.

7. In aerobic metabolism most of the high-energy phosphate is generated by:

A. glycolysis acting on glycogen and generating lactate
B. glycolysis acting on glucose and generating pyruvate
C. the pyruvate dehydrogenase reaction
D. the Krebs Tricarboxylic Acid Cycle
E. protons passing through the F_o - F_1 complex.

Ans E because by far the most ATP is made by the electron transport system, which generates a pH gradient which in turn drives the F_o - F_1 ATPase.

8. Under normal conditions, most regions of the inner mitochondrial membrane are impermeable to:

A. O_2
B. CO_2
C. H^+
D. H_2O
E. N_2

Ans C because the inner membrane is impermeable to ions except through specific carriers. Many gases and small molecules can pass through.

9. The immediate driving force that powers the phosphorylation of ADP to ATP by ATP synthase is:

A. high [ADP] in the mitochondrion
B. high [NADH] in the cytoplasm
C. gradient of ATP across the inner mitochondrial membrane
D. gradient of glucose across the plasma membrane
E. gradient of H^+ across the inner mitochondrial membrane.

Ans E: The driving force for phosphorylation is the potential energy of the pH gradient, produced by electron transfer.

10. The term "oxidative phosphorylation" refers to a process in which:

A. ATP energy is used to drive protons across the inner mitochondrial membrane.
B. A proton gradient provides power for electron transfer from substrate to oxygen.
C. ATP energy supplies power for transfer of electrons from substrate to oxygen.
D. Electron transfer from substrate to oxygen provides energy to create a proton gradient, which then powers synthesis of ATP.
E. Phosphorylation of ADP creates a pH gradient across the inner mitochondrial membrane.

Ans D: The sequence of causation is: oxidation of substrate, delivery of electrons to the ETS, electron transfer to O_2, production of a pH gradient, re-entry of protons through Complex V, phosphorylation.

11. Products of the reactions catalyzed by the pyruvate dehydrogenase complex include all of the following (A - D) EXCEPT: If all are included, no exception, choose E.

A. lactate
B. NADH
C. acetyl CoA
D. CO_2
E. all of the above without exception.

Ans A: Lactate is the product of the lactate dehydrogenase reaction, not PDH.

12. Starting materials for the reactions catalyzed by the pyruvate dehydrogenase complex include all of the following (A - D) EXCEPT: If all are included, no exception, choose E.

A. NAD^+
B. CoA
C. pyruvate
D. ADP
E. all of the above without exception.

Ans D: ADP is involved at the substrate level in glycolysis and the TCA Cycle, but not in the PDH complex.

13. All of the following (A - D) are coenzymes in BOTH the pyruvate dehydrogenase (complex) reaction AND the α-ketoglutarate dehydrogenase reaction EXCEPT: If all are, no exception, choose E.

A. NAD^+
B. FAD
C. lipoic acid
D. CoA
E. all of the above without exception.

Ans E: The two reactions have the same array of coenzymes, all those shown here and also thiamin pyrophosphate.

14. In mitochondria treated with the uncoupler DNP (dinitrophenol), the number of ATP derived from the oxidation of one NADH is:

A. 0
B. 1
C. 2
D. 3
E. 6

Ans A because in an uncoupled system, the proton gradient is short-circuited, allowing electron transfer to proceed without any phosphorylation.

15. The tricarboxylic acid cycle can do many wondrous things. One thing it can NOT do is:

A. be part of the pathway that converts carbons of glucose to the carbon skeleton of glutamate
B. metabolize the carbon skeleton of aspartate to CO_2
C. be part of the pathway that converts the carbon skeleton of glutamate to glucose
D. contribute to metabolizing a majority of the carbons of glucose to CO_2
E. provide part of the pathway for net synthesis of glucose from fat.

Ans E: Mammalian cells cannot make net amounts of glucose from fat because both C's of acetyl CoA are lost in passing through the TCA cycle. An exception could be argued for glycerol, which can enter gluconeogenesis, but this is a minor part of most fat molecules.

16. In the tricarboxylic acid cycle, four oxidation - reduction reactions produce reducing equivalents. The word that characterizes the name of the enzyme in all four cases is:

A. oxidase
B. reductase
C. dehydrogenase
D. isomerase
E. synthetase

Ans C: All dehydrogenases catalyze oxidations when they remove electrons from a substrate, and reductions when the electrons are passed to an acceptor.

17. Which one of the following is changed by the presence of an enzyme?

A. ΔG
B. $\Delta G^{o\prime}$
C. K_{eq}
D. the position of the equilibrium, expressed as concentration of products/reactants
E. the time required to reach equilibrium.

Ans E. An enzyme influences *only* the rate of a reaction (from a change in the activation energy), not the free energy or position of equilibrium.

18. A state of high "energy charge" would:

A. inhibit glycolysis, pyruvate dehydrogenase, and citrate synthase.
B. stimulate glycolysis and pyruvate dehydrogenase, but inhibit citrate synthase.
C. inhibit glycolysis and citrate synthase, but stimulate pyruvate dehydrogenase.
D. stimulate glycolysis, pyruvate dehydrogenase, and citrate synthase.
E. stimulate glycolysis, pyruvate dehydrogenase, citrate synthase, and the mitochondrial ATP synthase.

Ans A because high energy charge means the cell has plenty of energy, so those reactions that lead to production of more energy are inhibited.

19. After complete metabolism of one molecule of glucose to CO_2 and H_2O through 2 whole turns of the Krebs Cycle, the amount of oxaloacetate in the mitochondrion will be:

A. increased by two molecules
B. increased by one molecule
C. unchanged
D. decreased by one molecule
E. decreased by two molecules.

Ans C because the intermediates neither accumulate nor are they used up; during each turn of the cycle, 2 C's enter as acetyl (CoA) and 2 C's come out as CO_2.

20. Addition of cyanide to a tightly-coupled mitochondrial system causes:

A. increase in electron transport, inhibition of ATP formation
B. decrease in electron transport, increase in ATP formation
C. decrease in electron transport with a compensatory increase in the Krebs Cycle
D. decrease in the Krebs Cycle, electron transport, and ATP formation
E. decrease in electron transport only.

Ans D: If electrons cannot reach O_2, they back up, causing accumulation of the reduced forms of carriers, and all aerobic processes stop.

21. If the standard free energy change, $\Delta G°'$, for the transformation A $\rightleftharpoons$ B is positive, then:

A. The reaction will go to the right under all conditions.
B. The reaction will go to the right, starting from 1M concentrations each of A and B.
C. The equilibrium constant is greater than 1.
D. The reaction will go to the left, starting from standard conditions.
E. The equilibrium constant is negative.

Ans D: Standard free energy changes only predict with certainty how a reaction will proceed starting from standard conditions. Positive sign indicates left as written.

22. Given the process:

$$W \overset{+2}{\rightarrow} X \overset{-3}{\rightarrow} Y \overset{0}{\rightarrow} Z$$

with standard free energy changes in Kcal/mole as shown, what is the standard free energy change for the process Z $\rightarrow$ W?

A. −1 Kcal/mole
B. +1 Kcal/mole
C. 0 Kcal/mole
D. +2 Kcal/mole
E. −2 Kcal/mole.

Ans B: The over-all value is the arithmetic sum of the individual steps. Going from right to left reverses the sign.

DIRECTIONS: For each set of questions, choose the ONE BEST answer from the lettered list above it. An answer may be used one or more times, or not at all.

Questions 23 - 30:

A. ATP	E. CoQ
B. cytochrome (aa_3)	F. NADH dehydrogenase
C. cytochrome b	G. O_2
D. cytochrome c	H. NADH + H^+

23. Transfers electrons directly to oxygen.

Ans B: the same as cytochrome oxidase.

24. Accepts 2 hydrogens from the first complex of the electron transport system.

Ans E: CoQ is the carrier that transfers electrons from complexes I or II to complex III.

25. The terminal acceptor of electrons in aerobic respiration.

Ans G: All electrons end up on O_2 with the production of H_2O.

26. Its production by the electron transport system requires the presence of an electrochemical gradient of protons across the inner mitochondrial membrane.

Ans A: phosphorylation of ADP to ATP requires energy, derived from the potential energy of the proton gradient when H^+ pass back through ATP synthase.

27. Inhibited by cyanide.

Ans B: Cyanide binds very tightly to cytochrome oxidase and prevents transfer of electrons to oxygen.

28. Complex I.

Ans F: NADH dehydrogenase is an alternative name for Complex I.

29. Complex IV.

Ans B: Alternative names are cytochrome (aa_3) and cytochrome oxidase.

30. A component of Complex III.

Ans C: Complex III contains cytochromes b and c_1.

Questions 31 - 38: The reaction catalyzed by the enzyme:

A. succinate dehydrogenase
B. malate dehydrogenase
C. isocitrate dehydrogenase
D. α-ketoglutarate dehydrogenase
E. NADH dehydrogenase

F. citrate synthase
G. aconitase
H. fumarase
I. succinyl CoA synthetase

31. Produces FADH$_2$.

Ans A: Succinate dehydrogenase uses FAD as electron acceptor from succinate.

32. Has a mechanism and cofactor requirement very similar to those of pyruvate dehydrogenase.

Ans D: PDH and α-KG dehydrogenase are very similar in their complex structure, mechanism, and coenzyme requirements.

33. Is an oxidation-reduction reaction allosterically activated by ADP and inhibited by ATP.

Ans C: Isocitrate dehydrogenase is one of the control points of the TCA Cycle. So is citrate synthase, but that is not an oxidation-reduction reaction.

34. Results in phosphorylation at the substrate level.

Ans I: This enzyme, running in "reverse," allows the high energy of succinyl CoA to be conserved as a molecule of GTP.

35. Is the point of entry of an activated acetyl unit into the Tricarboxylic Acid Cycle.

Ans F: Citrate synthase is the "condensing enzyme" that joins acetyl (CoA) with oxaloacetate.

36. Is not part of the Tricarboxylic Acid Cycle.

Ans E: NADH dehydrogenase is the first complex of the electron transport system.

37. Is a reversible hydration-dehydration which interconverts two isomeric forms of the substrate.

Ans G: Aconitase interconverts citrate and isocitrate via the unsaturated acid *cis*-aconitate.

38. Regenerates oxaloacetate which can then enter into another turn of the citric acid cycle.

Ans B: Oxaloacetate is the product of malate dehydrogenase.

Questions 39 - 42:

A. $\Delta G < 0$	E. $-RT \ln K_{eq} < 0$
B. $\Delta G > 0$	F. $\Delta G^{\circ\prime} < 0$
C. $\Delta E_0 > 0$	G. $\Delta G^{\circ\prime} > 0$
D. $K_{eq} < 0$	H. $\Delta G^{\circ\prime} = 0$

39. Given the reaction A $\rightleftharpoons$ B; starting with [A] = 1 M and [B] = 0 M, the reaction is seen to proceed to the right. Then what do we definitely know about the reaction?

Ans A: Free-energy change is negative if a reaction proceeds. All the other criteria can be influenced by concentration or other effects.

40. If at equilibrium [A] = 0.75 M and [B] = .25 M, what do these new data tell us in addition?

Ans G because if equilibrium favors A, K_{eq} is <1 and standard free energy change is positive.

41. If the reaction is coupled to another: B $\rightleftharpoons$ C, and if at equilibrium [A] = 0.43 M, [B] = 0.14 M, and [C] = 0.43 M, then what can we conclude about the reaction A $\rightleftharpoons$ C?

Ans H: If equilibrium concentrations are equal, then K_{eq} = 1 and standard free energy change is zero.

42. Which of the alternatives can never occur under any conditions?

Ans D: K_{eq} is [products]/[reactants]; there is no such thing as a negative concentration.

5. AMINO ACID METABOLISM

A. M. Chandler

I. FUNCTIONS OF AMINO ACIDS IN MAN

Amino acids are ingested in large amounts as structural components of dietary proteins. Unlike carbohydrate and fat, there are no large reserve stores of protein in the body. Thus, a continuous intake is required if tissue breakdown is to be avoided.

Amino acids are:

   ♦ Precursors for the synthesis of proteins.
   ♦ A source of energy under certain conditions.
   ♦ Involved in the detoxification of drugs, chemicals and metabolic by-products.
   ♦ Involved as direct neurotransmitters or as precursors to neurotransmitters.
   ♦ Precursors to several peptide hormones and thyroid hormone.
   ♦ Precursors to histamine, NAD and miscellaneous compounds of biological importance.

II. ESSENTIAL AND NON-ESSENTIAL AMINO ACIDS

All twenty amino acids are essential for life. A lack of a sufficient amount of any one of them leads to severe metabolic disruption and ultimate death.

Most microorganisms and plants are able to synthesize all 20 from glucose or CO_2 and NH_3. Mammals, however, including man, have during the process of evolution lost the ability to synthesize the carbon skeletons for several of the amino acids. Therefore, it is <u>essential</u> that these particular amino acids be obtained through the diet. Those amino acids that are not synthesized at a sufficient rate to meet demand are termed the **essential amino acids** and for man number ten. A useful mnemonic to assist in remembering the essential amino acids is **PVT TIM HALL**:

P- phenylalanine	T- tryptophan	H- histidine*
V- valine	I- isoleucine	A- arginine *
T- threonine	M- methionine	L- lysine
		L- leucine

Note that histidine and arginine are marked with asterisks. These amino acids are undoubtedly required for the infant and growing child, but it is less clear that they are essential for the normal, healthy adult.

III. NITROGEN BALANCE

The greatest portion of N intake is in the form of amino acids in the protein of the diet. After digestion, absorption and metabolic processing, the excess N derived from the NH_2 not required for growth or maintenance is excreted in the urine in the form of urea, NH_3 and other nitrogenous compounds. The normal, healthy adult is in "**nitrogen balance**" or "**equilibrium**". That is, the amount of N ingested in the diet over a given period of time equals that excreted in the urine and feces as excretory products.

Positive and Negative Nitrogen Balance. During pregnancy, infancy, childhood and in the recovery phase from a severe illness or surgery, the amount of N taken in and retained exceeds that excreted. The organism is said to be in a state of **positive nitrogen balance**. On the other hand, during starvation, immediately following severe trauma, surgery or other acute stress such as infections, N excretion exceeds intake and retention and the organism is in a state of **negative nitrogen balance**. A gradual, prolonged negative N balance is associated with **senescence**. Nitrogen balance is humorally controlled. Positive N balance is associated with growth hormone, insulin and with testosterone and other anabolic steroids. Negative N balance is associated with glucocorticoid action in mobilizing amino acids from muscle tissue.

IV. PROTEIN QUALITY

The dietary source of protein is also important for maintaining N balance. Not all proteins have the same biological value (BV). Proteins derived from animal sources have a high BV because they contain all the essential amino acids in the proper proportions. Plant proteins, on the other hand, usually are in lower tissue concentration and are harder to digest. In general, plant proteins are deficient in one or more essential amino acids, primarily lysine, tryptophan or methionine. In some third-world countries animal proteins are almost non-existent and protein intake may be limited to only one or two plant sources. The lack of a single essential amino acid leads to severe growth retardation in children and in adults to negative N balance. Growing children are particularly vulnerable and this is evidenced by the prevalence of **kwashiorkor** (See Chapter 10). Strict vegetarians can do well if they plan a diet containing a mixture of vegetable proteins, each one compensating for a defect in the other.

V. PROTEIN DIGESTION

A. Gastric Digestion

The first phase of protein digestion takes place in the stomach. **Gastrin**, a polypeptide hormone, is secreted into the blood by the antral gastric mucosa upon stimulation by foods. **Ethanol** is a particularly strong stimulator of gastrin release. Gastrin stimulates **chief cells** of the gastric mucosa to secrete the inactive proenzyme, **pepsinogen**, the **parietal cells** to secrete HCL and the **epithelial cells** to secrete **mucroproteins**. Once in contact with the very acidic environment of the stomach (pH< 5.0), a peptide fragment is cleaved from the pepsinogen molecule yielding the active protease, **pepsin**. Pepsin can then activate more pepsinogen autocatalytically. In addition to pepsinogen, other zymogens are secreted which yield pepsins B, C and D.

Pepsins (pH optima of 2.5) hydrolyze ingested proteins at sites involving aromatic amino acids, leucine and acidic amino acids. Because of the relatively short residence time of the stomach contents, digestion is limited. The partially digested, relatively large polypeptides then enter the duodenum of the small intestine.

B. Intestinal Digestion

Pancreatic Zymogens. The pancreas secretes several proenzymes into the duodenum along with a slightly alkaline fluid buffered to the pH optima of the active forms. The proenzymes include **trypsinogen, chymotrypsinogens, procarboxypeptidases** and **proelastase.**

Activations. Upon stimulation by the entrance of food into the intestine the intestinal mucosa secretes the enzyme **enterokinase** which acts on trypsinogen converting it to trypsin. Trypsin in turn activates more trypsinogen and the other proenzymes.

$$\text{Trypsinogen} \xrightarrow{\text{enterokinase}} \text{trypsin + peptide}$$

$$\text{Chymotrypsinogens} \xrightarrow{\text{trypsin}} \text{chymotrypsins + peptides}$$

$$\text{Procarboxypeptidases} \xrightarrow{\text{trypsin}} \text{carboxypeptidases + peptides}$$

$$\text{Proelastase} \xrightarrow{\text{trypsin}} \text{elastase + peptide}$$

Other Enzymes. The following brush border enzymes are secreted into the intestinal lumen but also work intracellularly:

Aminopeptidases: broad specificity; hydrolyze N-terminal amino acids.
Dipeptidases: hydrolyze dipeptides such as glycylglycine.
Prolinase: hydrolyzes peptides containing proline at the N-terminus.

VI. AMINO ACID ABSORPTION

Amino acid absorption is very rapid in the small intestine and is carried out by active transport mechanisms and is, therefore, an energy-expending process. Several specific transport mechanisms have been identified involving different classes of amino acids. These include those for:

♦ small neutral amino acids
♦ large neutral amino acids
♦ basic amino acids
♦ acidic amino acids
♦ proline

Amino acids of the same class compete with one another for absorption sites.

The Gamma-Glutamyl Cycle (Figure 5-1). In addition to the transport mechanisms listed above, a general absorption mechanism involving all amino acids with a free amino group has been proposed, referred to as the **gamma-glutamyl cycle.** It provides a role for glutathione and explains the presence of 5-oxoproline in the urine. For every amino acid transported across the membrane by this proposed mechanism, three ATP's are consumed during the regeneration of glutathione.

In addition to amino acids, some small peptides are absorbed directly into the blood without hydrolysis. Very rarely, whole proteins are absorbed intact. The major route of entry into the blood is *via* the portal vein. Once in the blood amino acids are rapidly absorbed into cells. Liver and kidney take up the largest fraction. A blood-brain barrier exists for some amino acids, especially for glutamic acid.

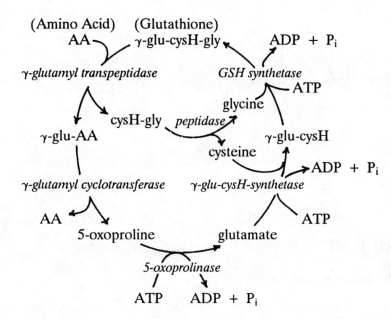

Figure 5-1. The Gamma-Glutamyl Cycle.

VII. AMINO ACID DEGRADATION

In the breakdown of amino acids the first task the cell must accomplish is the removal of the alpha-amino groups. The two major mechanisms by which this is accomplished are **transamination** and **oxidative deamination**.

A. Transamination

Twelve amino acids can undergo transamination. These are ala, arg, asN, asp, cys, ile, leu, lys, phe, trp, tyr, and val. (Mnemonic: **VAL AT CAPITAL**).

The general reaction involved is:

$$\underset{\text{R-CH-COOH}}{\overset{\overset{\displaystyle NH_2}{|}}{}} + \underset{\text{R}'\text{-C-COOH}}{\overset{\overset{\displaystyle O}{||}}{}} \rightleftharpoons \underset{\text{R}'\text{-CH-COOH}}{\overset{\overset{\displaystyle NH_2}{|}}{}} + \underset{\text{R-C-COOH}}{\overset{\overset{\displaystyle O}{||}}{}}$$

The enzymes involved are called **transaminases** or **aminotransferases**. The reactions catalyzed by transaminases are freely reversible with equilibrium constants approaching 1.0. They can be found both in the mitochondria and the cytosol. The transfer of amino groups from most amino acids to α-ketoglutarate to form glutamate takes place in the cytosol. The glutamate formed can then enter the mitochondria via a special transport mechanism where it can undergo oxidative deamination or else form aspartic acid which can then reenter the cytoplasm.

Mechanism of transamination. All transaminases share a common reaction mechanism and use the same cofactor, **pyridoxal phosphate (PLP)**, which is derived from the vitamin, **pyridoxine (vitamin B$_6$)**. PLP is covalently bound to the enzyme via a Schiff's base linkage to an epsilon amino group of a specific lysine located in the active site. During transamination, a Schiff's base forms between the amino acids and PLP.

Details of the reaction (Next page).

First stage:

$$\underset{AA_1}{\overset{\overset{\displaystyle NH_2}{|}}{R_1\text{-CH-COOH}}} + \underset{PLP\text{-}E}{\overset{\overset{\displaystyle HC=O}{|}}{E}} \underset{H_2O}{\overset{}{\rightleftharpoons}} \underset{Aldimine}{\overset{\overset{\displaystyle N=CH\text{-}E}{|}}{R_1\text{-CH-COOH}}} \longleftrightarrow \underset{Ketimine}{\overset{\overset{\displaystyle N\text{-}CH_2\text{-}E}{||}}{R_1\text{-C-COOH}}} \underset{H_2O}{\overset{}{\rightleftharpoons}} \underset{PMP\text{-}E}{\overset{\overset{\displaystyle CH_2\text{-}NH_2}{|}}{E}} + \underset{\alpha\text{-Ketoacid}_1}{\overset{\overset{\displaystyle O}{||}}{R_1\text{-C-COOH}}}$$

Second stage:

$$\underset{\alpha\text{-Ketoacid}_2}{\overset{\overset{\displaystyle O}{||}}{R_2\text{-C-COOH}}} + \underset{PMP\text{-}E}{\overset{\overset{\displaystyle CH_2\text{-}NH_2}{|}}{E}} \underset{H_2O}{\overset{}{\rightleftharpoons}} \underset{Ketimine}{\overset{\overset{\displaystyle N\text{-}CH_2\text{-}E}{||}}{R_2\text{-C-COOH}}} \longleftrightarrow \underset{Aldimine}{\overset{\overset{\displaystyle N=CH\text{-}E}{|}}{R_2\text{-CH-COOH}}} \underset{H_2O}{\overset{}{\rightleftharpoons}} \underset{PLP\text{-}E}{\overset{\overset{\displaystyle HC=O}{|}}{E}} + \underset{AA_2}{\overset{\overset{\displaystyle NH_2}{|}}{R_2\text{-CH-COOH}}}$$

B. Oxidative Deamination

L-glutamate is the ultimate product from the majority of the transaminations that occur. Glutamate then enters mitochondria where the following reaction occurs:

$$L\text{-glutamate} + NAD^+ + H_2O \underset{dehydrogenase}{\overset{glutamate}{\rightleftharpoons}} \alpha\text{-ketoglutarate} + NH_4^+ + NADH$$

The reaction is catalyzed by *glutamate dehydrogenase* found in both the mitochondria and cytoplasm. The enzyme can use both NAD^+ and $NADP^+$ but NAD^+ is preferred in the catabolic direction. The NADH formed within the mitochondrion can enter the electron transport system and yield 3 ATP.

Glutamate dehydrogenase is a very complex enzyme made up of six identical 56,000-dalton subunits. It is regulated allosterically by a number of effectors.

Inhibitors: ATP, GTP, NADH
Activators: ADP, GDP, specific amino acids
Hormones: thyroxine and steroid hormones

C. Amino Acid Oxidases

L-amino acid oxidase is found in liver, kidney and snake venom. It is involved primarily in the deamination of lysine (E = oxidase molecule):

$$L\text{-amino acid} + H_2O + E\text{-FMN} \longrightarrow \alpha\text{-ketoacid} + NH_3 + E\text{-FMNH}_2$$

D-amino acid oxidase:

$$D\text{-amino acid} + H_2O + E\text{-FAD} \longrightarrow \alpha\text{-ketoacid} + NH_3 + E\text{-FADH}_2$$

The flavin nucleotides (FMN and FAD) are derived from **riboflavin (vitamin B$_2$)**. The regeneration of the oxidized flavin nucleotides is carried out utilizing molecular oxygen yielding H_2O_2 which is rapidly degraded by *catalase*.

VIII. THE UREA CYCLE (KREBS-HENSELEIT CYCLE)

Ammonia is considered to be toxic to cells of the CNS in higher organisms and must be kept at low concentrations. In many aquatic animals like fish and tadpoles, NH_3 diffuses directly into the surrounding water through gills. These animals are called **ammonotelic**. Others such as birds and reptiles excrete the ammonia as <u>uric acid</u> (**uricotelic**). Man and most other terrestrial vertebrates excrete NH_3 as <u>urea</u> (**ureotelic**).

Steps in the production of urea are shown in Figure 5-2, and described in sections A - E (below and next page):

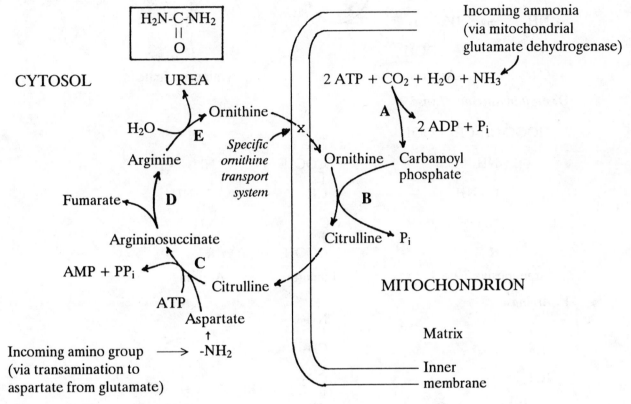

Figure 5-2. The Urea Cycle. Letters in boldface refer to enzymes described in sections A - E in text.

A. *Carbamoyl Phosphate Synthetase*.

The enzyme is mitochondrial, and responds to allosteric activation by N-acetylglutamate.

$$2\ ATP + CO_2 + NH_3 + H_2O \longrightarrow \underset{\text{Carbamoyl-P}}{H_2N\text{-}\overset{\overset{\displaystyle O}{\|}}{C}\text{-}O\text{-}PO_3} + 2\ ADP + P_i$$

B. *Ornithine Carbamoyl Transferase*

Portion within dotted box = **R**

Ornithine + Carbamoyl-P $\longrightarrow$ Citrulline + P_i

C. *Argininosuccinate Synthetase*

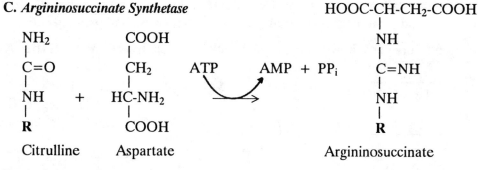

NH₂	COOH			HOOC-CH-CH₂-COOH

Citrulline   Aspartate   Argininosuccinate

D. *Argininosuccinate Lyase*

Argininosuccinate   Fumarate   Arginine

E. *Arginase*

Arginine   Ornithine   Urea

Overall Reaction:

$$2\,NH_3 + CO_2 + 3\,ATP + 3\,H_2O \longrightarrow Urea + 2\,ADP + AMP + 2\,P_i + PP_i$$

The urea cycle is intimately tied to the citric acid cycle by way of fumarate and aspartate. One N is derived from NH_3 and one from aspartate. Each of the enzymes of the urea cycle have been identified with specific genetic defects. Upon ingestion of proteins such patients show hyperammonemia, lethargy, vomiting and other signs of CNS disturbance.

IX.   EXCRETION OF FREE AMMONIA

In addition to excretion as urea, **NH₃** can be removed from cells by transferring it to the gamma carboxyl of glutamate to form **glutamine**. Glutamine is non-toxic and can pass through the blood-brain barrier.

$$Glutamate + ATP + NH_3 \xrightarrow{\text{glutamine synthetase}} glutamine + ADP + P_i$$

Glutamine is transported in the blood either to the liver where the amide group is removed and converted to urea, or else to the kidneys where it encounters in the tubules the enzyme, *glutaminase*. Glutaminase hydrolyzes the glutamine to glutamate and NH_3. The glutamate is reabsorbed by the tubules and the ammonia is excreted in the urine.

X.   DEGRADATION OF THE CARBON SKELETONS

After removal of the α-amino groups, the carbon skeletons of the 20 amino acids undergo a series of reactions that result in products that are members of the glycolytic pathway, the citric acid cycle or ketone bodies. There are only seven of these: pyruvate, acetyl CoA, acetoacetyl CoA, α-ketoglutarate, succinyl CoA, fumarate and oxaloacetate.

Table 5-1 outlines the catabolism of the 20 amino acids.

Amino Acid	Products	Number of Enzymatic Steps	Cofactors	Glycogenic or Lipogenic
Alanine	Pyruvate	1	PLP	G
Glycine	Pyruvate	2	N^5,N^{10}-methylene THFA	G
Serine	Pyruvate	1		G
Cysteine	Pyruvate	2	PLP, NADH	G
Threonine	Pyruvate	3		G
Aspartic Acid	Oxaloacetate	1	PLP	G
Asparagine	Oxaloacetate	2	PLP	G
Histidine	α-ketoglutarate	5	THFA, PLP	G
Glutamic acid	α-ketoglutarate	1	PLP	G
Glutamine	α-ketoglutarate	2	PLP	G
Arginine	α-ketoglutarate	4	PLP, NAD	G
Proline	α-ketoglutarate	4	O_2, PLP	G
Methionine	succinyl CoA	9	ATP, CoA, NAD, biotin, Vit B_{12}	G
Valine	succinyl CoA	10	PLP, NAD, CoA, Vit B_{12}	G
Isoleucine	succinyl CoA	9	PLP, NAD, CoA, FAD, biotin, Vit B_{12}	G,L
Leucine	acetyl CoA, acetoacetyl CoA	6	thiamin PP, lipoic acid, biotin, Vit B_{12}, PLP, CoA, NAD, FAD	L
Phenylalanine	acetoacetyl CoA, fumarate	7	O_2, NADPH, tetrahydrobiopterin	G,L
Tyrosine	acetoacetyl CoA, fumarate	6	O_2,NADPH, tetrahydrobiopterin	G,L
Tryptophan	acetoacetyl CoA, alanine	9	O_2,NADPH,NAD	G,L
Lysine	acetoacetyl CoA	9	NADPH, NAD, NADP, PLP, CoA, FAD	G,L

Table 5-1. Catabolism of the 20 Amino Acids

XI. ONE-CARBON FRAGMENT METABOLISM

In the degradation and synthesis of amino acids (and many other compounds), there is often a need to remove, add or rearrange 1-C units. With the exception of the decarboxylases which remove 1-C units as CO_2 or HCO_3^-, all 1-carbon transfers require the participation of cofactors. The precursors of these cofactors must be supplied by the diet, either as vitamins or as essential amino acids.

A. Tetrahydrofolate (THFA) (Tetrahydropteroylglutamate)

Structure. This compound is the most versatile of the carriers of 1-C fragments and carries them at several levels of oxidation, from the most reduced, methyl, to the most oxidized, methenyl:

$-CH_3$	methyl
$-CH_2-$	methylene
$-CH=O$	formyl
$-CH=NH$	formimino
$-CH=$	methenyl

Tetrahydrofolate is composed of three units: (a) a pteridine derivative, (b) p-aminobenzoic acid (PABA) and (c) glutamic acid. The 1-C fragments are carried either on N^5 or N^{10} or as a bridge between both (Figure 5-3). The **sulfa drugs** act by inhibiting the bacterial enzymes responsible for coupling the PABA moiety to the other components to form the THFA molecule. The bacteria are then inhibited in their ability to carry out 1-C transfers. The sulfa drugs are structural analogues of the PABA moiety. Other drugs such as **methotrexate** also act through interference with the action of tetrahydrofolate. (See Chapter 11)

Metabolism. The several forms of 1-C fragments carried by THFA are interconvertible one to the other (Figure 5-3). Each reaction is reversible, therefore, 1-C fragments can be donated or accepted by each species, creating an equilibrium mixture of 1-C fragments in different oxidation states.

THFA is involved in transferring a methyl group to Vitamin B_{12} in the reactions converting homocysteine to methionine. This is the only methyl transfer involving THFA. All other methylations involve S-adenosylmethionine.

B. S-Adenosylmethionine (SAM)

SAM is the cofactor involved in transmethylations. The general reaction is:

$$SAM + R(acc) \xrightarrow{\text{methyl transferase}} RCH_3 + \text{S-adenosylhomocysteine}$$

The structure of SAM and its formation and metabolism are shown in Figure 5-4.

C. Biotin

All carboxylations via transcarboxylases use **biotin** as a cofactor. Each carboxylation reaction involves a specific transcarboxylase.

$$E\text{-biotin} + ATP + HCO_3^- \longrightarrow E\text{-biotin-}CO_2 + ADP + P_i$$
$$E = \text{transcarboxylase}$$

The biotin is attached to the transcarboxylase through the ϵ-amino group of a lysine in the active center. When "activated" the CO_2 is attached to a nitrogen of the biotin molecule.

Biotin is a vitamin. It is bound tightly and irreversibly by **avidin,** a protein found in significant quantities in raw egg white.

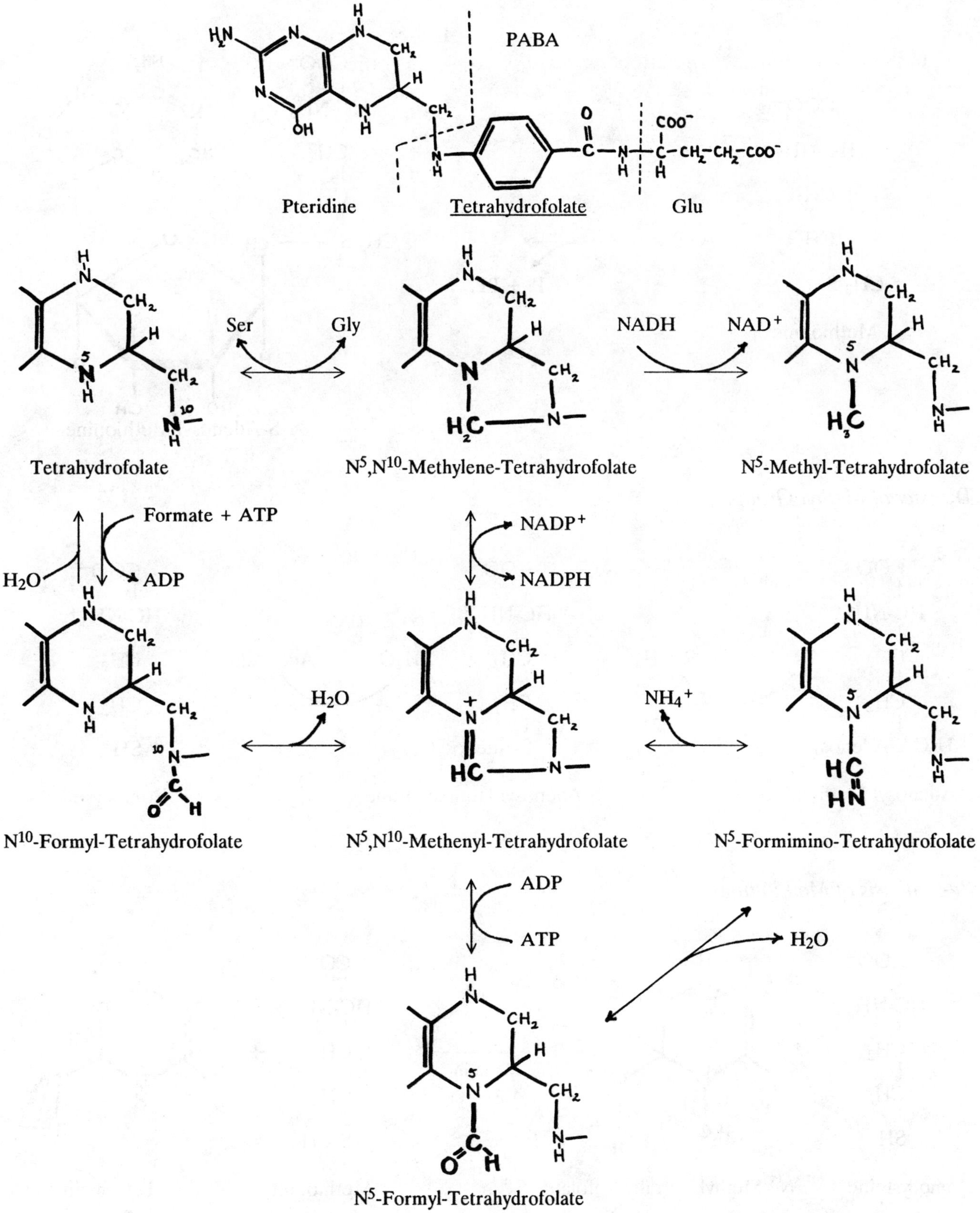

Figure 5-3. Tetrahydrofolic Acid and the Metabolism of the Various C-1-Carrying Forms.

Synthesis of SAM from Methionine:

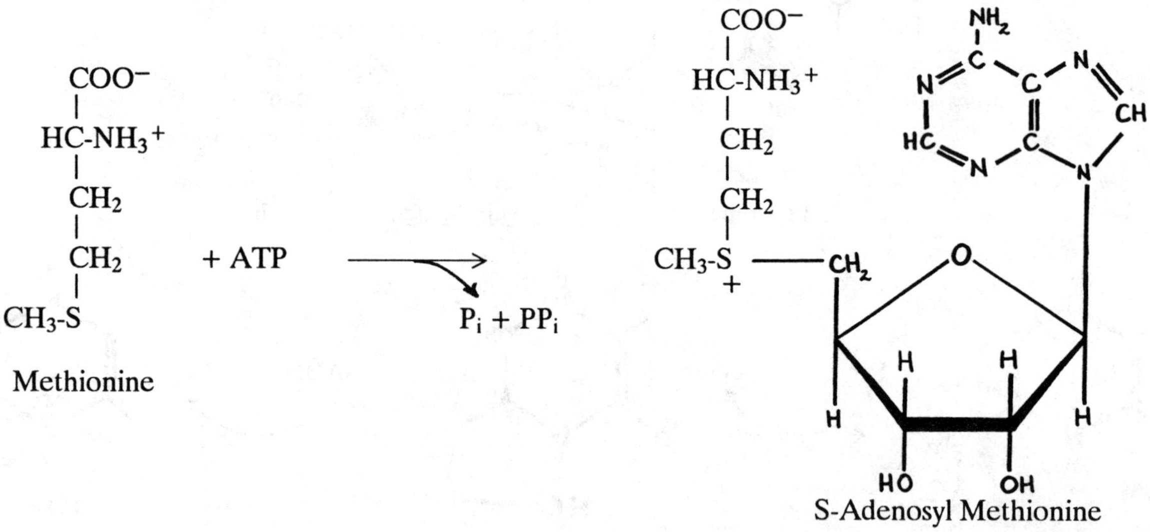

Transfer of Methyl Group:

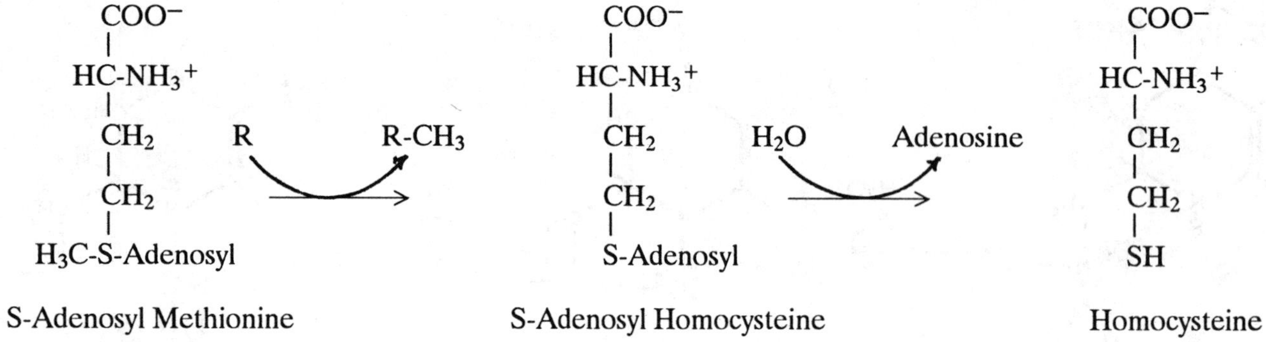

Re-synthesis of Methionine:

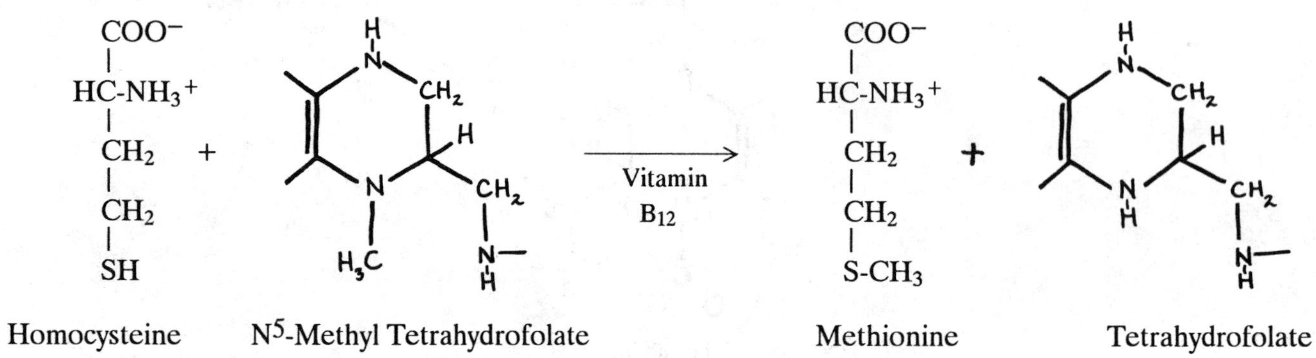

Figure 5-4. S-Adenosyl Methionine: Structure, Formation and Metabolism.

D. Cyanocobalamin (Vitamin B$_{12}$)

Vitamin B$_{12}$ (cyanocobalamin) has a complex ring structure and has as an essential component an atom of the trace metal, **cobalt**. It participates in the transfer of methyl groups from N^5-methyltetrahydrofolate to homocysteine to regenerate methionine. It is also involved in certain one-carbon rearrangement reactions as in the conversion of methylmalonyl CoA to succinyl CoA in the latter stages of the degradation of Met, Val and Ile. An inability to obtain enough Vitamin B$_{12}$ in the diet or an inability to absorb it in the intestinal tract leads to the condition of **pernicious anemia**. The appearance of methylmalonic acid in the blood and urine can also be caused by vitamin B$_{12}$ deficiency.

XII.   METABOLISM OF PHENYLALANINE AND TYROSINE

Several genetic defects exist involving the enzymes of phenylalanine and tyrosine degradation. Tyrosine is also the precursor to several neurohormones and to melanin. Therefore, the metabolism of these amino acids will be covered in slightly more detail.

A. Degradation of Phenylalanine and Tyrosine

The scheme for the degradation of phenylalanine and tyrosine is shown in Figure 5-5.

Phenylketonuria. The most common genetic disturbance in the metabolism of these amino acids occurs at the first step, the oxidation of phenylalanine to tyrosine by *phenylalanine hydroxylase* (*phenylalanine 4-monooxygenase*). The absence of this enzyme or a defect in its functioning caused by defects in other components of the system leads to the accumulation of **phenylpyruvate** causing the condition known as **phenylketonuria**. This condition has a frequency of about 1/15,000 and if untreated is characterized by mental retardation, CNS damage and hypopigmentation. If diagnosed early, some affected children may be spared from the major damaging effects of phenylpyruvate accumulation by placing them on a diet low in phenylalanine, substituting the phenylalanine with tyrosine.

An essential cofactor for phenylalanine hydroxylase is **tetrahydrobiopterin** which is also oxidized, during the oxidation of phenylalanine to tyrosine, to **dihydrobiopterin**. The dihydro form must be reduced back to the tetrahydro form to sustain the overall conversion reaction. The enzyme carrying out this reduction uses NADPH as a cofactor. Some cases of phenylketonuria are the result of defects in this enzymatic step or in steps related to the synthesis of tetrahydrobiopterin.

Alcaptonuria. A second defect in tyrosine catabolism has been observed in persons with the condition **alcaptonuria**. A diminished activity of the enzyme converting **homogentisic acid** to 4-maleylacetoacetate leads to the accumulation of homogentisic acid in the urine which is spontaneously oxidized by atmospheric oxygen to produce a black or dark-colored urine. This condition is relatively benign.

B. Conversion of Tyrosine to Neurohormones and Melanin

Figure 5-6 shows the conversion of tyrosine to DOPA, dopamine, norepinephrine, epinephrine and melanin.

DOPA (dihydroxyphenylalanine) is a key intermediate in these pathways. Genetic defects in the conversion of DOPA to **melanin** causes **albinism**, a condition of very severe hypopigmentation. **Dopamine** and **norepinephrine** are neurotransmitters. **Epinephrine** is synthesized primarily in the adrenal medulla. Disturbances in the functioning of CNS pathways using dopamine as a transmitter have been linked to the condition of **schizophrenia**. A lack of sufficient dopamine production in certain brain structures like the *substantia nigra* or the destruction of this structure by toxic compounds leads to **Parkinson's Disease** or the

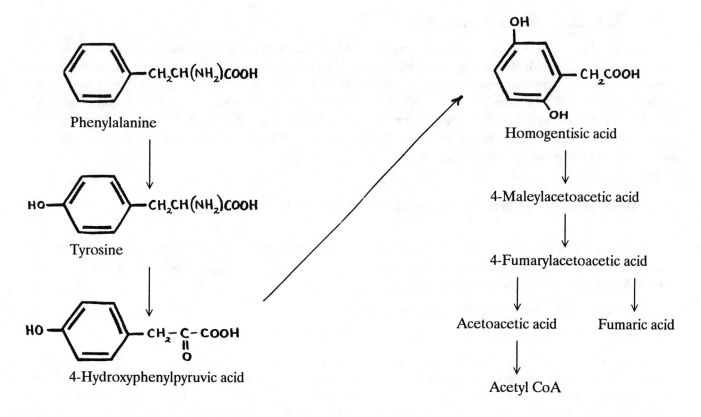

Figure 5-5. Degradation of Phenylalanine and Tyrosine.

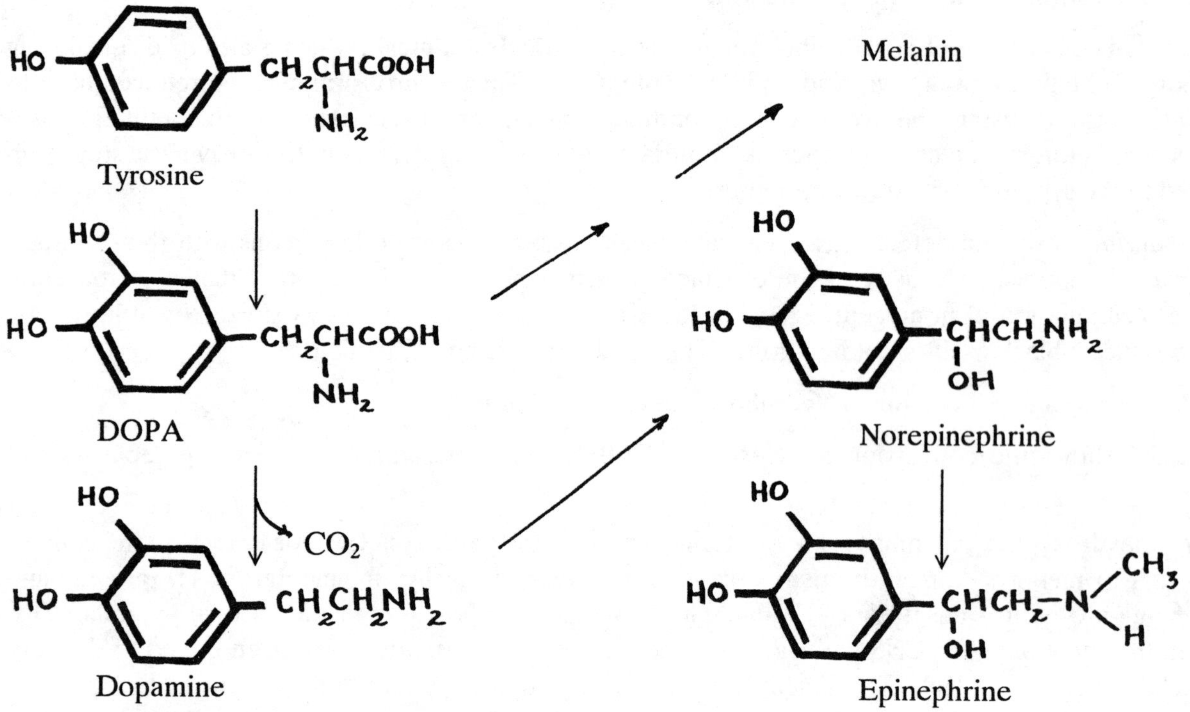

Figure 5-6. Metabolism of Tyrosine.

Parkinsonian syndrome. Administration of large quantities of L-DOPA can reverse many of the symptoms in some cases.

Norepinephrine is found primarily at nerve endings of the adrenergic (sympathetic) nervous system.

XIII.   GENERAL PRECURSOR FUNCTIONS OF AMINO ACIDS

In addition to serving as precursors to proteins, amino acids also act as precursors to many other compounds of biological importance. Table 5-2 lists a few of these compounds and the amino acid(s) from which they are derived.

Compound	Amino Acid Precursor(s)
Neurotransmitters	
many amino acids serve directly as neurotransmitters	
dopamine, epinephrine,norepinephrine	Phe, Tyr
serotonin (5-hydroxytryptamine)	Trp
GABA (γ-aminobutyric acid)	Glu
acetylcholine (via ethanolamine)	Ser
Miscellaneous compounds	
indole acetic acid (plant hormone)	Trp
creatine and phosphocreatine	Met, Gly, Arg
spermine, spermidine	Met, Arg (Orn)
histamine	His
thyroxine	Tyr
heme	Gly
polypeptide hormones	all 20 amino acids
NAD	Trp
taurine	Cys
carnitine	Lys
purine bases	Gly
carnosine, anserine	His
Coenzyme A	Cys

Table 5-2. Important Derivatives of Amino Acids

XIV. REVIEW QUESTIONS ON AMINO ACID METABOLISM

> **DIRECTIONS:**   For each of the following multiple-choice questions (1 - 31), choose the ONE BEST answer.

1. A man is in negative nitrogen balance when his:

A. dietary nitrogen drops below the recommended daily allowance
B. fecal nitrogen excretion exceeds his urinary nitrogen excretion
C. diet contains more nonessential amino acids than essential amino acids
D. urinary nitrogen excretion exceeds his dietary nitrogen intake
E. dietary nitrogen intake exceeds his urinary nitrogen excretion.

Ans D: Negative N balance is a sign of prolonged protein deprivation or muscle protein breakdown caused by glucocorticoid action (stress).

2. If a mature animal in nitrogen balance is placed on a diet deficient only in phenylalanine, which of the following conditions is most likely to occur?

A. Nitrogen balance will become negative and remain that way so long as the deficiency exists.
B. Nitrogen balance will become negative temporarily, but the individual will adapt and N balance will gradually return to zero.
C. Nitrogen intake will continue to equal nitrogen excretion (balance = 0).
D. Nitrogen balance will become positive and remain that way so long as the deficiency exists.
E. Nitrogen balance will become positive temporarily, but the individual will adapt and N balance will gradually return to zero.

Ans A: Protein synthesis can occur only if all 20 amino acids are present. If one is lacking, other tissues are catabolized to supply the missing essential amino acid to the more critical cells and organs

3. Which of the following enzyme pairs is involved in the conversion of amino acid nitrogen into two compounds that directly provide the urea nitrogen?

A. glutamic-oxaloacetic transaminase and diamine oxidase
B. L-amino acid oxidase and racemase
C. serine dehydratase and glutamate dehydrogenase
D. carbamoyl phosphate synthetase and glutamic-oxaloacetic transaminase
E. glutamine synthetase and glutaminase.

Ans D: These enzymes are the only ones mentioned that are members of the urea cycle.

4. Cofactors involved in the regeneration of methionine from homocysteine include:

A. lipoic acid
B. retinoic acid
C. biotin and thiamin pyrophosphate
D. cofactors derived from tetrahydrofolic acid and vitamin B_{12}
E. cofactors derived from vitamins E and K.

Ans D: The enzyme, which contains vit B_{12}, transfers the methyl group from methyl-THFA to B_{12} and then to homocysteine to form methionine.

5. The rate-limiting reaction in the urea cycle is that catalyzed by:

A. argininosuccinase
B. argininosuccinate synthetase
C. arginase
D. ornithine transcarbamoylase
E. carbamoyl phosphate synthetase.

Ans E: Carbamoyl phosphate synthetase is the key rate limiting enzyme in the cycle and is located in mitochondria.

6. The utilization of ammonia for synthesis of the α-amino group of non-essential amino acids:

A. is dependent on the action of glutamate dehydrogenase
B. is achieved by reversal of the urea cycle
C. is mediated by carbamoyl phosphate
D. is dependent on intestinal bacteria
E. requires the participation of glutamine.

Ans A: Glutamate dehydrogenase can, under conditions of abundant energy, add NH_3 to α-ketoglutarate to form glutamate. NADPH is coenzyme.

7. Which one of the following is an essential amino acid?

A. asparagine
B. glycine
C. glutamic acid
D. methionine
E. serine.

Ans D: Met is the "M" in PVT TIM HALL.

8. The combination of which of the following enzymatic activities provides the major route of flow of nitrogen from amino acids to ammonia in man?

A. amino acid oxidases and glutamate dehydrogenase
B. glutamate dehydrogenase and glutaminase
C. transaminases and glutamate dehydrogenase
D. transaminases and glutaminase
E. glutaminase and amino acid oxidases.

Ans C: Transaminases mobilize the α-amino groups eventually to α-ketoglutarate to form glutamate, which is then oxidized and deaminated to α-ketoglutarate again and NH_3. NADH is also formed.

9. An amino acid which is both ketogenic and glucogenic is:

A. tyrosine
B. alanine
C. leucine
D. glutamic acid
E. histidine.

Ans A: The aromatic amino acids, Phe and Tyr, are catabolized to fumarate (glycogenic) and acetoacetic acid (ketogenic).

10. Dopamine is a neurohormone in brain. Its precursor is:

A. cysteine
B. phenylalanine
C. asparagine
D. tryptophan
E. lysine.

Ans B: Dopamine is derived from DOPA which is an oxidative products of Tyr. Tyrosine in turn is formed from phenylalanine.

11. Transport of amino acids by the gamma-glutamyl cycle involves direct participation of which one of the following components?

A. GTP
B. aspartic acid
C. glutathione
D. glutamine
E. glucose.

Ans C: The tripeptide, glutathione, is the key component of the γ-glutamyl cycle.

12. Phenylketonuria is a genetic defect caused by absence of:

A. α-keto acid decarboxylase
B. tyrosinase
C. homogentisic acid oxidase
D. phenylalanine hydroxylase
E. alanine transaminase.

Ans D: The majority of cases of phenylketonuria is caused by a missing or defective phenylalanine hydroxylase resulting in the failure to oxidize Phe to Tyr.

13. The enzyme enterokinase is important in the intestinal digestion of dietary protein because it converts:

A. pepsinogen to pepsin
B. procarboxypeptidase A to carboxypeptidase
C. trypsinogen to trypsin
D. procarboxypeptidase B to carboxypeptidase B
E. chymotrypsinogen to chymotrypsin.

Ans C: Enterokinase, secreted by cells of the intestinal mucosa, makes the initial conversion of trypsinogen to trypsin. Trypsin then converts the other zymogens to their active forms.

14. Excess lysine added to a diet might impair the absorption of:

A. proline
B. arginine, histidine
C. phenylalanine, tyrosine
D. aspartic acid, glutamic acid
E. glycine, alanine.

Ans B: Arg, His and Lys are all basic amino acids. They will compete with one another for the same transport mechanisms.

15. A diet that consists exclusively of corn and no other source of protein, will lead to a state of:

A. nitrogen equilibrium
B. positive nitrogen balance
C. negative nitrogen balance
D. hyperammonemia
E. obesity.

Ans C: Corn is deficient in the essential amino acids lysine and tryptophan.

16. A reaction type in which S-adenosylmethionine participates is:

A. transmethylation
B. transformylation
C. transadenylation
D. transamination
E. transcription.

Ans A: Nearly all transmethylases use S-adenosylmethionine as a cofactor.

17. Which one of the following intermediates directly links the urea cycle with the TCA cycle?

A. acetyl CoA
B. arginine
C. fumarate
D. glutamate
E. pyruvate.

Ans C: The TCA and urea cycles are linked by the intermediates fumarate and oxaloacetate. Oxaloacetate is easily converted to aspartate by transamination.

18. Which one of the following groups consists entirely of amino acids which are nutritionally essential for humans?

A. Valine, isoleucine, tyrosine, arginine
B. Leucine, methionine, isoleucine, alanine
C. Glutamic acid, arginine, cysteine, tryptophan
D. Valine, isoleucine, tyrosine, lysine
E. Lysine, tryptophan, phenylalanine, threonine.

Ans E: Only Lys, Trp Phe and Thr are in the mnemonic PVT TIM HALL.

19. The coenzyme involved in transaminations and many other amino acid transformations is derived from:

A. niacin
B. pyridoxine
C. flavins
D. thiamine
E. vitamin B_{12}.

Ans B: The cofactors for trans-aminases are pyridoxal phosphate and pyridoxamine phosphate, both derived from pyridoxine (Vit B_6).

20. The major nitrogen-containing compound(s) excreted in urine is:

A. amino acids
B. uric acid
C. creatinine
D. urea
E. NH_4^+.

Ans D: About 80% of excreted N is in the form of urea. About 20% in the form of NH_3 or NH_4^+.

21. The major source of urinary ammonia produced by the kidney is:

A. leucine
B. glycine
C. glutamine
D. asparagine
E. glutamic acid.

Ans C: The kidney tubules contain high activity of glutaminase which hydrolyzes glutamine to Glu + NH_3. The NH_3 is excreted into the urine.

22. The phosphorylated form of creatine is:

A. a source of high energy phosphate for ATP formation in muscle
B. a component of the urea cycle
C. excreted by the kidney
D. an important hormone
E. none of the above.

Ans A: Creatine phosphate is at high levels in muscle and acts as a high energy phosphate donor to ADP during muscle activity to help maintain ATP levels for continued contractions.

23. All of the following occur in humans EXCEPT:

A. serine → cysteine
B. phenylalanine → tyrosine
C. glutamate → proline
D. oxaloacetate → lysine
E. homocysteine → methionine.

Ans D: Lysine is one of the essential amino acids and cannot be synthesized by mammals.

24. The primary site of urea synthesis is in the:

A. kidney
B. skeletal muscles
C. liver
D. small intestine
E. brain.

Ans C: The liver is the only organ having the urea cycle.

25. Pathways for the synthesis of pyrimidines, urea, and citrulline have in common a requirement for:

A. acetate
B. carbamoyl phosphate
C. tetrahydrofolate
D. propionate
E. pyruvate.

Ans B: Carbamoyl phosphate is a common precursor of pyrimidines, urea and citrulline. Different forms of carbamoyl phosphate synthase are required, however.

26. In maple syrup urine disease (now called branched chain aminoaciduria), the defective metabolic step involves:

A. an oxidative decarboxylation
B. an amino acid transaminase
C. a methionine deficiency in the diet
D. an amino acid hydroxylase
E. fixation of amino groups to carbon skeletons.

Ans A: Deficient oxidative decarboxylation of branched chain amino acids leads to their accumulation in blood and urine, and of their keto-acid byproducts. The mixture of these ketones in urine has the odor of maple syrup.

27. Which one of the following amino acids is non-essential as a human nutrient?

A. lysine
B. phenylalanine
C. valine
D. threonine
E. proline.

Ans E: Proline can be synthesized from TCA cycle intermediates in mammalian tissues.

28. Hormones made from amino acid precursors include those of the:

A. pancreas
B. thyroid gland
C. adrenal medulla
D. parathyroid glands
E. all of the above.

Ans E: All of the tissues mentioned secrete hormones that are derived from amino acid precursors.

29. Glutamic acid dehydrogenase carries out the oxidative deamination of glutamate in mitochondria. The products of this reaction when the cell is in a low energy state are:

A. glutamate, NH_4^+, NADH
B. α-ketoglutarate, NH_4^+, NAD^+
C. α-ketoglutarate, NH_4^+, NADH
D. α-ketoglutarate, NH_4^+, ATP
E. α-ketoglutarate, NH_4^+, $NADP^+$.

Ans C: Glutamate dehydrogenase, operating in the catabolic direction, yields NH_4^+, NADH and α-ketoglutarate as products.

30. Degradation of serine, alanine and cysteine is likely to produce:

A. α-ketoglutarate
B. pyruvate
C. fumarate
D. succinate
E. phenylpyruvate.

Ans B: The major degradation products of these two and three carbon amino acids all enter the glycolytic pathway at the level of pyruvate.

31. Release of gastrin would be most strongly induced by ingestion of:

A. ethanol
B. fats
C. glucose
D. protein
E. DNA.

Ans A: Ethanol is the most potent stimulator of gastrin release from the antral portion of the stomach.

Questions 32 - 35: Metabolism of aromatic amino acids involves the following pathways:

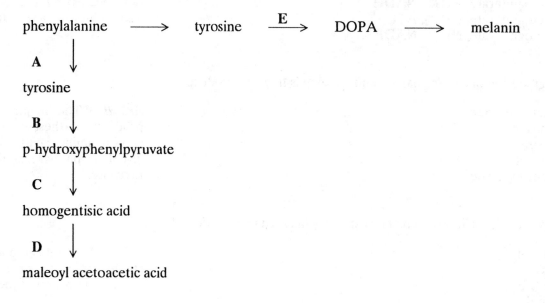

Match the condition below with an enzymatic block (lettered in diagram above).

32. Tyrosinosis.

Ans B: Lack of tyrosine aminotransferase leads to a build up of tyrosine and metabolic byproducts causing a group of symptoms known as tyrosinosis.

33. Albinism.

Ans E: Lack of tyrosine hydroxylase or other enzymes in the melanogenic pathway leads to failure to form melanin in skin and other tissues (albinism).

34. Alkaptonuria.

Ans D: Lack of homogentisate oxidase causes accumulation of homogentisic acid in blood and urine. Air oxidation of urinary homogentisic acid forms dark pigments yielding the condition called alkaptonuria.

35. Phenylketonuria.

Ans A: The most common cause of phenylketonuria is the lack of sufficient phenylalanine hydroxylase activity in the tissues.

Questions 36 - 43: Which of the following is most directly involved as a precursor in the biosynthesis of the metabolites in questions 36 - 43?

        A. tyrosine
        B. tryptophan
        C. serine
        D. glutamic acid
        E. glycine

36. Porphyrins.

Ans E: Glycine is one of the substrates of δ-amino levulinate synthetase, the first enzyme in porphyrin biosynthesis.

37. γ-Aminobutyric acid.

Ans D: GABA (γ-aminobutyric acid), an important neurohormone, is formed by the decarboxylation of glutamic acid.

38. Serotonin.

Ans B: The important indole neurohormone, serotonin, is derived from tryptophan.

39. Epinephrine.

Ans A: Epinephrine, the neurotransmitter for the adrenergic pathway, is derived from tyrosine.

40. Acetylcholine.

Ans C: Serine contributes methyl groups for the synthesis of the choline portion of acetyl choline, an important neurotransmitter for the parasympathetic pathways.

41. Thyroxine.

Ans A: Tyrosine is iodinated in the thyroid gland to form thyroxine and related derivatives, hormones necessary for proper metabolism.

42. Niacin.

Ans B: Niacin can be synthesized in limited amounts by humans using tryptophan as a precursor. It is an important component of NAD and NADP.

43. Creatine.

Ans C: Serine accepts a guanidino group from arginine to form guanidinoacetic acid. Methylation of this compound then yields creatine.

Questions 44 - 46:

      A. α-ketoglutaric acid
      B. glutamic acid
      C. aspartic acid
      D. acetyl CoA
      E. pyruvic acid

44. Can be deaminated enzymatically to form oxaloacetic acid.

Ans C: Transamination of aspartic acid yields oxaloacetic acid, a member of the TCA cycle.

45. Is formed directly from alanine by transamination.

Ans E: Both alanine and pyruvate are three carbon compounds readily interconvertible by transamination.

46. Deamination directly yields α-ketoglutarate.

Ans B: Both glutamate and α-ketoglutarate are five-carbon compounds readily interconvertible by transamination.

Questions 47 - 50:

      A. enterokinase
      B. chymotrypsinogen
      C. carboxypeptidase
      D. pepsin
      E. trypsin

47. An exopeptidase.

Ans C: Carboxypeptidases are exopeptidases; all others listed are endopeptidases.

48. A protease with a pH optimum less than 3.0.

Ans D: The gastric protease, pepsin, has is optimal pH at 3.0 or less.

49. A precursor of an endopeptidase.

Ans B: Chymotrypsinogen is the precursor of chymotrypsin, an endopeptidase.

50. An intestinal enzyme which initiates zymogen activation.

Ans A: Enterokinase is a protease released from the intestinal mucosa which activates trypsinogen to trypsin. Trypsin then activates more of itself autocatalytically and also all the other zymogens released by the pancreas.

Questions 51 - 54:

        A. pyridoxal phosphate
        B. tetrahydrofolate
        C. adenosine 5'-triphosphate
        D. a flavin nucleotide
        E. NADH

51. The interconversion of serine and glycine.

Ans B: Serine has three carbons, glycine, two. One carbon can be passed back and forth via N^5,N^{10}-THFA.

52. Transaminations.

Ans A: All transaminases use pyridoxal phosphate or pyridoxamine phosphate as coenzymes.

53. The formation of S-adenosyl methionine.

Ans C: The adenosyl group of S-adenosylmethionine is donated by ATP.

54. The oxidation of amines.

Ans D: Certain D and L-amino acids and other amines can be oxidized in liver and kidney by enzymes using FMN or FAD as coenzymes.

6. PORPHYRINS

Thomas Briggs

I. STRUCTURE AND CHEMISTRY

Porphyrins: cyclic tetrapyrroles found in **all aerobic cells**. With a system of conjugated double bonds, they absorb light strongly and are intensely colored, as in blood.

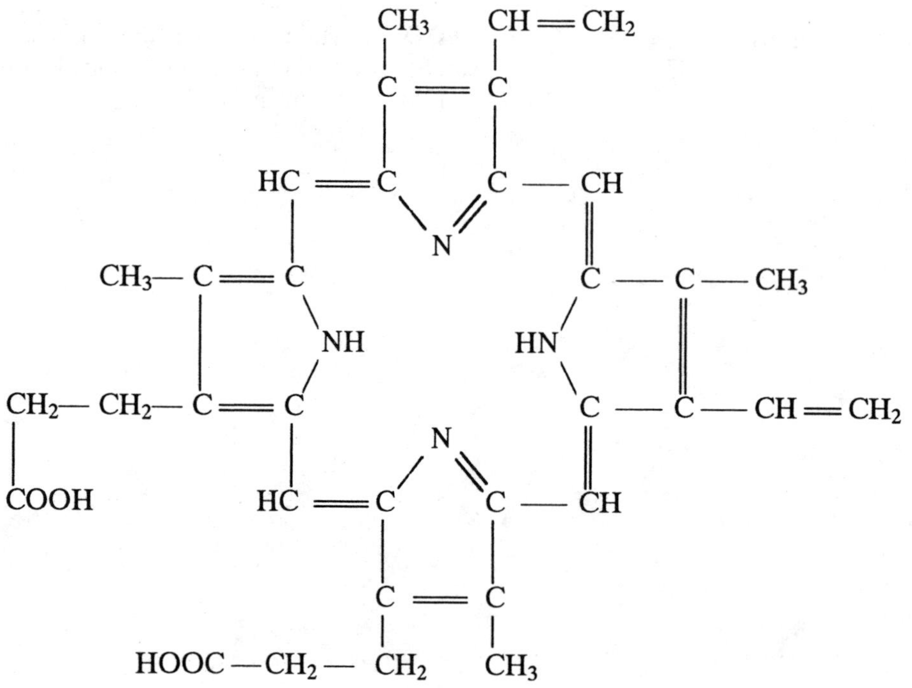

Figure 6-1. Protoporphyrin IX.

Porphyrins vary in the type and sequence of side-chains on the tetrapyrrole ring. Many are derivatives of **Protoporphyrin IX**, which has the particular sequence shown above. Biologically active forms usually have a **metal ion** in the center, are associated with a **protein**, and take part in enzymic activities.

Heme: **iron protoporphyrin**, isomer **IX**. It occurs in:

> ♦ *hemoglobin* and *myoglobin* (O_2 transport and storage)
>
> ♦ *cytochromes* (electron transport)
>
> ♦ *catalase* (simple breakdown of H_2O_2 to H_2O and O_2)
>
> ♦ *peroxidases* (use H_2O_2 to oxidize an organic substrate).

Hemin, Hematin: terms which refer to the oxidized (Fe^{+++}) form of heme.

Tetrapyrroles (not heme) are also part of:

> ♦ chlorophyll
>
> ♦ vitamin B_{12}.

II. METABOLISM

A. Synthesis

ALA synthase is the first enzyme in the chain of heme synthesis:

$$
\begin{array}{llll}
\text{COOH} \quad \text{O} & \text{COOH} & \text{pyridoxal} & \text{COOH} \quad \text{O} \qquad + \text{ CoASH} \\
| \qquad\quad \| & | & \text{phosphate} & | \qquad\quad \| \\
\text{CH}_2\text{-CH}_2\text{- C-SCoA} \ + & \text{CH}_2\text{-NH}_2 & \xrightarrow{\text{\textit{ALA}}} & \text{CH}_2\text{-CH}_2\text{- C - CH}_2\text{-NH}_2 \ + \ \text{CO}_2 \\
& & \text{\textit{Synthase}} & \delta\text{-amino-} \\
\text{Succinyl CoA} & \text{Glycine} & & \text{levulinic acid (ALA)}
\end{array}
$$

An important control point, *ALA synthase* is rate-limiting and its production is repressed by the end product of the sequence, heme (or derivatives such as hemin [ferriheme]). Heme (and hemoglobin) synthesis is stimulated by **erythropoietin**, a glycoprotein.

Two ALA are then condensed to **porphobilinogen** (PBG), a monopyrrole, by *ALA dehydratase*, a Zn-containing enzyme which is sensitive to inhibition by lead. Four PBG are joined, under the influence of two enzymes, (1) *PBG deaminase* (older term, *Uroporphyrinogen III synthetase*), producing a linear tetrapyrrole, hydroxymethylbilane, and (2) *Uroporphyrinogen III cosynthase*, producing one particular isomeric cyclic tetrapyrrole, **Uroporphyrinogen III**, the first of a series of colorless porphyrinogens. This is then trimmed through several decarboxylations, giving rise to the specific sequence of methyl, vinyl, and propionic acid side-groups that is to be found in heme. The still colorless product is next oxidized with NADPH and molecular oxygen to produce the colored protoporphyrin. Finally, **iron** is inserted by *ferrochelatase*, another enzyme which is inhibited by lead.

Heme synthesis occurs in all aerobic cells, but especially in blood-forming tissues. The process is partly mitochondrial and partly cytoplasmic. In humans 80% of the body's heme is in **hemoglobin** of erythrocytes. Paradoxically, these cells do not synthesize heme or hemoglobin after maturity, because they lack nuclei, mitochondria, and many of the necessary enzymes. A mammalian erythrocyte's entire supply of heme and hemoglobin is produced prior to maturation. Hemoglobin reversibly binds and transports **oxygen** (see Chapter 1). Carbon monoxide competes strongly at the binding site with O_2, accounting for its toxicity.

B. Degradation and Excretion

Heme is degraded in the reticuloendothelial system of liver, spleen and bone marrow. The metabolism involves a number of stages:

1. *Ring opening* by the complex enzyme heme oxygenase (with NADPH), to form **biliverdin**, a linear tetrapyrrole;

2. *Reduction* of biliverdin to **bilirubin**, an orange-yellow pigment which is insoluble in water;

3. *Transport* of the insoluble bilirubin as a complex with **serum albumin**;

4. *Uptake* by liver cells as a complex with **ligandin**, a binding protein;

5. *Conjugation* with **glucuronic acid**; this is obligatory for bilirubin to be secreted into bile:

$$\text{Bilirubin} \ + \ \text{UDP Glucuronic acid} \ \xrightarrow{\substack{\textit{UDP-glucuronyl} \\ \textit{transferase}}} \ \text{Bilirubin diglucuronide}$$

6. *Excretion* of the diglucuronide via bile through the intestinal tract.

Bilirubin is the major **bile pigment** in humans. Though much less soluble in water than biliverdin, evolution may have favored its production in mammals because it is an effective anti-oxidant.

Bacterial metabolism of bilirubin in the bowel leads to formation of a complex mixture of pigments, including **urobilinogen**, a colorless substance which can readily be oxidized to **urobilin**, a pigment which produces much of the color of feces. Abnormally light-colored feces ("clay-colored stools") is a sign that biliary flow is greatly diminished or absent. Some of these pigments also undergo an enterohepatic circulation and may be found in urine. Thus urinary urobilin can be a useful indicator that some degree of biliary excretion of bilirubin is occurring.

III. DISORDERS OF PORPHYRIN AND HEME METABOLISM

A. Synthesis

These rare diseases, called **porphyrias**, can be life-threatening. They usually take the form of a partial block in one of the enzymes of heme synthesis, leading to inadequate levels of heme for feedback control of ALA synthase, followed by overproduction and excretion of precursors (ALA, PBG) or porphyrin intermediates. Urine may be red. The conditions are usually inherited, but can sometimes be caused by toxic agents such as lead or some pesticides. They may be accompanied by photosensitivity of the skin, abdominal pain, psychic disturbances, and muscular paralysis all occurring in intermittent attacks. ("Mad King George III" was not mad, but probably had one of the heritable porphyrias, see Sec IV).

It is important for patients with porphyria to avoid alcohol and certain drugs, especially barbiturates, which induce cytochrome P-450, depleting intracellular heme and leading to derepression of ALA synthase. Disease symptoms are induced or exacerbated due to excessive activity of the porphyrin synthetic pathway and accumulation of porphyrins and precursors. Administration of hematin reverses these effects because it inhibits and/or represses ALA synthase, restoring feedback control. A high carbohydrate diet is also useful in this regard (reason unclear).

B. Metabolism

Problems can occur at many points; build-up of bilirubin in blood and tissues causes a visible yellow color called **icterus** or **jaundice** (Table 6-1). In contrast to the rare porphyrias, jaundice is relatively common. A mild form often occurs in the newborn, due to a temporary underdevelopment of the conjugation and excretion process. Phototherapy, exposure to visible light, is useful because it causes a physical change in the bilirubin molecule which allows it to be readily excreted even when unconjugated. Severe and prolonged accumulation of unconjugated bilirubin in the newborn is dangerous because unconjugated bilirubin can penetrate nervous tissue, concentrate in brain cells (**kernicterus**), and cause mental retardation.

Bilirubin in blood can be quantified by the Van den Bergh diazo reaction. In this test, conjugated bilirubin reacts immediately ("**Direct**"); unconjugated bilirubin reacts only after addition of methanol ("**Indirect**"). This information as to the type of bilirubin is readily obtainable and is useful in diagnosis: conjugated bilirubin in blood ("regurgitation" hyperbilirubinemia) is caused by failure of the liver to excrete bilirubin into bile, even though that organ's conjugating ability may be normal. Unconjugated bilirubin in blood ("retention" hyperbilirubinemia) is caused by overproduction of bilirubin (hemolytic jaundice) or by failure of hepatic uptake or conjugation (Table 6-1).

Only conjugated bilirubin can pass into urine. Therefore, bilirubin in urine (**choluric jaundice**) signifies regurgitation hyperbilirubinemia; jaundice without urinary bilirubin (**acholuric jaundice**) signifies retention hyperbilirubinemia, a raised level of unconjugated bilirubin only.

Area of Defect in Bilirubin Metabolism	Cause	Main Form of Bilirubin Found	Bilirubin in Urine?	Urobilinogen in Urine
overproduction	hemolytic disorders	Unconjugated	No	↑
uptake, conjugation and secretion by liver	underdevelopment in neonatal hyperbilirubinemia	Unconjugated	No	↓
general hepatic deficiency	hepatitis, cirrhosis	Both (variable)	Yes	↓
biliary transport	blockage of bile duct by gallstone or cancer of pancreas	Conjugated	Yes	0
various	heritable defects in metabolism and transport, see below			

Table 6-1. Major Causes of Jaundice.

IV. GENETIC DISEASES

A. The Porphyrias

Erythropoietic protoporphyria (EPP)
Molecular defect: Deficient *ferrochelatase*.
Pathway affected: Last step of heme synthesis.
Expression: Increased protoporphyrin (plasma and feces).
Genetics: Autosomal dominant.

Acute intermittent porphyria (AIP)
Molecular defect: Deficient *porphobilinogen deaminase*.
Pathway affected: porphobilinogen → hydroxymethylbilane.
Expression: Increased δ-aminolevulinic acid and PBG.
Genetics: Autosomal dominant.

Other Hereditary Porphyrias
>   Hereditary coproporphyria (HCP)
>   Porphyria cutanea tarda (PCT)
>   Variegate porphyria

B. Bilirubin Metabolism: Hyperbilirubinemia

Crigler-Najjar Syndrome, Type 1
>   Molecular defect: absent hepatic *bilirubin : UDP-glucuronyl transferase.*
>   Pathway affected: bilirubin → bilirubin glucuronide.
>   Expression: Increased unconjugated bilirubin.
>   Genetics: Autosomal recessive.
>   Treatment: Exchange transfusion.

Gilbert Syndrome (benign condition, occurring as a heterogeneous group)
>   Molecular defect: impaired hepatic uptake and conjugation of bilirubin.
>   Pathway affected: bilirubin → bilirubin glucuronide.
>   Expression: mild unconjugated hyperbilirubinemia.
>   Genetics: Autosomal dominant or heterogeneous ?

Dubin - Johnson Syndrome
>   Molecular defect: impaired hepatic secretion of conjugated bilirubin into bile.
>   Expression: conjugated hyperbilirubinemia.
>   Genetics: Autosomal recessive.

V. REVIEW QUESTIONS ON PORPHYRINS

> *DIRECTIONS:* For each of the following multiple-choice questions (1 - 17), choose the ONE BEST answer.

1. Bilirubin diglucuronide is

A. elevated in neonatal hyperbilirubinemia
B. usually found in the bile duct
C. lipid-soluble
D. elevated in hemolytic jaundice
E. normally excreted mainly in the urine.

Ans B: A, C, and D pertain to the unconjugated form; though conjugated BR may appear in urine in obstructive jaundice, it is not normally there.

2. All of the following statements (A - D) about catalase are correct EXCEPT: If all are correct, no exception, choose E. Catalase:

A. is a heme protein
B. is part of the electron transport system
C. breaks down hydrogen peroxide
D. produces oxygen as a product
E. all of the above are correct without exception.

Ans B: Catalase is one of the enzymes that break down H_2O_2; it has nothing to do with the ETS.

3. Unconjugated bilirubin is

A. elevated in neonatal hyperbilirubinemia
B. usually found in the bile duct
C. normally excreted in the urine
D. water-soluble
E. not involved in the development of kernicterus.

Ans A: It is only the conjugated form that is water-soluble and excreted in bile; Unconjugated BR is elevated in neonatal hyperbilirubinemia; it *can* pass into brain cells and cause damage.

4. The enzyme responsible for conjugation of bilirubin, prior to secretion into bile, is

A. ALA synthase
B. PBG deaminase
C. uroporphyrinogen cosynthase
D. glucuronyl transferase
E. ferrochelatase.

Ans D: This is the enzyme that transfers glucuronyl units from UDPG to bilirubin.

5. All of the following (A - D) are consequences of bile duct obstruction EXCEPT: If all are, no exception, choose E.

A. increased conjugated bile pigments in the serum
B. decreased excretion of cholesterol
C. decreased bilirubin glucuronides in the feces
D. increased levels of fecal lipids
E. all of the above without exception.

Ans E: With bile flow obstructed, conjugated BR will accumulate in blood instead of being excreted in feces; since bile acid excretion is blocked, B and D will also occur.

6. In porphyrin synthesis, the committed step is

A. condensation of 2 PBG (porphobilinogen)
B. condensation of glycine and succinyl CoA
C. isomerization of isopentenyl-PP to dimethylallyl-PP
D. condensation of 2 ALA (δ-aminolevulinic acid)
E. formation of uroporphyrinogen III.

Ans B because this is the first, rate-limiting, and controlling step in the synthetic pathway.

7. The function of erythropoietin is to

A. promote synthesis of hemoglobin
B. prevent pernicious anemia
C. prevent thalassemia
D. promote binding of oxygen to hemoglobin
E. regulate reduction of methemoglobin.

Ans A: This is the hormone that regulates production of red cells and therefore of hemoglobin.

8. All of the following statements (A - D) about peroxidase are correct EXCEPT: If all are correct, no exception, choose E. Peroxidase:

A. is a heme protein
B. is part of the respiratory chain
C. breaks down hydrogen peroxide
D. oxidizes organic substrates
E. all of the above are correct without exception.

Ans B: Peroxidase breaks down H_2O_2 using organic substrates as electron acceptors. It has nothing to do with the ETS.

9. Porphyrins are made from:

A. succinic acid and glycine
B. lysine and proline
C. alanine and glutamic acid
D. acetyl CoA and oxaloacetate
E. glycerol and fatty acid.

Ans A: Condensation of these two is catalyzed by ALA synthase, the first enzyme in the pathway. There are no other starting materials.

10. The first step in heme synthesis requires

A. thiamin
B. riboflavin
C. niacin
D. pyridoxal phosphate
E. cobalamin.

Ans D because this is the cofactor required by ALA synthase.

11. In porphyria, administration of hematin is beneficial because hematin:

A. induces cytochrome P-450, leading to increased metabolism of porphyrin intermediates
B. induces greater activity of ALA synthase
C. restores feedback control of overactive ALA synthase
D. causes increased conversion of porphyrins to bilirubin
E. represses excessive activity of ferrochelatase.

Ans C: Porphyrias result, in part, from an overactive ALA synthase due to a failure of feedback control by end-product. Hematin represses ALA synthase.

12. Which of the following does NOT contain an iron-porphyrin?

A. carboxyhemoglobin
B. catalase
C. peroxidase
D. myoglobin
E. ferritin.

Ans E: This is the protein for storage of iron atoms; it contains no heme.

13. Porphyria is associated with

A. phenylalanine-ammonia lyase deficiency
B. hypoxanthine-guanine phosphoribosyl transferase deficiency
C. overproduction of δ-aminolevulinic acid
D. bilirubin glucuronyl transferase deficiency
E. ALA synthetase deficiency.

Ans C: The porphyrias are in part due to a failure of regulation of ALA synthase, leading to excess production of ALA.

14. Bile pigments can be produced from:

A. cholic acid
B. heme
C. triglycerides
D. cholesterol
E. phospholipids.

Ans B: The bile pigments, which include bilirubin as the major member in humans, are made from breakdown of the porphyrin moiety of heme.

15. Urobilinogen can NOT be derived from

A. hemoglobin
B. catalase
C. flavoprotein
D. cytochrome c
E. peroxidase.

Ans C: Flavoprotein contains no porphyrin, so cannot lead to production of bilirubin or its bacterial product, urobilinogen.

16. Urobilinogen is formed in the

A. spleen
B. liver
C. bone marrow
D. bowel
E. kidney.

Ans D because urobilinogen is a product made by intestinal bacteria from bilirubin.

17. The last step in heme synthesis is catalyzed by:

A. Ferrochelatase
B. Uroporphyrinogen cosynthase
C. ALA synthase
D. UDP-glucuronyl transferase
E. heme oxygenase

Ans A: Insertion of iron comes last in the synthetic pathway.

DIRECTIONS: For the following set of questions, choose the ONE BEST answer from the lettered list. An answer may be used one or more times, or not at all.

Questions 18 - 25:

   A. "Direct" bilirubin in serum is elevated
   B. "Indirect" bilirubin in serum is elevated
   C. BOTH may be elevated
   D. NEITHER is elevated

18. Hepatitis.

Ans C: Both the hepatic conjugating and secreting ability may be affected.

19. Neonatal jaundice.

Ans B: Hepatic uptake and conjugation are undeveloped, leading to build-up of unconjugated BR.

20. Acute intermittent porphyria.

Ans D: Porphyrias have to do with abnormal production of porphyrins, not bilirubin metabolism.

21. Cholelithiasis.

Ans A: Blockage of bile duct leads to back-up of conjugated BR, which leaks out into bloodstream.

22. Crigler-Najjar Syndrome.

Ans B: Molecular defect is in the conjugating enzyme, causing accumulation of unconjugated BR.

23. Cancer of head of pancreas.

Ans A because bile flow can be obstructed, with result as in question 21.

24. Gilbert syndrome.

Ans B: With compromised hepatic uptake and conjugation of bilirubin, it is the unconjugated form which accumulates.

25. Rh blood disease.

Ans B because hemolysis leads to increased release of unconjugated BR. Even though the liver's excreting ability may be normal, production is more than it can handle.

7. LIPIDS

Chi-Sun Wang

I. CLASSIFICATION OF LIPIDS

A. Major Lipid Classes

Fatty acids: long-chain monocarboxylic acids.

Acylglycerols: triacylglycerols, diacylglycerols and monoacylglycerols.

Phospholipids: phosphoglycerides and sphingomyelin.

Glycosphingolipids: cerebroside (ceramide monosaccharide), cerebroside sulfate, ceramide oligosaccharide and ganglioside. The common structural component of glycosphingolipids is ceramide (N-acylsphingosine). They are similar to sphingomyelin as derivatives of a ceramide, but do not contain phosphorus and the additional nitrogenous base.

Other lipids: sterols, terpenes, waxes, aliphatic alcohols.

Lipoproteins: lipids combined with other classes of compounds.

B. Derived Lipids

These are molecules, soluble in lipid solvents, that are produced by hydrolysis of natural lipids.

II. FUNCTIONS OF LIPIDS

A. Structural Component

All cellular membranes, including myelin, consist of a lipid bilayer. Membranes function as permeability barriers and in many other ways. See Chapter 9.

B. Enzyme Cofactors

Several enzymes require lipids for their activity. Examples include phospholipid in the blood clotting cascade, coenzyme A, etc.

C. Energy Storage

Energy storage is the major function of triacylglycerols which are largely found in adipose tissue. These lipids do not require hydration and yield 9 Kcal/g of energy on complete combustion, compared to 4 Kcal/g for carbohydrates and 4 Kcal/g for proteins. This is because the C-H / O ratio is higher.

D. Hormones and Vitamins

Prostaglandins: arachidonic acid is a precursor for the biosynthesis of prostaglandins.

Steroid Hormones: estradiol, testosterone, cortisol, and many other gonadal and adrenal cortical hormones produced from cholesterol.

Fat-soluble Vitamins: A, D, E, K.

E. Aid in Fat Digestion: *Components of Bile*

F. Insulation: *Subcutaneous Adipose Tissue*

G. Thermogenesis: *Brown Adipose Tissue*

III. DIGESTION AND ABSORPTION OF LIPIDS

Most dietary lipid is in the form of triacylglycerols and is digested in the small intestine. The emulsification of lipid droplets by **bile salts** allows a large increase of the surface area of the droplets. Lipase produced by the pancreas catalyzes the hydrolysis of triacylglycerols to fatty acids and monoacylglycerols, which form **micelles.** These micelles also contain bile salts, cholesterol and fat-soluble vitamins. The micelles then migrate by diffusion to intestinal mucosa and enter the intestinal epithelial cells, where free fatty acids and monoacylglycerols are re-esterified to form triacylglycerols. Triacylglycerol, cholesterol, cholesterol ester, phospholipid, and specific proteins are assembled into **chylomicron** particles which subsequently enter the lymphatics for transport to the thoracic duct and the bloodstream.

IV. FATTY ACIDS

A. Chemistry of Fatty Acids

Saturated	Nbr of Carbons	Structure
Acetic	2	CH_3COOH
Propionic	3	CH_3CH_2COOH
Butyric	4	$CH_3(CH_2)_2COOH$
Hexanoic	6	$CH_3(CH_2)_4COOH$
Octanoic	8	$CH_3(CH_2)_6COOH$
Decanoic	10	$CH_3(CH_2)_8COOH$
Lauric	12	$CH_3(CH_2)_{10}COOH$
Myristic	14	$CH_3(CH_2)_{12}COOH$
Palmitic	16	$CH_3(CH_2)_{14}COOH$
Stearic	18	$CH_3(CH_2)_{16}COOH$
Unsaturated	Nbr of Double Bonds	Structure
Oleic	1	18 C, *cis*-Δ^9
Linoleic	2	18 C, *cis,cis*-$\Delta^{9,12}$
Linolenic	3	18 C, all *cis*-$\Delta^{9,12,15}$
Arachidonic	4	20 C, all *cis*-$\Delta^{5,8,11,14}$

Table 7-1. Some Common Fatty Acids.

Comments on Fatty Acids. Cis-Δ^9 for oleic acid means that the double bond is between the 9th and 10th carbons (numbering the carbons starting with the carboxyl carbon as 1) and the double bond has the *cis* geometric configuration:

$$\begin{array}{ccc}
H & & H \\
\backslash & & / \\
C & = & C \\
/ & & \backslash \\
CH_3(CH_2)_7 & & (CH_2)_7COOH
\end{array}$$

The most common fatty acids found in mammals are straight-chain, unsubstituted, monocarboxylic acids and have an even number of carbons, usually 12-20 (Table 7-1). When a fatty acid contains more than one double bond, it is polyunsaturated. The polyunsaturated fatty acids linoleic and linolenic are essential in the diet of mammals.

B. β-Oxidation of Fatty Acids

Steps in β-Oxidation (Figure 7-1). After initial **activation** of a fatty acid in the cytoplasm (enzymatic step 1 in the figure), the entrance of acyl-CoA into mitochondria requires the **carnitine transport** system. The β-oxidation of acyl-CoA then consists of a repeated series of four steps: **dehydrogenation, hydration, dehydrogenation**, and **thiolytic cleavage** (steps 2-5 in the figure). The last step in each series releases a 2-carbon fragment as acetyl CoA.

Cofactor Requirement. ATP (for activation), carnitine (for entrance into mitochondria), NAD^+, FAD, and CoA.

Oxidation of unsaturated fatty acids requires *cis-Δ^3 trans-Δ^2-enoyl-CoA isomerase.*

Stoichiometry. The stoichiometry for the oxidation of palmitate ($C_{16}H_{32}O_2$) can be calculated as shown below:

Palmitate

7 turns of β-oxidation (7 NAD^+, 7 FAD)	+ 35 ATP
8 acetyl groups	+ 96 ATP
1 ATP used for initial activation	− 2 ATP
(but cleavage of PP_i costs an extra ~P)	

Net: 129 ATP

C. Catabolism of Odd-numbered-carbon Fatty Acids.

Oxidation of odd-numbered-carbon fatty acids yields successive molecules of acetyl-CoA, ending with 1 molecule of **propionyl-CoA**. The major pathway for the metabolism of propionyl-CoA is summarized in the three reactions shown in Figure 7-2 (page 117). In these enzyme reactions, **biotin** is a cofactor for propionyl-CoA carboxylase and **vitamin B$_{12}$** is a cofactor for methylmalonyl-CoA mutase. Succinyl-CoA can then enter into the TCA cycle.

D. Biosynthesis of Saturated Fatty Acids.

Acetyl-CoA is the major building block for fatty acid synthesis.

Lipogenesis from carbohydrate requires the participation of **the mitochondrion.** (Figure 7-3).

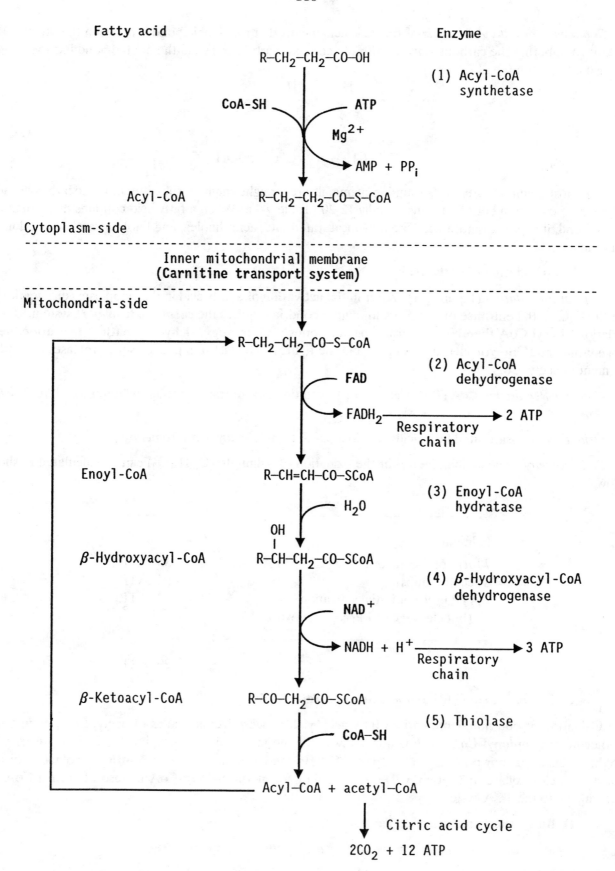

Figure 7-1. β-Oxidation of Fatty Acids in Mitochondria.

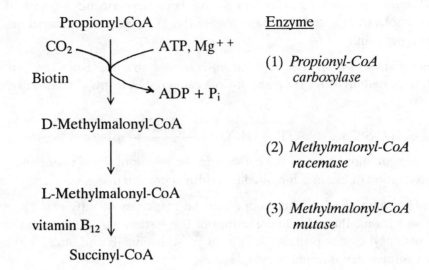

Enzyme

(1) *Propionyl-CoA carboxylase*

(2) *Methylmalonyl-CoA racemase*

(3) *Methylmalonyl-CoA mutase*

Figure 7-2. Metabolism of Propionyl CoA.

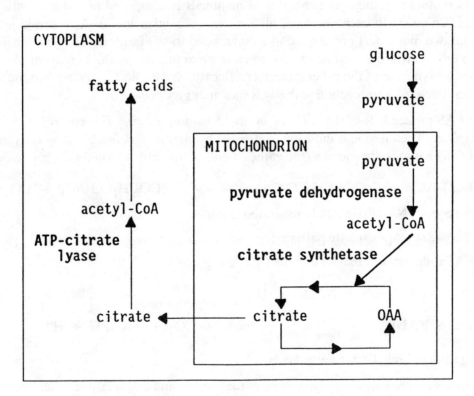

Figure 7.3. Involvement of Mitochondrion in Lipogenesis from Carbohydrate.

Transport of acetyl-CoA to the cytoplasm is mediated via **citrate**, whose formation in mitochondria is catalyzed by *citrate synthase*. Citrate is then converted to acetyl-CoA in cytoplasm by *ATP-citrate lyase*:

$$\text{Citrate} + \text{ATP} + \text{CoA} \xrightarrow[\textit{lyase}]{\textit{ATP-citrate}} \text{acetyl-CoA} + \text{ADP} + \text{P}_i + \text{Oxaloacetate}$$

The synthesis of this inducible enzyme is increased in caloric excess. The acetyl-CoA is then utilized for biosynthesis of fatty acids.

Reducing equivalents are needed, generated from the hexose monophosphate pathway and "malic enzyme."

Animals cannot convert fatty acids into glucose because pyruvate → acetyl-CoA is irreversible. Also, though the 2 carbons from acetyl CoA can traverse the TCA cycle to oxaloacetate, 2 carbons are also lost, so there can be no net gain.

Formation of malonyl-CoA. The major *de novo* route is in cytoplasm. The first step of fatty acid synthesis is the formation of **malonyl-CoA** from acetyl-CoA catalyzed by *acetyl-CoA carboxylase*:

$$CH_3\text{-}\overset{\overset{\displaystyle O}{\|}}{C}\text{-S-CoA} + ATP + HCO_3^- \xrightarrow{\text{Biotin}} \text{Malonyl-CoA} + ADP + P_i$$

The enzyme contains biotin as the cofactor in the reaction. Biotin-containing enzymes cannot function if intestinal absorption of biotin is inhibited by **avidin**, a protein in egg white.

Fatty acid synthase (FAS). There appear to be two types of fatty acid synthase systems. In bacteria, plants, and lower forms, the individual enzymes of the system may be separate and the acyl radicals are in combination with acyl carrier protein (ACP). In yeast, mammals and birds, the enzymes are combined as a **multienzyme complex** that is found in cytoplasm.

The fatty acid synthase multienzyme complex of mammals is composed of a dimer with a monomeric molecular weight of 267,000. In each monomer all 7 enzyme activities and an ACP reside in a single polypeptide chain. The two monomers are aligned in a cyclic head-to-tail fashion. The thiol group in 4′-phosphopantetheine (Pan) of ACP on one monomer is in close proximity to the thiol group of a cysteine residue attached to β-ketoacyl synthase of the other monomer. Because both thiols participate in the synthase reaction, only the dimer is active. The reaction pathway is shown in Figure 7-4.

Synthesis by FAS proceeds for fatty acids of up to 16 carbon atoms. The control step is catalyzed by *acetyl-CoA carboxylase*. The enzyme is inducible, **activated by citrate**, and **inhibited by fatty acyl CoA**. The resulting malonyl-CoA carbons are added to the carboxyl end of the fatty acid and CO_2 is released:

$$R\text{-CO-ACP} + {}^-OOC\overset{*}{C}H_2\overset{*}{C}O\text{-CoA} \longrightarrow RCO\overset{*}{C}H_2\overset{*}{C}O\text{-ACP} + CO_2$$

The synthesis requires **NADPH** which is generated from:

1. The **hexose monophosphate pathway**;

2. An NADP-dependent **malate enzyme** ("malic enzyme"):

$$\text{Oxaloacetate (cytosol)} + NADH + H^+ \xrightarrow[\text{dehydrogenase}]{\text{malate}} \text{malate} + NAD^+$$

$$\text{Malate} + NADP^+ \xrightarrow[\text{enzyme}]{\text{malic}} \text{pyruvate} + CO_2 + NADPH + H^+$$

E. Elongation of Long Chain Fatty Acids.

Elongation occurs in three areas — cytoplasm, mitochondria and endoplasmic reticulum:

> *Cytoplasm* contains a system for fatty acid elongation which is similar to the one used for *de novo* synthesis. Elongation also requires NADPH.

> *Mitochondria* use acetyl-CoA and CoA esters, similar to a reversal of β-oxidation, but require both NADH and NADPH.

> *Endoplasmic reticulum (microsomes)* use malonyl CoA and CoA ester, not ACP-esters. NADPH is required.

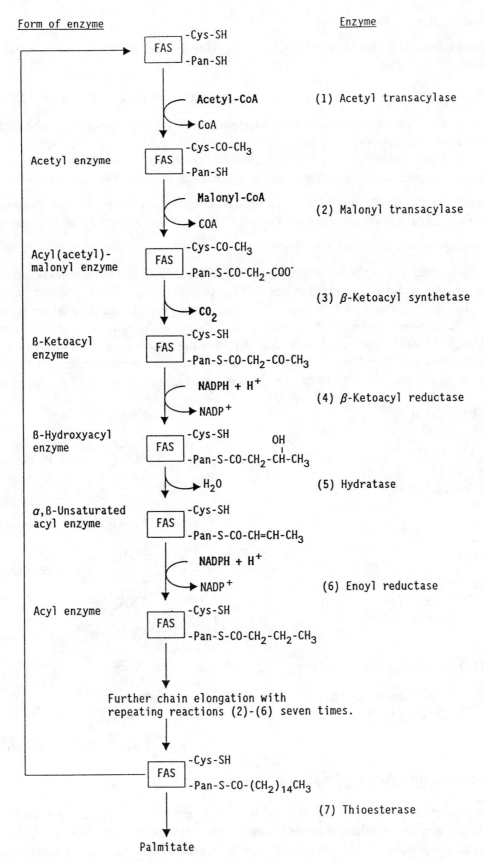

Figure 7-4. Reactions Carried Out by Fatty Acid Synthase.

F. Desaturation of Fatty Acids.

Desaturation occurs in endoplasmic reticulum. *NADPH is required.* Both the fatty acid and NADPH are oxidized by molecular oxygen:

$$\text{Fatty acyl-CoA} + \text{NADPH} + \text{H}^+ + \text{O}_2 \longrightarrow \text{Unsaturated fatty acyl-CoA} + \text{NADP}^+ + 2\,\text{H}_2\text{O}$$

All desaturations in mammalian systems are **farther than 6 carbon atoms** from the methyl end of the fatty acid. Therefore, any fatty acid having a double bond closer to the methyl end than 6 carbons must either be obtained in the diet or be produced from a dietary fatty acid. Thus by a combination of desaturation and elongation, linoleic acid can be converted to arachidonic acid for conversion to prostaglandins.

Double bonds are all cis, although *trans* are present occasionally in fats from plant sources. Partially hydrogenated margarine may also contain *trans* fatty acids which result from the manufacturing process.

G. Prostaglandins

The polyunsaturated C-20 fatty acids give rise to physiologically and pharmacologically active acid derivatives known as **eicosanoids**: prostaglandins (PG), prostacyclins (PGI), thromboxanes (TX), and leukotrienes (LT). Arachidonate and the other C-20 precursors usually are derived from the 2-position of phospholipids in the plasma membrane as a result of phospholipase A_2 activity. **Anti-inflammatory corticosteroids are thought to inhibit phospholipase A_2 activity through induction of an inhibitory protein named lipocortin.** Eicosanoids can be put into three groups, based on the polyunsaturated acid from which they are derived. The groups, their biosynthetic origins, and some examples are shown below (Figure 7-5).

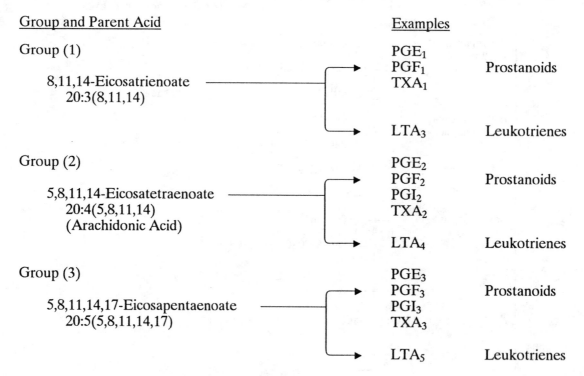

Figure 7-5. Relationships Among the Eicosanoids.

The major classes of prostaglandins (as they are known collectively) are designated with a numerical subscript which denotes the **number of double bonds** in the molecule. Because of a cyclization reaction, prostanoids have 2 fewer double bonds than the parent fatty acid, while leukotrienes have the same number.

Two major pathways are followed in the synthesis of the eicosanoids:

* The **cyclooxygenase pathway** produces the prostanoids. It includes prostaglandin endoperoxide which has two enzyme activities, *cyclooxygenase* and *peroxidase*, and consumes two molecules of O_2. **Aspirin** was found to inhibit the cyclooxygenase reaction. By this pathway, **thromboxanes** are synthesized in platelets and upon release cause **vasoconstriction** and **platelet aggregation**. **Prostacyclins** (e.g., **PGI_2**) are produced by blood vessel walls and are potent **inhibitors of platelet aggregation**.

* The **lipoxygenase pathway** produces **leukotrienes**, a family of conjugated trienes. The slow-reacting substrate of anaphylaxis is a mixture of LTC_4, LTD_4, LTE_4. These leukotrienes also cause vascular permeability and are important in **inflammatory** and immediate **hypersensitivity reactions**.

V. KETONE BODIES

A. Synthesis of Ketone Bodies

Ketogenesis is a hepatic process (Figure 7-6). In liver mitochondria, fatty acids are oxidized to **acetyl CoA**, a portion of which is oxidized via the TCA cycle to CO_2 and a portion of which is converted via **3-hydroxy-3-methylglutaryl-CoA** (HMG-CoA) to the ketone bodies acetoacetate and β-hydroxybutyrate.

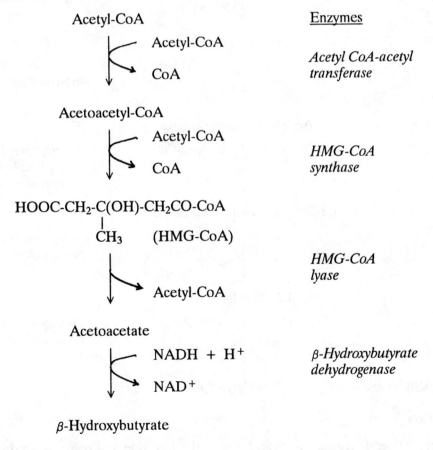

Figure 7-6. Hepatic Ketogenesis.

B. Metabolism and Regulation.

Utilization. Ketone bodies are utilized primarily by muscle.

Starvation. Liver glycogen in the rat is depleted by about 70% after a 12-hour fast and is almost gone after 24 hours of starvation. Acetyl-CoA derived from fatty acid β-oxidation is channeled into ketogenesis. The synthetic pathways for carbohydrates and fatty acids are inhibited. Normally, adult brain depends entirely on glucose for its energy. Under conditions of long-term starvation (more than 3 days), the brain derives a considerable part of its energy from ketone bodies.

Insulin deficiency. In regard to lipid metabolism, diabetes and starvation resemble each other. Fatty acids mobilized from adipose tissue raise the level of acetyl-CoA in the liver and promote **ketogenesis**.

Consequences of ketosis. Ketone bodies are excreted in large part as the sodium salts. Depletion of Na^+ and other cations leads to **acidosis** (ketoacidosis).

VI. TRIACYLGLYCEROLS

A. Synthesis *de novo* is mainly in liver and adipose tissue (Figure 7-7).

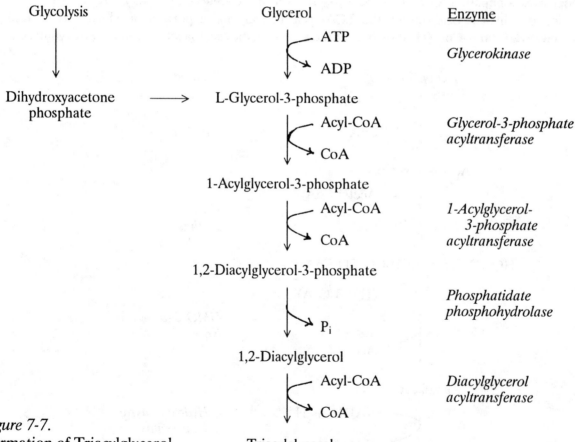

Figure 7-7.
Formation of Triacylglycerol.

B. Utilization

Digestion. Dietary triacylglycerols are hydrolyzed in the intestine and resynthesized by intestinal cells, then transferred to the lymph as **chylomicrons**. Secretion of chylomicrons requires intestinal protein synthesis.

Transport. Triacylglycerols produced in liver are released as **very low density lipoproteins** (VLDL).

Lipoprotein lipase is an enzyme bound to capillary endothelium. Peripheral tissues require it in order to utilize the triacylglycerol from triacylglycerol-containing lipoproteins. (See also section IX F, page 126).

Hormone-sensitive lipase (HSL) hydrolyzes triacylglycerols to fatty acids and glycerol during fat mobilization in adipose tissue. (See also section IX G, page 126). Because adipose tissue has no glycerol kinase, glycerol is not re-utilized there but is metabolized predominantly by the liver.

Regulation. Fatty acid taken up by the liver undergoes (1) oxidation, or (2) esterification to triacylglycerols. In the normal fed condition, the liver has plenty of carbohydrate to be oxidized for energy, therefore (2) > (1). In the fasting state, fatty acid oxidation proceeds at a higher rate, therefore (1) > (2).

The esterification pathway (2) is never saturated, therefore the shift to oxidation in fasting is not because esterification has reached a maximum; it is because the oxidative pathway is turned on.

VII. PHOSPHOLIPIDS

A. Phosphoglycerides

Structure. Phosphoglycerides have long-chain fatty acids esterified to glycerol in the 1 and 2 positions, a phosphate in position 3, and a nitrogen-containing base, such as choline, esterified to the phosphate.

Phospholipases. There are several phospholipases which hydrolyze phospholipids at specific places in the molecule. They are useful in investigations requiring highly specific cleavage of phospholipids and phosphate esters.

Plasmalogens. These phosphoglycerols contain an α,β-unsaturated alcohol in ether linkage. Thus the plasmalogens containing ethanolamine have the general structure shown below (Figure 7-8).

Synthesis. The synthetic pathway for phosphoglycerides is shown in Figure 7-9.

$$H_2C\text{-}O\text{-}CH = CHR_1$$
$$|$$
$$R_2COOCH \quad O^-$$

L-Phosphatidalethanolamine
$$|\qquad\quad|$$
$$H_2C\text{-}O\text{-}P\text{-}OCH_2CH_2NH_2$$
$$||$$
$$O$$

$$OH$$
$$|$$
$$CH_3(CH_2)_{12}\text{-} CH = CH\text{-}CH\text{-}CH\text{-}CH_2OH$$

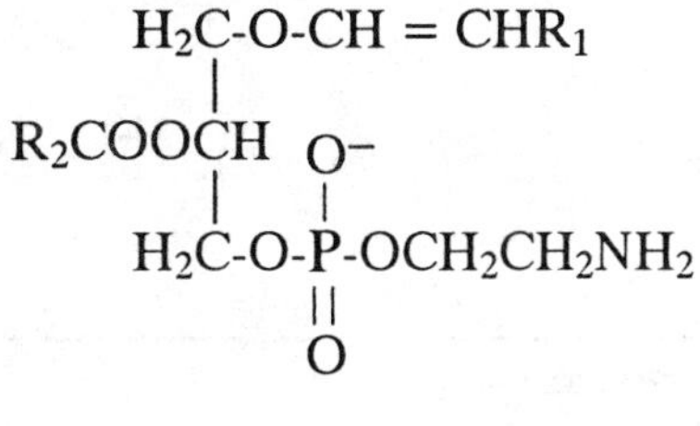

Ceramide (N-Acyl-Sphingosine)
$$|$$
$$HN\text{-}C\text{-}R$$
$$||$$
$$O$$

$$OH \qquad\qquad O^-$$
$$|\qquad\qquad\quad |$$
$$CH_3(CH_2)_{12}\text{-}CH = CH\text{-}CH\text{-}CH\text{-}CH_2\text{-}O\text{-}P\text{-}O\text{-}CH_2\text{-}CH_2\text{-}N^+(CH_3)_3$$

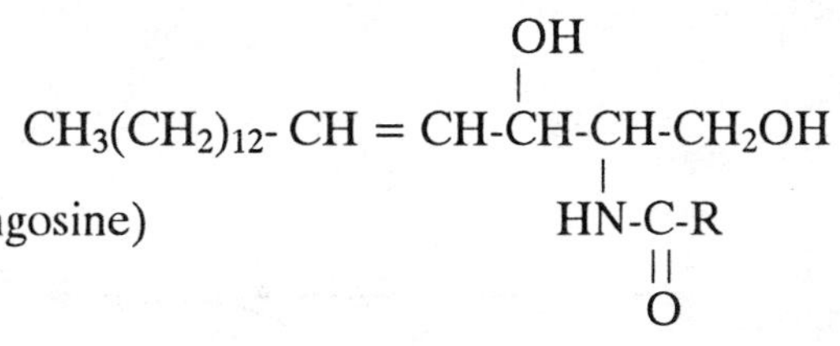

Sphingomyelin
$$|\qquad\qquad || $$
$$HN\text{-}C\text{-}R \quad O$$
$$||$$
$$O$$

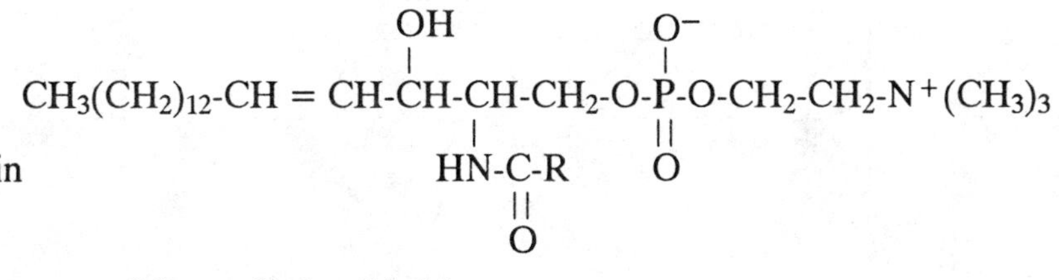

Figure 7-8. Structures of Some Polar Lipids.

Figure 7-9.
Synthetic Pathway for Phosphoglycerides.

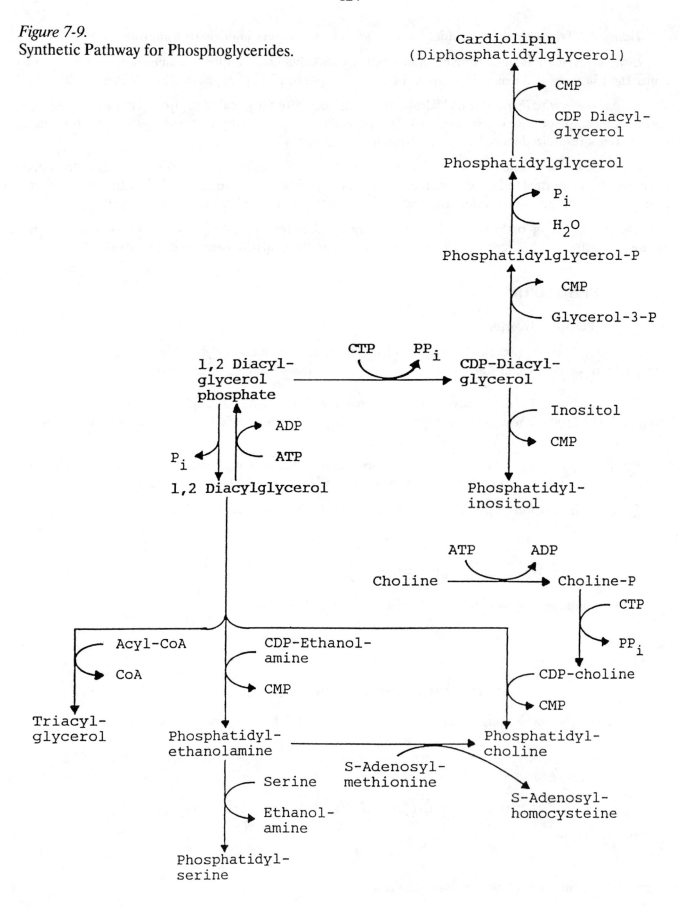

B. Sphingomyelin

Structure: Ceramide and choline in a phosphodiester linkage (Figure 7-8).

Synthesis:

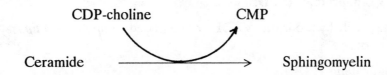

VIII. GLYCOSPHINGOLIPIDS (GLYCOLIPIDS).

Glycosphingolipids are the same as sphingomyelin in that they contain ceramide but no glycerol. They do not contain phosphate.

A. Cerebroside

Cerebroside (ceramide-hexose), found primarily in brain, consists of a hexose linked to ceramide. On complete hydrolysis it yields one molecule of sphingosine, a fatty acid, and a hexose which is most frequently D-galactose, less commonly D-glucose.

B. Cerebroside Sulfatides

These are sulfate esters of galactocerebroside.

C. Ceramide Oligosaccharides

Designated by the terms ceramide disaccharide, ceramide trisaccharide, etc.

D. Gangliosides

Gangliosides are ceramide oligosaccharides that contain at least one residue of **sialic acid** (N-acetyl-neuraminic acid).

E. Synthesis

Biosynthesis of glycosphingolipids requires **sugar nucleotides**. These are: UDP-Glc, UDP-Gal, UDP-GlcNAc, UDP-GalNAc and CMP-sialic acid.

F. Metabolic Blocks in Glycolipid Catabolism

Several **lysosomal disorders** exist, in which various glycolipids accumulate ("Lipid storage diseases," see Section X, pages 127 - 128, and Figure 7-10, page 129).

IX. LIPOPROTEINS

A. Lipid Transport in Blood

Lipids are insoluble in water and so must be transported as lipoproteins. Lipoproteins are classified according to increasing density: chylomicrons, very low density lipoproteins, low density lipoproteins, and high density lipoproteins. Nine apolipoproteins (ApoA-I, A-II, A-IV, B, C-I, C-II, C-III, D and E) have been isolated and characterized.

B. Chylomicrons

Chylomicrons are the largest of the lipoprotein particles which transport dietary triacylglycerols, cholesterol and other lipids from the intestine to adipose tissue and the liver. The triacylglycerols in chylomicrons are hydrolyzed within a few minutes by *lipoprotein lipase*, located on walls of capillaries. The cholesterol-rich residual particles, known as **remnants**, retain ApoE but not ApoC, and are taken up by the liver by a receptor specific for ApoE. Chylomicrons are especially abundant in blood after a fat-rich meal.

C. Very Low Density Lipoproteins (VLDL)

VLDL are synthesized primarily by the liver. VLDL are degraded by *lipoprotein lipase*, resulting in the production of low density lipoproteins.

D. Low Density Lipoproteins (LDL)

Most LDL appears to be formed from VLDL, but there is evidence for some direct production by the liver. These lipoproteins are rich in **cholesterol ester** and are **atherogenic**.

Cells outside the liver and intestine obtain cholesterol from plasma LDL rather than by synthesizing it *de novo*. Uptake from blood is by a **receptor-mediated** process specific for ApoB. The steps in the uptake are: (1) LDL binds to regions of plasma membranes called "**coated pits**." (2) The receptor-LDL complexes are internalized by **endocytosis**. (3) These vesicles subsequently fuse with lysosomes, are degraded, and the cholesterol is released within the cell.

The cholesterol content of cells that have an active pathway for uptake of LDL is regulated in two ways. First, the released cholesterol **suppresses** the activity of 3-hydroxy-3-methylglutaryl-CoA reductase (***HMG-CoA reductase***), thereby inhibiting *de novo* synthesis of cholesterol. Second, the number of LDL receptors is regulated with a feedback mechanism. When cholesterol is abundant inside the cell, new **LDL receptors are not synthesized** and the uptake of additional cholesterol from plasma LDL decreases.

The genetic defect in most cases of **familial hypercholesterolemia** is due to an absence or deficiency of the normal receptor for LDL.

E. High Density Lipoproteins (HDL)

HDL are produced by the liver. A major function of HDL is to act as a repository for apolipoproteins C and E that are required in the metabolism of chylomicrons and VLDL. HDL also are important in the transport of cholesterol from the periphery to the liver. Plasma LCAT (*lecithin : cholesterol acyl transferase*) is activated by ApoA-I of HDL and brings about esterification of cholesterol on the surface of HDL, resulting in its accumulation in the interior of the lipoprotein particles. This can create a concentration gradient and draw cholesterol from extrahepatic tissues for reverse cholesterol transport; thus **HDL is anti-atherogenic**.

F. Lipoprotein Lipase (LPL)

Triacylglycerols in chylomicrons and VLDL cannot be taken up intact by tissues but must first undergo hydrolysis by LPL, an enzyme situated on the capillary endothelium of extrahepatic tissues. The enzyme requires apolipoprotein C-II as cofactor.

G. Mobilization of Fat by Hormone-Sensitive Lipase (HSL)

The fat reserves of mammals are stored in **adipose tissue**. Fatty acids in triacylglycerols must first be hydrolyzed to free fatty acids before they can be transported from adipose tissue to other tissues. The hydrolysis of the triacylglycerols within adipose tissue is catalyzed by *hormone-sensitive lipase*. The released free fatty acids are transported in blood by the plasma protein, **albumin**.

The lipolysis of adipose tissue fat (fat mobilization) is stimulated by **glucagon and epinephrine**. Lipolysis is inhibited by **insulin**, which stimulates glucose uptake for lipogenesis, decreasing cAMP in fat cells.

H. Hypo- and Hyperlipoproteinemias

A number of heritable disorders of lipoprotein metabolism occur. (See Section X, next page).

I. Fatty Liver.

Two main categories occur: The first is associated with a raised level of plasma free fatty acids. The second is usually due to a metabolic block in the production of plasma lipoproteins, which may be caused by a deficiency of choline. Orotic acid also may produce fatty liver, perhaps by inhibiting glycosylation of ApoB and thus interfering with the release of VLDL into plasma.

J. Hypolipidemic Drugs.

Mevastatin and *lovastatin*: inhibit the HMG-CoA reductase reaction.

Clofibrate and *gemfibrozil*: reduce hepatic synthesis of triacylglycerol by diverting the influx of fatty acid to oxidation, rather than to esterification.

Cholestyramine: a resin that binds bile salts and prevents reabsorption.

Probucol: enhances catabolism of LDL.

Nicotinic acid: reduces lipolysis in adipose tissue.

X. GENETIC DISEASES

A. Lipoprotein metabolism

Familial *lipoprotein lipase* deficiency
    Molecular defect: Deficient lipoprotein lipase.
    Pathway affected: Hydrolysis of glyceride ester bonds.
    Diagnosis: Accumulation of chylomicron triglyceride.
    Genetics: Autosomal recessive.

Familial hypercholesterolemia
    Molecular defect: Abnormal LDL receptors; 3 mutations:
        1. no receptor binding of LDL
        2. reduced receptor binding of LDL
        3. no internalization of bound LDL
    Pathway affected: Cellular uptake and delivery of LDL to lysosome.
    Diagnosis: High cholesterol in plasma.
    Genetics: Autosomal dominant.

Familial *lecithin : cholesterol acyltransferase* (LCAT) deficiency
    Molecular defect: Deficient LCAT.
    Pathway affected: Cholesterol → cholesteryl ester.
    Diagnosis: Abnormal plasma lipoprotein pattern.
    Genetics: Autosomal recessive.

B. Sphingolipidoses (Lipid Storage Diseases, Figure 7-10, page 129)

Farber
    Molecular defect: Deficient ceramidase.
    Pathway affected: Ceramide → sphingosine + fatty acid.
    Diagnosis: Deficient enzyme activity in leukocytes.
    Genetics: Autosomal recessive.

Fabry

> Molecular defect: Deficient α-*galactosidase A*.
> Pathway affected: Gal-Gal-Glc-Cer → Gal-Glc-Cer + Gal.
> Diagnosis: Increased tissue Gal-Gal-Glc-Cer; deficient α-galactosidase A (plasma).
> Genetics: X-linked.

Gaucher, types 1, 2 and 3

> Molecular defect: Deficient β-*glucocerebrosidase*.
> Pathway affected: Glucocerebroside → Glc.
> Diagnosis: Reduced enzyme activity in leukocytes.
> Genetics: Autosomal recessive (all three types).

G_{M1}-gangliosidosis types 1, 2 and 3

> Molecular defect: Mutated β-*gangliosidase A*.
> Pathway affected: Ganglioside G_{M1} → G_{M2}
> Diagnosis: Reduced enzyme activity in leukocytes.
> Genetics: Autosomal recessive.

Krabbe, globoid cell leukodystrophy

> Molecular defect: Deficient *galactocerebroside β-galactosidase*.
> Pathway affected: galactocerebroside → ceramide + Gal.
> Diagnosis: deficient enzyme activity in leukocytes.
> Genetics: Autosomal recessive.

Metachromatic leukodystrophy

> Molecular defect: Deficient *arylsulfatase A*.
> Pathway affected: Catabolism of sulfatides and sulfogalactoglycerolipids.
> Diagnosis: Deficient enzyme activity in leukocytes.
> Genetics: Autosomal recessive.

Niemann-Pick, types I and II

> Molecular defect: Deficient *sphingomyelinase*.
> Pathway affected: Sphingomyelin → ceramide + phosphorylcholine.
> Diagnosis: Deficient enzyme activity in leukocytes.
> Genetics: Autosomal recessive.

Sandhoff, G_{M2}-gangliosidosis type 2

> Molecular defect: Deficient *Hex A and B*; mutated β chain.
> Pathway affected: G_{M2} → N-acetyl-β-D-galactosamine, or
> globoside → N-acetyl-β-D-galactosamine.
> Diagnosis: Deficient activities of Hex A and Hex B in serum.
> Genetics: Autosomal recessive.

Tay-Sachs, G_{M2}-gangliosidosis type 1

> Molecular defect: Deficient *hexosaminidase A*; mutated α chain.
> Pathway affected: G_{M2} → N-acetyl-β-D-galactosamine.
> Diagnosis: Deficient Hex A in serum and leukocytes.
> Genetics: Autosomal recessive.

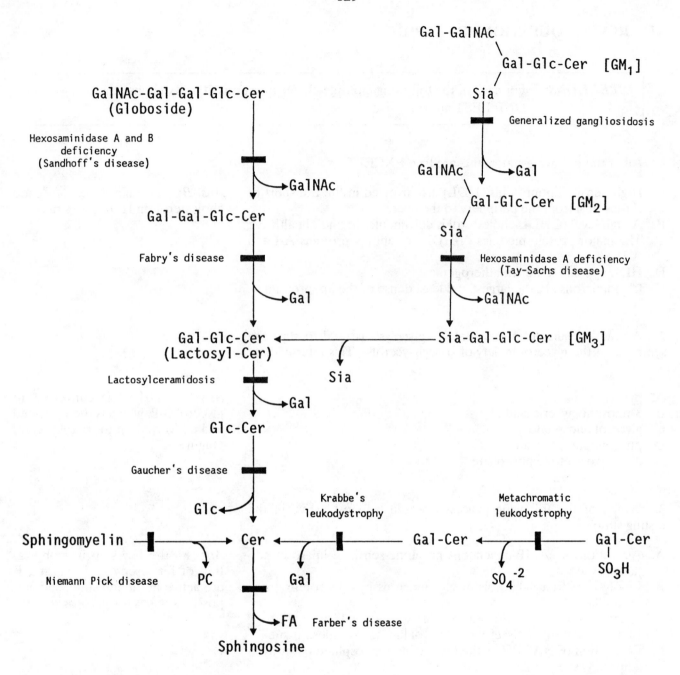

Figure 7-10. Catabolism of typical glycosphingolipids by human lysosomal glycosidases, sphingolipase, and sulfatase. Diseases associated with a hydrolase deficiency are indicated, and the block in catabolism in each is indicated by a solid bar. PC is phosphorylcholine and FA is fatty acid.

XI. REVIEW QUESTIONS ON LIPIDS

> **DIRECTIONS:** For each of the following multiple-choice questions (1 - 48), choose the ONE BEST answer.

1. All of the following statements are true EXCEPT:

A. High density lipoproteins (HDL) are involved in the transport of cholesterol from the periphery to the liver
B. A high level of HDL-cholesterol is detrimental to one's health
C. The major protein moieties of HDL are apolipoproteins A-I and A-II
D. HDL appears to be antiatherogenic
E. Chylomicrons are the largest and least dense of the lipoproteins.

Ans B: A high level of "good cholesterol" in HDL is desirable.

2. A readily-available intermediate in glycolysis is used to start the synthesis of the glycerol moiety of triacylglycerols. This intermediate is:

A. pyruvate
B. 3-phosphoglyceric acid
C. glycerol phosphate
D. phosphoenolpyruvate
E. dihydroxyacetone phosphate.

Ans E: DHAP is converted to glycerol-3-P, which is then acylated and converted on to triacylglycerol (Figure 7-7).

3. Which of the following processes would be most active in the fasting state?

A. Activation of cAMP-dependent hormone-sensitive lipase in adipose tissue
B. Lipolysis of triacylglycerols of chylomicrons by lipoprotein lipase in capillaries
C. Activation of acetyl CoA carboxylase by citrate in liver
D. Phosphorylation of glycerol by glycerol kinase in adipose tissue
E. Production of NADPH by the hexose monophosphate pathway in adipose tissue.

Ans A, the key step in mobilization of fat reserves. B, C, and E are active in the fed state, and adipose tissue has no glycerol kinase.

4. In the fed state, when fatty acids are being synthesized in liver they cannot immediately undergo β-oxidation. This is because:

A. cAMP and fatty acyl CoA depress the activity of acetyl CoA carboxylase
B. they are utilized in the esterification pathway
C. fatty acyl CoA inhibits mitochondrial citrate synthase
D. insulin leads to increased uptake of glucose into cells and an increased rate of glycolysis
E. gluconeogenesis takes precedence over fatty acid oxidation.

Ans B because the two processes are reciprocally regulated and during synthesis esterification is active (see VI. B).

5. In the control of fatty acid metabolism

A. when fuel is abundant fatty acid oxidation is enhanced
B. high carbohydrate intake causes increased fatty acid synthesis
C. high carbohydrate intake causes decreased fatty acid synthesis
D. palmitoyl CoA has no effect on fatty acid synthesis
E. ketone bodies enhance fatty acid synthesis.

Ans B because excess carbohydrate can be converted to fat (Figure 7-3).

6. Increased urinary acetoacetate might result when liver

A. glycogen is normal
B. glycogen is depleted
C. NADPH concentration is high
D. acetyl CoA concentration is low
E. lactate concentration is high.

Ans B: Glycogen depletion causes the switch to utilize fatty acid as energy source.

7. The role of hormone-sensitive lipase is to

A. hydrolyze the ester bonds in hormones
B. hydrolyze dietary fat and the enzyme is stimulated by epinephrine
C. mobilize fat from adipose tissue
D. hydrolyze triacylglycerols in liver
E. hydrolyze triacylglycerols in brain.

Ans C: This enzyme is the first step in utilizing fat reserves.

8. Acetyl-CoA carboxylase and other biotin-containing enzymes have lowered activity after ingestion of dietary

A. citrate
B. carnitine
C. avidin
D. lactalbumin
E. cyanide.

Ans C: Avidin has a high binding affinity for biotin, preventing its absorption.

9. In the lipogenesis of fatty acids from pyruvate, which of the following enzyme reactions occurs in cytoplasm?

A. pyruvate dehydrogenase
B. citrate synthase
C. propionyl-CoA carboxylase
D. succinate dehydrogenase
E. ATP-citrate lyase.

Ans E: The others are mitochondrial. See Figure 7-3.

10. About how many grams of stearate must be metabolized to give an amount of energy equivalent to 50 g of glycogen?

A. 5
B. 20
C. 40
D. 80
E. 94

Ans B: 50 × 4(Kcal/g for carbohydrate) = 200 Kcal; 200 ÷ 9(Kcal/g for fat) = about 20 Kcal.

11. Mitochondrial membranes can transport

A. fatty acid
B. fatty acyl-CoA
C. malonyl-CoA
D. acetyl-CoA
E. acyl-carnitine.

Ans E: The membrane is impermeable to A - D, but fatty acids can (and must) use the carnitine transport system (Figure 7-1).

12. All glycerol-containing lipids are synthesized from

A. triacylglycerol
B. cephalin
C. phosphatidic acid
D. phosphatidyl glycerol
E. monoacylglycerol.

Ans C: Phosphatidate (1,2-Diacyl-glycerol phosphate) is the key intermediate (Figure 7-9).

13. Ketosis is the consequence of increased blood levels of

A. acetoacetate
B. acetyl-CoA
C. β-hydroxy-β-methyl-glutarate
D. pyruvate
E. acetone.

Ans A: Acetoacetate and β-hydroxybutyrate are the major ketone bodies; acetone is produced in minor amounts from decarboxylation of acetoacetate.

14. The cofactors common to both oxidation and synthesis of fatty acids include

A. NAD, NADP
B. FAD, NADP
C. NAD, CoA
D. CoA, ATP
E. NAD, ATP.

Ans D: In oxidation, ATP is used for activation; in synthesis, it is used by citrate lyase and acetyl CoA carboxylase. CoA is used repeatedly in both processes (Figures 7-1 and 7-4).

15. Glucose catabolism through the pentose phosphate pathway stimulates fatty acid synthesis because it increases

A. acetyl-CoA
B. NADH
C. NADPH
D. ATP
E. glycogen

Ans C: Though A and D are also needed for fatty acid synthesis, NADPH is the rate-limiting cofactor.

16. Which of the following is NOT on the major route to β-hydroxy-butyrate in liver?

A. acetyl-CoA
B. malonyl-CoA
C. acetoacetyl-CoA
D. acetoacetate
E. hydroxymethyl glutaryl-CoA.

Ans B: Malonyl CoA is utilized for fatty acid synthesis.

17. The complete oxidation of palmitic acid to carbon dioxide and water leads to the formation of a net total of high energy bonds of

A. 39 ATP
B. 69 ATP
C. 99 ATP
D. 129 ATP
E. 131 ATP.

Ans D: See IV. B for the computation.

18. Essential fatty acids and precursors of prostaglandins include:

A. palmitic acid
B. lignoceric acid
C. linolenic acid
D. oleic acid
E. myristic acid.

Ans C: This is the only essential fatty acid among the five.

19. Triacylglycerols of chylomicrons are:

A. intact triacylglycerols of intestinal lumen
B. newly synthesized by intestinal cells
C. transported from the liver
D. synthesized by adipocytes
E. synthesized by muscle cells.

Ans B because chylomicrons are produced from dietary fat by intestinal cells.

20. An isomerase is needed for:

A. biosynthesis of saturated fatty acids
B. biosynthesis of steroids
C. oxidation of some unsaturated fatty acids
D. oxidation of saturated fatty acids
E. catabolism of steroids.

Ans C: β-Oxidation of some unsaturated fatty acids requires *cis*-Δ^3 → *trans*-Δ^2-enoyl CoA isomerase (See IV. B).

21. Substances requiring bile salt for transport into intestinal cells include:

A. vitamin A
B. vitamin B
C. vitamin C
D. glucose
E. amino acids.

Ans A because Vitamin A is the only water-insoluble compound among the five.

22. During fatty acid synthesis, $^{14}CO_2$ is incorporated into the carboxyl of malonyl-CoA. At the condensation step, this $^{14}CO_2$ is:

A. 50% incorporated into fatty acyl-CoA
B. 100% incorporated into acyl-CoA
C. eliminated as $^{14}CO_2$
D. transferred to biotin
E. transferred to succinyl CoA.

Ans C: The same C that enters into malonyl CoA is the one lost in the condensation step. See IV. D.

23. In metabolism of fatty acids in adipose tissue,

A. insulin promotes fatty acid synthesis and triacylglycerol synthesis
B. catecholamine promotes synthesis of triacylglycerol
C. glucagon inhibits hormone-sensitive lipase
D. insulin activates hormone-sensitive lipase
E. catecholamine promotes synthesis of fatty acids.

Ans A because in the fed state, insulin promotes the process of fat storage.

24. In the biosynthesis of phosphoglycerides, which of the following is required for activation of diacylglycerol phosphate?

A. ATP
B. dATP
C. GTP
D. UTP
E. CTP.

Ans E (Figure 7-9).

25. A fatty acid with an odd number of carbon atoms will enter the citric acid cycle in part as

A. citrate
B. isocitrate
C. α-ketoglutarate
D. succinate
E. malate.

Ans D: The remainder from β-oxidation is propionyl CoA, which is converted to succinyl CoA (Figure 7-2).

26. Which of the following carries 2-carbon units from mitochondria to cytosol for fatty acid synthesis?

A. oxaloacetate
B. malate
C. pyruvate
D. carnitine
E. citrate.

Ans E because citrate is the precursor of cytosolic acetyl CoA (Figure 7-3).

27. The *de novo* biosynthesis of triacylglycerols occurs mainly in

A. liver and brain
B. brain and muscle
C. adipose tissue and muscle
D. muscle and liver
E. liver and adipose tissue.

Ans E: See VI. A.

28. The major energy source for the brain is normally

A. blood glucose
B. blood amino acid
C. blood ketone bodies
D. blood fatty acids
E. blood lactic acid.

Ans A: Brain normally used glucose; it can adapt to using ketone bodies, but only after 2-3 days' starvation.

29. Oxidation of fatty acids occurs in or on the:

A. mitochondrial matrix
B. outer mitochondrial membrane
C. cytosol
D. intermembrane space
E. Golgi apparatus.

Ans A: Many oxidative processes occur in mitochondria.

30. Gangliosides contain all of the following EXCEPT

A. neuraminic acid
B. galactose
C. fatty acid
D. sphingosine
E. phosphate.

Ans E: Gangliosides are a glycolipid, not a phospholipid (see VIII).

31. Carnitine acyltransferase is necessary for:

A. fatty acid oxidation
B. fatty acid synthesis
C. cholesterol synthesis
D. cholesterol oxidation
E. digestion of triacylglycerols.

Ans A because the carnitine transport system is needed for fatty acids to enter mitochondria for oxidation.

32. The role of lipoprotein lipase is to:

A. digest dietary lipoproteins in intestinal lumen
B. mobilize adipose tissue fat
C. carry out intracellular lipolysis of lipoproteins
D. hydrolyze triacylglycerols of plasma lipoproteins for transport of the released fatty acids into tissues
E. hydrolyze cholesterol esters of plasma lipoproteins.

Ans D: LPL is required to hydrolyze triacylglycerols of chylomicrons and VLDL.

33. Fatty acids in transport from adipose tissue to energy-utilizing tissues like muscle occur in the blood as:

A. chylomicrons
B. fatty acids bound to albumin
C. VLDL
D. LDL
E. HDL.

Ans B (See IX. G).

34. The rate-controlling step in the synthesis of fatty acids is catalyzed by:

A. acetyl CoA carboxylase
B. fatty acid synthase
C. HMG CoA synthase
D. hormone-sensitive lipase
E. acyl transacylase.

Ans A: This is the first and committed step for the pathway of fatty acid synthesis.

35. Mitochondria do NOT contain:

A. cytochrome oxidase
B. citrate synthase
C. HMG CoA synthase
D. β-ketoacyl CoA thiolase
E. acetyl CoA carboxylase

Ans E: Acetyl-CoA carboxylase is a cytosolic enzyme for fatty acid synthesis.

36. Activation of fatty acids requires:

A. ATP
B. dATP
C. GTP
D. UTP
E. CTP

Ans A: FA + CoA + ATP → F Acyl CoA + AMP + PP_i (Figure 7-1).

37. The immediate precursor of prostaglandins is:

A. myristate
B. palmitate
C. linoleate
D. linolenate
E. arachidonate

Ans E: Prostaglandins are made from C-20 fatty acids.

38. Hydrolysis of sphingomyelin would yield all of the following EXCEPT:

A. phosphate
B. choline
C. fatty acid
D. sphingosine
E. glycerol

Ans E: Sphingolipids contain no glycerol (Figure 7-8).

39. Cholestyramine exerts its hypocholesterolemic effect by interaction with:

A. cholesterol
B. triacylglycerols
C. lipoproteins
D. bile salts
E. cholesterol esters.

Ans D: By binding bile salts, cholestyramine promotes their excretion, causing the liver to convert more cholesterol to new bile salts.

40. Acetyl-CoA carboxylase

A. is inhibited by acyl-CoA
B. catalyzes the non-equilibrium step in the catabolism of odd-numbered carbon fatty acids
C. contains avidin
D. is inhibited by ATP
E. is inhibited by citrate.

Ans A because acyl-CoA is the end product that exerts feedback regulation.

41. Ketone bodies

A. can be a source of energy for muscle in fasting state
B. can be a source of energy for brain in feeding state
C. are made during periods of rapid glycolysis
D. can be a precursor of gluconeogenesis
E. are required for triacylglycerol synthesis.

Ans A: Ketone bodies, made during periods of rapid lipolysis, can be an important source of energy during starvation.

42. Aspirin inhibits the synthesis of

A. prostaglandins
B. steroid hormones
C. phosphatidic acid
D. lipoproteins
E. epinephrine.

Ans A: Aspirin inhibits the cyclooxygenase pathway in the synthesis of prostaglandins.

43. Conversion of diacylglycerol to lecithin requires:

A. CMP-choline
B. CDP-choline
C. CTP-choline
D. CTP-ethanolamine
E. CDP-ethanolamine.

Ans B: (See Figure 7-9).

44. The lipoprotein that accumulates in blood when receptors on membranes of cells of peripheral tissues are defective, preventing endocytosis, is:

A. chylomicrons
B. HDL
C. IDL
D. LDL
E. VLDL.

Ans D, the so-called "bad cholesterol."

45. The last step in a cycle of β-oxidation is catalyzed by:

A. 3-*cis* → 2-*trans* enoyl CoA isomerase
B. acyl CoA dehydrogenase
C. acyl CoA thiolase
D. hydroxymethyl glutaryl (HMG) CoA synthase
E. hydroxyacyl CoA dehydrogenase.

Ans C (Figure 7-1).

46. Intermediates in the degradation of fatty acids are linked to:

A. acyl carrier protein
B. coenzyme A
C. CDP
D. carnitine
E. pyrophosphate.

Ans B (Figure 7-1).

47. Hydrolysis of lecithin would yield all of the following EXCEPT:

A. phosphate
B. choline
C. fatty acid
D. sphingosine
E. glycerol.

Ans D: Lecithin is the common name for diacylphosphatidyl choline.

48. Lovastatin exerts its hypolipemic effect by

A. promoting β-oxidation
B. inhibiting HMG-CoA reductase
C. promoting fat mobilization in adipose tissue
D. inhibiting ketone bodies production by inhibition of acetyl CoA-acetyl transferase
E. lowering acetyl-CoA production by inhibition of pyruvate dehy drogenase.

Ans B: Lovastatin is one of a family of compounds that inhibit the first step in cholesterol synthesis.

DIRECTIONS: For each set of questions, choose the ONE BEST answer from the lettered list above it. An answer may be used one or more times, or not at all.

Questions 49 - 54: For each of the genetic diseases in the following set of questions, choose the ONE most appropriate answer from the following list of enzyme deficiencies:

A. α-galactosidase
B. β-galactosidase
C. sphingomyelinase
D. ceramidase
E. α-glucosidase

F. β-glucosidase
G. hexosaminidase A
H. lipoprotein lipase (LPL)
I. lecithin:cholesterol acyltransferase (LCAT)

49. Tay-Sachs disease.

50. Farber's disease.

51. Gaucher's disease.

52. Niemann-Pick disease.

53. Krabbe's disease.

54. Fabry's disease.

Answers:
49. G
50. D
51. F

52. C
53. B
54. A

The location of the block in catabolism associated with each disease is shown in Figure 7-10.

8. STEROIDS

Thomas Briggs

I. CHOLESTEROL: STRUCTURE AND CHEMISTRY

Carbon skeleton. The full system in cholesterol consists of four fused rings, two "angular methyl groups," and an 8-carbon side-chain, 27 carbons in all (Figure 8-1). In many steroid derivatives the side-chain is reduced or absent.

Substitution. Saturated carbons of the ring system may bear substituents which project on either side of the plane of the ring. In structural formulas, these are designated as follows:

$\cdots$ (dotted line) $= \alpha$ (alpha, <u>a</u>way from observer).

—— (solid line) $= \beta$ (<u>b</u>eta, toward, or <u>b</u>y observer).

This notation is the same as that used for the hydroxyl group on the anomeric carbon of sugars (Chapter 3, page 32).

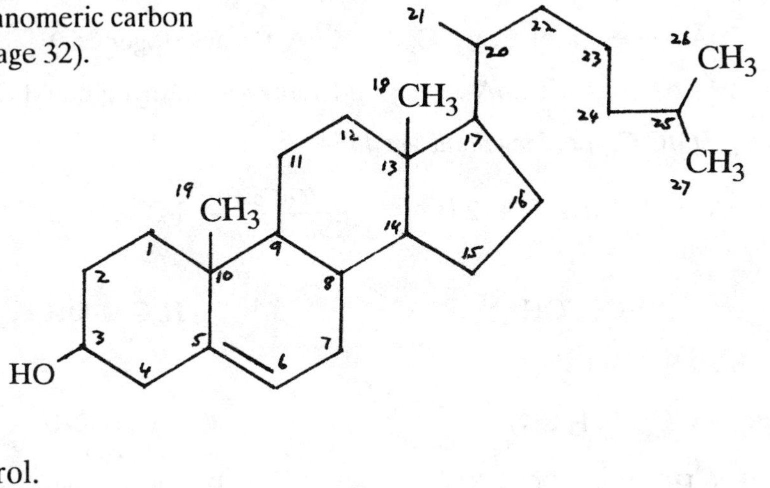

Figure 8-1. Cholesterol.

The **3β-hydroxyl** group, though hydrophilic, cannot overcome the non-polarity of the rest of the molecule, hence cholesterol is **insoluble in water.**

The $\Delta^{5\text{-}6}$ double bond can be reduced in two ways:

5α-H: 5α-cholestan-3β-ol; new hydrogen *trans* to the angular methyl group (C-19); rings A/B *trans*.

5β-H: 5β-cholestan-3β-ol; new hydrogen *cis* to C-19; rings A/B *cis*. This is mainly a bacterial product (coprosterol in older usage, from Gr. $\kappa o \pi \rho o \sigma$, meaning feces, where it is found). But the 5β-configuration occurs in some derivatives such as the bile acids and certain metabolites of steroid hormones.

II. OCCURRENCE AND FUNCTION

The name means **"bile-solid-alcohol."** It is found in bile, (a frequent component of gallstones, where it was first discovered), and is a crystalline solid and an alcohol.

Cholesterol occurs in **all cells** and tissues of higher organisms, but is especially abundant in nervous tissue and egg yolk. Related sterols are found in plants and higher microorganisms; these are poorly absorbed from the digestive tract. For storage, sterols are often **esterified** with unsaturated fatty acids. The total amount in a human averages about 180 grams. A typical value for blood cholesterol in developed countries is 180 mg/100 ml. Note that on a weight basis this is twice as high as the level of blood glucose.

Cholesterol has a universal function as a **component of membranes**, up to 25% (w/w), with a complex effect on fluidity (see chapter 9). Also it is a **precursor** of many important biological substances (see **bile acids, hormones**). In human disease, its insolubility makes deposits troublesome, especially in **atherosclerotic plaques** and **gallstones**.

In blood, cholesterol is carried as a complex with **lipoproteins**, partly esterified with a fatty acid, and partly with the 3-OH free. (See Chapter 7, section IX). The cholesterol in blood can exchange readily with liver cholesterol, but some cholesterol pools, especially that in brain, exchange very slowly.

III. BIOSYNTHESIS OF CHOLESTEROL

A. Acetate to Squalene

♦ *Formation of β-hydroxy-β-methyl-glutaryl CoA* (HMG CoA)

$$2 \text{ acetyl CoA} \rightleftharpoons \text{ acetoacetyl CoA} + \text{CoA} \text{ (as in ketogenesis)}$$

$$\text{acetoacetyl CoA} + \text{acetyl CoA} \rightleftharpoons \beta\text{-hydroxy-}\beta\text{-methyl-glutaryl CoA, or HMG CoA}$$

♦ *Reduction of HMG CoA to Mevalonic Acid*

$$\text{HMG CoA} + 2 \text{ NADPH} + 2 \text{ H}^+ \xrightarrow{\text{HMG CoA Reductase}} \text{MVA} + 2 \text{ NADP}^+ + \text{HSCoA}$$

```
      H3C   OH                          H3C   OH
        \  /                              \  /
         C                                 C
        / \                               / \
      CH2  CH2                          CH2  CH2
       |    |                            |    |
     HOOC  CO~SCoA                     HOOC  CH2OH
```

HMG CoA   **Mevalonic Acid** (MVA)

This reaction is the first committed, largely irreversible step in cholesterol biosynthesis, and is an **important control point**. The enzyme is rate-limiting for the entire pathway, and is inhibited by the eventual end-product, cholesterol.

♦ *Then mevalonic acid is converted* to **isopentenyl pyrophosphate**, which is the "active isoprene unit": the precursor of all isoprenoid compounds such as terpenes; vitamins A, D, E, K; sterols; rubber.

$$MVA \quad \xrightarrow[\text{3 times}]{ATP, \ Mg^{++}} \quad H_2C = C\text{-}CH_2\text{-}CH_2OPP \qquad \begin{array}{l} \textbf{Isopentenyl} \\ \textbf{Pyrophosphate} \end{array}$$
$$\hspace{8.5cm} | \hspace{3.5cm}$$
$$\hspace{8.5cm} CH_3 \hspace{2.5cm} (CO_2 \text{ is lost})$$

♦ *Isopentenyl-PP can isomerize* to **dimethylallyl-PP**, which condenses with another isopentenyl-PP to form **geranyl-PP**, a monoterpene (C-10):

$$H_2C = C\text{-}CH_2\text{-}CH_2OPP \rightleftharpoons H_3C\text{-}C = CH\text{-}CH_2OPP \longrightarrow H_3C\text{-}C = CH\text{-}CH_2CH_2C = CH\text{-}CH_2OPP$$
$$\hspace{0.8cm}| \hspace{5cm} | \hspace{5.5cm} | \hspace{2.5cm} |$$
$$\hspace{0.8cm}CH_3 \hspace{4.7cm} CH_3 \hspace{5.2cm} CH_3 \hspace{2cm} CH_3$$

♦ *Geranyl-PP condenses* with an *i*-pentenyl-PP to make **farnesyl-PP**, a sesquiterpene (C-15).

♦ *Two farnesyl-PP condense* head-to-head (NADPH required) to **squalene**, a triterpene (C-30). Farnesyl-PP is also the precursor of other polyisoprenoids such as dolichol and ubiquinone.

B. Squalene to Cholesterol *(Figure 8-2)*

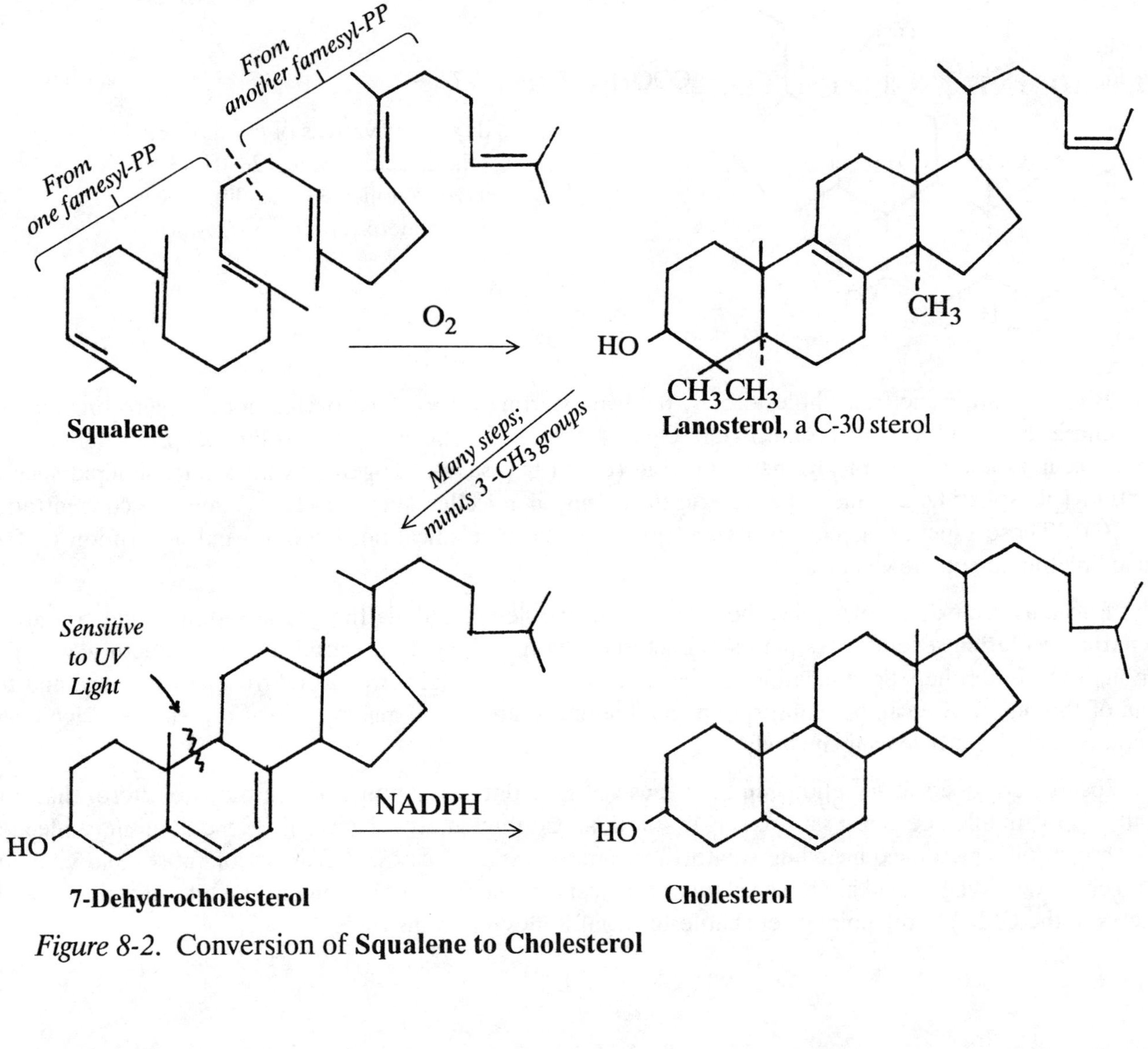

Figure 8-2. Conversion of **Squalene to Cholesterol**

Ring B of 7-dehydrocholesterol can be split by UV light (at arrow, Figure 8-2) to form vitamin D_3. The presence of 7-dehydrocholesterol in human skin explains how sunlight leads to the formation of vitamin D. In the body, vitamin D (cholecalciferol) is transported to the liver and hydroxylated at C-25, then to the kidney and hydroxylated at 1α, to form **1α,25-dihydroxycholecalciferol**, or **calcitriol**. *This is the biologically active form.* Vitamin D can be regarded as a hormone because it acts on target cells by the same mechanism as the steroid hormones. The active form promotes calcium absorption in the intestine by inducing the synthesis of a **calcium binding protein**.

IV. METABOLISM OF CHOLESTEROL

A. Secretion

Some cholesterol is simply **secreted as is** into intestine. A little free cholesterol also occurs in bile.

B. Conversion to Bile Acids *(Figure 8-3)*

Quantitatively this is the major route for the excretion of cholesterol in humans.

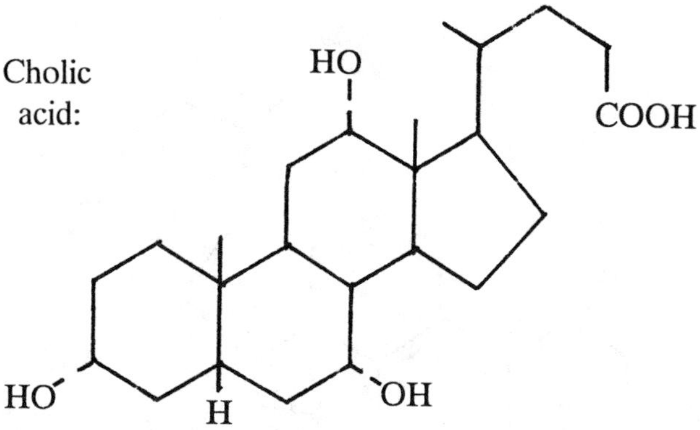

Figure 8-3.

Bile acids are derivatives of cholanoic acid (C-24):
Cholic acid: 3α,7α,12α-triol
Deoxycholic: 3α,12α-diol
Chenodeoxycholic: 3α,7α-diol

Bile acids are made from cholesterol by the liver, starting with 7α-hydroxylation by a microsomal monooxygenase that employs cytochrome P-450, O_2, and NADPH. This is the controlling step. In humans bile acids occur as conjugates of **glycine** or **taurine** (called bile salts). Together with a phospholipid such as lecithin (phosphatidyl choline), they aggregate as **mixed micelles** above a critical micellar concentration (CMC). These function as detergents and promote the emulsification, lipolysis, and absorption of fats, including the fat-soluble vitamins.

Bile salts are extensively reabsorbed (>95%) in the ileum, and via the portal system, undergo **enterohepatic circulation** several times per day. Bacterial action in the gut may result in some structural changes. Though the enterohepatic circulation is efficient, the bile salt pool turns over 6-10 times per day, and the sum of the bile salts escaping reabsorption can amount to about 500 mg/day, which represents a significant amount of cholesterol leaving the body.

Insufficient bile salt secretion can be a cause of gallstone formation because the cholesterol that normally occurs in bile needs bile salts to stay in solution. Continued disturbance of BS metabolism can lead to malabsorption syndromes, including steatorrhea, and in extreme cases, deficiency of fat-soluble vitamins. Oral chenodeoxycholic acid has been useful as replacement therapy to supplement the bile acid pool (well in excess of the CMC) to the point where cholesterol gallstones may redissolve.

C. Steroid Hormones

Production of hormones from cholesterol is quantitatively minor, but of major importance physiologically.

Major types: The examples shown in Figure 8-4 are the principal members of each major category (gonadal and adrenal) secreted in the human.

Formation from Cholesterol: All steroid-producing tissues cleave the side-chain of cholesterol between carbons 20 and 22 to form **pregnenolone** (enzyme: *20,22-lyase*, or *desmolase*). This is the rate-limiting step. In most cases, pregnenolone is then converted to **progesterone**.

♦ Gonads: The *corpus luteum* stops at progesterone. The testis, after 17-hydroxylation, cleaves the remaining side-chain (enzyme: *17,20-lyase*) to form **19-carbon** steroids (androstenedione, then testosterone). In many tissues, a 5α-reductase converts testosterone to 5α-dihydrotestosterone which is the active hormone in those tissues. The ovary also makes testosterone but does not release it. Instead, it aromatizes ring A (resulting in the loss of carbon 19) to form **18-carbon** steroids (estradiol, estrone).

Gonadal steroids

i. **progestational** ii. **androgens** iii. **estrogens**

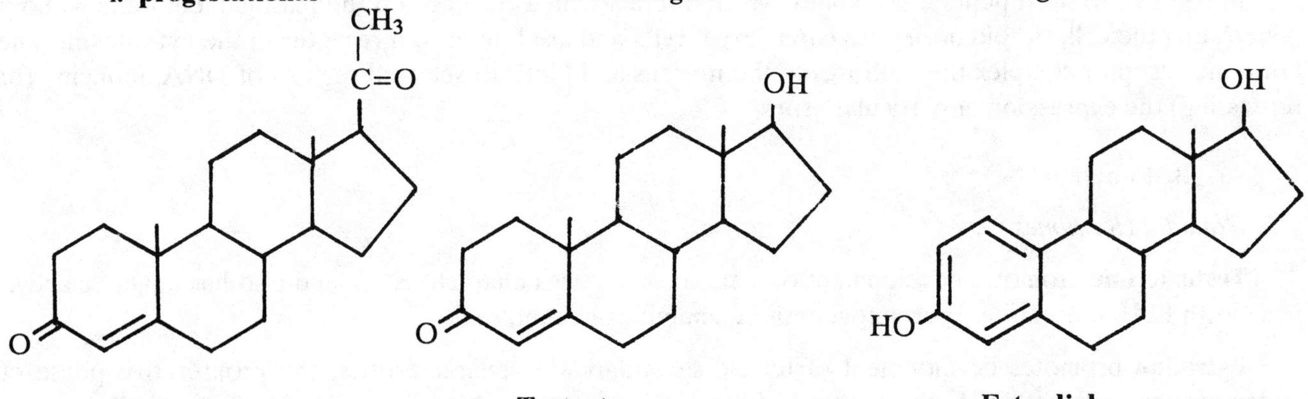

Progesterone Testosterone Estradiol

Adrenal steroids (corticosteroids)

i. **glucocorticoids** ii. **mineralocorticoids**

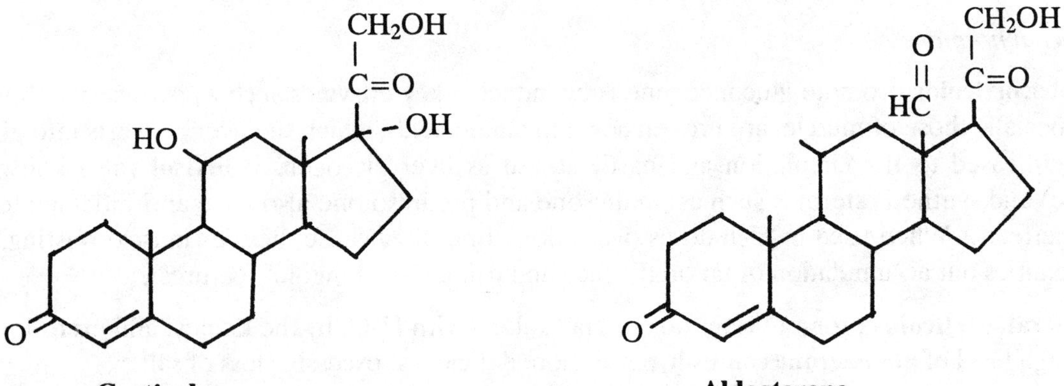

Cortisol Aldosterone

Figure 8-4. Representative Steroid Hormones in the Human

♦ Adrenal cortex: hydroxylations occur in sequence at positions **17, 21, 11, 18**. Except: the *zona fasciculata*, which makes cortisol, has no 18-hydroxylase, and the *zona glomerulosa*, which makes aldosterone, has no 17-hydroxylase. Some conversion of pregnenolone to androgens (especially **dehydroepiandrosterone**, or **DHEA**) and estrogens may also occur. In a patient with Cushing's syndrome, there is excess production of glucocorticoids which cause protein wasting and hyperglycemia (due to gluconeogenesis); adrenal androgens may cause signs of virilization such as hirsutism.

Transport: Testosterone and estradiol in the circulation are bound to **sex hormone-binding globulin (SHBG)**. Cortisol is bound to **transcortin** or **corticosteroid-binding globulin (CBG)**. In each case, the bound form is in equilibrium with a small amount of the free hormone. The unbound fraction is the biologically active form. Progesterone also binds to CBG. Aldosterone has no specific carrier and is present at a much smaller total level, which is entirely in the active (free) form.

Metabolism: Reductions occur involving ring A, and the products are excreted in bile and/or urine as conjugates of **sulfate** or **glucuronic acid**. Androgens are oxidized at C-17 to ketones of the class known as **17-ketosteroids**. Urinary 17-KS are indicative of androgen metabolism of *both* testicular and adrenal origin.

V. ACTION OF STEROID HORMONES

A. Mechanism

In contrast to most peptide hormones, which interact with a receptor on the plasma membrane without penetrating the cell, steroid hormones **enter target cells** and are bound by a **receptor** in the **cytoplasm**. The hormone-receptor complex then diffuses to the nucleus and binds to selected regions of DNA, inducing (or repressing) the **expression** of particular **genes**.

B. Function

Gonadal Hormones

Testosterone promotes development of male secondary sex characteristics, and also has anabolic activities. With FSH, it promotes spermatogenesis by seminiferous tubules.

Estradiol promotes development of female secondary sex characteristics, the proliferative phase of endometrium, and, with FSH, development of the ovarian follicle and finally, ovulation. Estrogens also seem to be critical for bone maturation and normal bone development in the male.

Progesterone, as the name implies, promotes gestation. It supports the secretory phase of endometrium, the luteal phase of ovary, and inhibits further ovulation.

Adrenal hormones

Glucocorticoids promote gluconeogenesis by inducing key enzymes such as pyruvate carboxylase. Proteins, especially those of muscle, are broken down to amino acids, which the liver converts into glucose which is partly released to the circulation and partly stored as liver glycogen. **Cortisol** (also known as hydrocortisone) and synthetic steroids such as prednisone and prednisolone also have anti-inflammatory and anti-immune effects. When used in high doses over a long time they cause, besides muscle wasting, lipolysis on the extremities but accumulation of fat on the face and trunk ("Cushingoid" features).

Mineralocorticoids promote retention of Na^+, along with H_2O, by the kidney, and excretion of H^+ and K^+. A high level of **aldosterone** causes hypertension; deficiency, excessive loss of salt.

C. Regulation

Gonadal hormones: Trophic hormones are the same in both sexes. The hypothalamic **Gonadotropin Releasing Hormone (GnRH)**, which is the same as **Luteinizing Hormone Releasing Hormone (LHRH)**, stimulates the pituitary to release both **Follicle Stimulating Hormone** (FSH, formerly known as ICSH) and **Luteinizing Hormone (LH)**.

In the male, FSH induces spermatogenesis in seminiferous tubules; feedback is by a glycoprotein, **inhibin**. LH stimulates production of **testosterone** by Leydig cells. Feedback is on both pituitary and hypothalamus.

In the female, FSH induces follicular development and production of **estradiol**. At mid-cycle, there is a positive feedback effect by estrogen, causing a surge of both FSH and LH production by the pituitary. Luteinization ensues, with production of **progesterone** by the *corpus luteum*.

Adrenal hormones

The hypothalamus puts out **Corticotropin Releasing Hormone** (CRH) which stimulates the pituitary to release **Adrenocorticotropic Hormone** (ACTH), which then signals the adrenal cortex to secrete **cortisol**. Negative feedback by cortisol occurs on both pituitary and hypothalamus. In extreme cases, such as when steroid drugs are used in high doses for a long time, adrenal **atrophy** may occur due to prolonged inhibition of ACTH production. Conversely, lack of cortisol production, as in an enzymatic defect in the adrenal, may lead to adrenal **hyperplasia** due to excessive ACTH production in an attempt by the pituitary to compensate for the lack of steroid hormone.

Production of **aldosterone** is regulated largely by the renin-angiotensin system. In response to a perceived drop in perfusion pressure, the kidney produces **renin**, which converts **angiotensinogen**, a peptide made by the liver, to angiotensin I. This is converted by an enzyme in lung tissue to **angiotensin II**, which induces production of aldosterone in the *zona glomerulosa* of the adrenal. Potassium also induces formation of aldosterone. Thus an excess of renin or aldosterone causes hypertension; deficiency results in salt loss.

VI.   CHOLESTEROL LEVELS

Because of its insolubility in an aqueous medium, cholesterol must be "packaged" as part of a lipoprotein in order to be transported in blood (see chapter on Lipids). But a high level of blood cholesterol, especially that contained in LDL, is a risk factor in atherosclerotic heart disease. Total cholesterol of 180 - 200 mg/100mL is borderline high, but a more meaningful figure is one that incorporates the relative values of HDL and LDL, since HDL is protective. The ratio of Total/HDL does this: for males it should be <4.5; females can have the ratio up to 5.0.

But since many people have a high ratio which puts them at risk, much attention is being given to factors that lower cholesterol levels:

♦ *Diet*: Cholesterol itself can be avoided by substituting vegetable products for meat and dairy products. In addition, consuming more polyunsaturated fats and fewer saturated fats has a cholesterol-lowering effect. Fats containing monounsaturates seem to be benign. There are exceptions to these generalizations.

♦ *Exercise*: Physical activity seems to increase the HDL/LDL ratio, which correlates negatively with atherosclerosis.

♦ *Drugs*: Some that have been helpful are: **nicotinic acid** and fibric acid derivatives such as **clofibrate**, which lower cholesterol levels by decreasing hepatic secretion of VLDL; **compactin**, **mevinolin**, and **lovastatin**, which inhibit HMG CoA reductase, the controlling step in cholesterol synthesis; and **cholestyramine**, a resin which, taken orally, binds bile acids and promotes their excretion rather than enterohepatic circulation, causing the liver to replace them by new synthesis from cholesterol, some of which comes from the blood.

VII. DISORDERS OF CHOLESTEROL AND STEROID METABOLISM

A. Familial Hypercholesterolemias

This group of conditions results from a heritable lack of functional receptors for LDL. Since endocytosis of LDL cannot occur, intracellular HMG CoA reductase becomes more active, while at the same time blood levels of LDL reach very high levels. (These are discussed more fully in the chapter on Lipids). Diet and drugs may help the individual heterozygous for the condition; little can be done for the homozygote.

B. Inborn Errors of Steroid Metabolism

Congenital Adrenal Hyperplasia (female pseudohermaphroditism, adrenogenital syndrome)

> *21-Hydroxylase* deficiency
>> Molecular defect: Deficient 21-Hydroxylase; several mutant types.
>> Pathway affected: C-21 hydroxylation.
>> Diagnosis: Increased 17-ketosteroids and 21-deoxysteroids (blood and urine).
>> Genetics: Autosomal recessive.

Other enzyme defects causing congenital adrenal hyperplasia

Deficient Enzyme	Compound	Genetics
11β-Hydroxylase	17-Ketosteroids (urine)	Autosomal recessive.
3β-Hydroxysteroid Dehydrogenase	DHEA (blood)	Autosomal recessive.

Male Pseudohermaphroditism, androgen deficiency

> *5α-Reductase* deficiency
>> Molecular defect: Deficient enzyme from more than one mutation.
>> Pathway affected: Testosterone → Dihydrotestosterone (DHT).
>> Diagnosis: Elevated ratio of testosterone/DHT.
>> Genetics: Autosomal recessive.

Other enzyme defects

Deficient Enzyme	Compound	Genetics
20,22-Lyase	All steroids	Autosomal recessive
17-Hydroxylase	Testosterone low	Autosomal recessive
17,20-Lyase	Pregnanetriolone	
17-Dehydrogenase	Δ^4-androstenedione / testosterone ratio	

Androgen receptor defects: complete and incomplete testicular feminization; Reifenstein syndrome and the infertile male syndrome.

>> Molecular defect: Abnormal or absent androgen receptor protein.
>> Pathway affected: Binding of androgens to receptors.
>> Diagnosis: Absent binding of DHT to receptors.
>> Genetics: X-linked recessive.

VIII.   CHOLESTEROL IN PERSPECTIVE

Cholesterol has many roles in health and disease, both in its own right, and as a precursor of a variety of biologically-important substances (Figure 8 - 5):

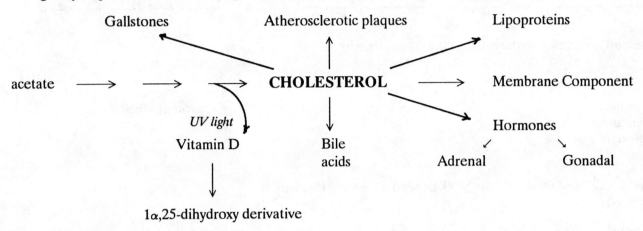

Figure 8-5. The many functions of cholesterol.

IX.   REVIEW QUESTIONS ON STEROIDS

> **DIRECTIONS:**   For each of the following multiple-choice questions (1 - 20), choose the ONE BEST answer.

1.   Estriol is excreted in the urine as the conjugate with

A. glucuronic acid
B. cysteine
C. glycine
D. glutamine
E. protein.

Ans A: Steroid hormone metabolites are usually excreted as the glucuronates or sulfates.

2.   Which of the following does NOT depend on micellar activity for absorption?

A. glycine
B. vitamin E
C. cholesterol
D. vitamin A
E. stearic acid.

Ans A because all the others are hydrophobic and need bile salts for absorption.

3.   All of the following are likely to be found in bile salt micelles EXCEPT: If all are found, no exception, choose E.

A. sodium glycocholate
B. phosphatidyl choline
C. sodium taurocholate
D. 1,25-dihydroxycholecalciferol
E. All the above are found without exception.

Ans D: A, B, and C are normal components of human bile salt micelles; D is not a dietary constituent nor is it secreted into bile by the liver.

4.   The mechanism of action of progesterone involves all of the following (A - D) EXCEPT: If all are involved, no exception, choose E.

A. interaction with a cytoplasmic receptor
B. penetration of the hormone into the cell
C. activation of adenylate cyclase
D. altered genetic expression
E. all of the above without exception.

Ans C because this is what many peptide hormones do, but not steroids.

5.   Important derivatives of cholesterol include all of the following (A - D) EXCEPT: If all are included, no exception, choose E).

A. aldosterone
B. cortisol
C. estradiol
D. taurocholate
E. all of the above.

Ans E: Cholesterol is converted to steroid hormones and bile salts.

6. An intermediate on the way to cholesterol, that is convertible to cholecalciferol by ultraviolet light, is:

A. squalene
B. lanosterol
C. 7-dehydrocholesterol
D. lathosterol
E. calcitriol.

Ans C: This is the $\Delta^{5,7}$-diene that is cleaved by UV light to cholecalciferol (Vitamin D_3).

7. All of the following statements about cholesterol are true EXCEPT: If all are true, no exception, choose E.

A. Cholesterol often occurs in gallstones.
B. Cholesterol is a solid at room temperature.
C. Cholesterol is an alcohol.
D. Cholesterol is insoluble in water.
E. All the above are true without exception.

Ans E: The name means "bile-solid-alcohol"; the one polar –OH group cannot overcome the hydrophobicity of all the rest of the molecule, so it's insoluble in H_2O.

8. You have had your total cholesterol checked at a health fair, and the result is 190 mg/100 mL. What should you do next?

A. Nothing; 190 is a safe level.
B. Change your eating habits so as to avoid foods such as eggs that contain significant amounts of cholesterol.
C. Go on a stringent diet that eliminates all saturated fat.
D. Get the test repeated; it might not have been accurate the first time.
E. Pay a little extra to get the whole lipid profile determined; this would be much more informative.

Ans E: Though B, C, and D contain an element of truth, you need the whole profile done because you might have a high level of HDL, which would be OK.

9. Some virilization may be seen in a patient with Cushing's syndrome, but not in a patient on high doses of prednisone because:

A. prednisone has weak glucocorticoid activity and prolonged use does not cause adrenal atrophy
B. prednisone has weak mineralocorticoid activity and is a strong glucocorticoid
C. prednisone cannot produce a Cushingoid toxicity syndrome
D. virilization in the patient with Cushing's syndrome is caused by adrenal androgens
E. Cushing's syndrome is an autosomal dominant trait.

Ans D: Prednisone is a strong glucocorticoid but is not an androgen. In Cushing's syndrome there is excess adrenal cortical function and release of adrenal androgens such as DHEA.

10. The concentration of plasma cholesterol can be reduced by all of the following (A - D) EXCEPT: If all reduce the plasma cholesterol, no exception, choose E.

A. inhibition of the enzyme, HMG-CoA reductase
B. increased conversion to bile acids by the liver
C. a low fat, low cholesterol diet
D. reduction in the number of LDL receptor sites on cell membranes
E. all the above reduce plasma cholesterol without exception.

Ans D because this would reduce endocytosis of LDL, causing accumulation of LDL in the circulation.

11. All of the following are convertible to BOTH ketone bodies AND cholesterol EXCEPT:

A. mevalonic acid
B. HMG CoA
C. acetoacetyl CoA
D. acetyl CoA
E. palmitic acid

Ans A: MVA occurs after the branch-point; even palmitate can be cleaved to acetyl CoA which can go to both cholesterol and KB.

12. Quantitatively, the major way for cholesterol to be eliminated from the body is by biliary excretion after conversion to:

A. bilirubin
B. bile acids
C. cortisol glucuronide
D. cholesteryl ester
E. LDL.

Ans B: Though metabolites of steroid hormones are excreted, and a little cholesterol itself, bile acids are the major end-product of cholesterol metabolism.

13. Biosynthesis of cholesterol involves all of the following EXCEPT:

A. dimethylallyl pyrophosphate
B. succinyl CoA
C. isopentenyl pyrophosphate
D. squalene
E. lanosterol.

Ans B: Succinyl CoA is an intermediate in the TCA cycle.

14. In humans, cholesterol

A. is converted to glycocholic acid by intestinal bacteria
B. is a part of cell membranes
C. can be catabolized mainly to acetyl CoA
D. is a precursor of squalene
E. is excreted mainly as the glucuronide.

Ans B: Cholesterol has a vital function in its own right as component of membranes, stiffening the membrane at higher temperatures and preserving fluidity at low temperatures.

15. High doses of hydrocortisone over 2 months can cause

A. a decrease in liver glycogen
B. larger skeletal muscles
C. atrophy of the adrenal cortex
D. a rise in blood ACTH
E. decreased gluconeogenesis.

Ans C: This would cause a Cushing-like condition, with muscle wasting, increased gluconeogenesis and blood glucose levels, and a negative feedback on the adrenal, which would even atrophy in time.

16. A lack of receptors for LDL will probably induce

A. a low level of blood LDL
B. a low activity of HMG CoA reductase
C. decreased plasma cholesterol
D. decreased endocytosis of LDL
E. increased activity of ACAT (cholesterol esterifying enzyme).

Ans D: Lack of receptors will prevent endocytosis of LDL, causing decreased cellular esterification (by ACAT), increased cellular synthesis, and increased blood LDL cholesterol.

17. The surge of gonadotropins around the time of ovulation

A. is caused by the effect of GnRH on the ovary
B. is caused in part by a negative feedback, acting on the hypothalamus, of estradiol secreted during the luteal phase
C. starts the development of follicles that release estradiol during the luteal phase
D. starts the development of the *corpus luteum* that releases progesterone during the luteal phase
E. stimulates the adrenal cortex to release cortisol.

Ans D: Gonadotropins are released by the pituitary; the surge occurs in mid-cycle and starts the luteal phase.

18. All of the following statements (A - D) concerning Vitamin D are true EXCEPT: If all are true, no exception, choose E.

A. It is not itself active in stimulating calcium transport by intestine and calcium mobilization from bone *in vivo*.
B. It can be produced by ultraviolet light acting on $\Delta^{5,7}$ sterols.
C. It is converted to the 1,25-dihydroxy-derivative which stimulates the intestinal mucosa to transport calcium.
D. It is acted upon by liver and then by kidney.
E. All the above are true without exception.

Ans E: Cholecalciferol, the product of UV light, is converted by liver, then kidney, to the 5,7-dihydroxy derivative, which is calcitriol, the most active form.

19. Cholestyramine lowers blood cholesterol levels because it:

A. interrupts the efficient enterohepatic circulation of bile salts
B. inhibits HMG CoA reductase
C. depresses triglyceride synthesis, thereby lowering levels of lipoproteins in blood
D. induces the production of receptors for LDL
E. depresses the production of VLDL by liver.

Ans A: This resin binds bile acids, preventing their reabsorption. Then the liver has to convert more cholesterol to new bile salts.

20. In mammals, which of the following can NOT take place?

A. estrone → estradiol
B. estrone → dihydrotestosterone
C. 17-hydroxypregnenolone → testosterone
D. progesterone → estrogen
E. testosterone → estradiol.

Ans B because this would require the C-19 angular methyl group to be put back on, which cannot occur.

MATCHING: For each set of questions, choose the ONE BEST answer from the list of lettered options above it. An answer may be used one or more times, or not at all.

Questions 21 - 29:

A. Lanosterol
B. 7-Dehydrocholesterol
C. Calcitriol
D. Sodium taurocholate
E. Pregnenolone
F. Progesterone
G. Testosterone
H. Dihydrotestosterone
I. Estradiol
J. Cortisol
K. Aldosterone
L. Dehydroepiandrosterone

21. Androgen produced by action of 5α-reductase in some target tissues.

Ans H: This is the product of the reductase acting on testosterone.

22. Most potent glucocorticoid in humans.

Ans J: This is the only glucocorticoid in the list.

23. Deficient in a patient with rickets.

Ans C: Rickets is caused by deficiency of Vitamin D, of which calcitriol is the most active form.

24. Typically an adrenal androgen and precursor of elevated urinary 17-ketosteroids in a patient with an adrenal tumor.

Ans L: DHEA is often elevated in patients with adrenal tumors. Urinary 17-KS are made from all androgens, including testosterone, which is, however, not typically of adrenal origin.

25. Produced by aromatization of ring A of testosterone.

Ans I: Estrogens result from removal of the C-19 angular methyl group when ring A is made aromatic.

26. A common bile salt.

Ans D which is the result of conjugation with taurine of cholic acid, the most common bile acid.

27. Can be converted to adrenal androgens without involving progesterone.

Ans E: Pregnenolone can be converted directly to DHEA.

28. Direct precursor of Vitamin D.

Ans B: The UV-sensitive diene.

29. Synonym for hydrocortisone.

Ans J: Also the most potent glucocorticoid in humans.

Questions 30 - 39:

A.	Progesterone	F.	Estradiol
B.	Testosterone	G.	Cholic acid
C.	Glucagon	H.	Cholecalciferol
D.	Aldosterone	I.	HMG CoA
E.	Calcitriol	J.	Cortisol

30. Has a carbonyl group at C-18.

Ans D: C-18 is the aldehyde group of aldosterone.

31. Causes activation of adenylate cyclase in liver.

Ans C: Glucagon is the only peptide hormone in the list.

32. Induces production of pyruvate kinase.

Ans J because pyruvate kinase is a key enzyme of gluconeogenesis, induced by glucocorticoids.

33. A hormone that does NOT enter the cell.

Ans C: Glucagon's action involves cAMP while the hormone stays outside; the steroids enter cells.

34. A steroid ketone that promotes gestation.

Ans A: The name itself defines the function.

35. Produced by the action of UV light.

Ans H: a synonym for Vitamin D_3, the product of the UV-induced cleavage of 7-dehydrocholesterol.

36. Use of the drug cholestyramine to increase its excretion is one way to lower cholesterol levels.

Ans G: The drug binds bile acids and prevents reabsorption into the enterohepatic circulation.

37. Use of a drug such as lovastatin to inhibit its reduction is another way to lower cholesterol levels.

Ans I: Lovastatin is an inhibitor of HMG CoA reductase, the rate-limiting enzyme in cholesterol synthesis.

38. Promotes retention of Na^+.

Ans D: Aldosterone is the most potent mineralocorticoid in humans; one effect is to promote reabsorption of Na^+ by kidney.

39. Has an aromatic ring.

Ans F: Aromatic rings are found only in estrogens, in which the A ring is aromatic.

9. MEMBRANES
Thomas Briggs

I. OVERVIEW

A. What Membranes Are

A membrane is a thermodynamically **stable, non-covalent assembly** of lipid and protein; often these major components also have carbohydrate residues attached to them. It is arranged as a **bilayered sheet**, with each layer being termed a "**leaflet**." Because interactions among components are non-covalent, each leaflet is **fluid** in that individual molecules are free to move within the plane of that leaflet. A membrane also is **asymmetric** in that one leaflet is different from the other, and many **directional functions** are carried out by the membrane.

B. What Membranes Do

♦ *Permeability barriers.* A membrane forms the boundary between a cell and its environment, and also forms compartments within a cell. But a membrane is highly **selective** in what it allows to pass through it. Some molecules cross a membrane easily; other hardly at all, or only with the aid of a transport mechanism. The presence of transport mechanisms enables a membrane to selectively regulate what ions and molecules may pass through it.

♦ *Fuse.* Membranes may fuse with each other on close approach. Fusion is an integral part of cell functions such as endocytosis, exocytosis, budding of vesicles from the Golgi, and fusion of sperm and egg. A reversal of fusion occurs on the separation of membranes during cell division.

♦ *Create and maintain gradients.* Membranes not only regulate what may diffuse through, but also contain energy-dependent devices for creation of specific **gradients**, and for maintenance of different concentrations of a substance on each side.

♦ *Regulate flow of information.* A membrane can have **receptors** on one side which are specific for a particular informational entity such as a hormone. Binding of the hormone leads to **transmission of a signal** across the membrane, with diverse physiological effects.

♦ *Convert energy.* A membrane often has structural elements which convert one form of chemical energy to another. Examples: use of ATP energy to produce a gradient (as in **active transport**); use of a gradient to generate ATP (as in **oxidative phosphorylation**).

II. CHEMISTRY AND STRUCTURE

A. Lipid Components

Lipids of membranes are **amphipathic**, that is, they have both hydrophobic and hydrophilic characteristics. A typical amphipathic lipid has a polar "head group" containing a charge or other polar group, and a non-polar "tail" such as a long fatty acid chain or other hydrocarbon entity. The polar head-groups are attracted to water molecules, while the non-polar portions are repelled by water, and instead experience many weak interactions among themselves. The result (see below) can be a closed spherical structure such as a **micelle**, with hydrophobic portions sequestered in the interior, or a **bilayered sheet**, as occurs in membranes, with non-polar portions facing each other in the interior, and polar portions facing water on either side.

The total of many weak, non-covalent interactions confers sufficient stability on membrane-like structures that they can even self-assemble from constituent molecules. In this case, however, the asymmetry and directionality of natural membranes are lacking.

There are several types of amphipathic lipids:

- *Phospholipids.* Most phospholipids are **phosphoglycerides**, i.e., they are derivatives of phosphatidate (see Chapter 7). Examples are **phosphatidyl choline** (lecithin), and the phosphatidyl derivatives of glycerol, ethanolamine, and inositol. Diphosphatidyl glycerol (cardiolipin) is prominent in mitochondrial membranes.

  One phospholipid that is not a derivative of phosphatidate is the sphingolipid, **sphingomyelin**. It is structurally very similar to lecithin, but consists of phosphoryl choline attached to ceramide, and thus has no glycerol.

- *Glycolipids.* Other sphingolipids have carbohydrate residues as their polar head-groups. These may be one or several glucose, galactose, etc. units attached to ceramide, and are termed **cerebrosides**. If one or more of the sugars is an acid sugar (N-acetyl-neuraminic acid, or NANA) the lipid is a **ganglioside**.

- *Cholesterol.* Included in spaces between hydrophobic fatty acid chains, cholesterol is oriented with the -OH group facing the water phase. See below (Section C-2) for its effect on a membrane's properties.

B. Proteins

Membranes contain 25 - 75% protein. **Intrinsic proteins** have much interaction with the membrane. They are stably embedded in the basic lipid bilayer and can only be removed with difficulty and by disrupting the integrity of the membrane. **Peripheral proteins** are weakly attached to the membrane, or to other proteins, and can be readily removed. Proteins are responsible for most of a membrane's functions.

Proteins of membranes typically have **domains** enriched in hydrophobic or in hydrophilic amino acid residues; these determine how the protein interacts with the membrane. For instance, an amino acid sequence with much leucine, valine, etc. is apt to coil into an α-helix which then becomes a membrane-spanning region of the protein. The non-polar R groups interact stably with the non-polar interior of the membrane. Amino acid sequences with many charged and polar R groups are likely to be part of those domains that are located in the aqueous medium on either side of the bilayer.

Membrane proteins may be **glycosylated**. The carbohydrate residues are often N-acetyl derivatives of **glucose** or **galactose**, attached to the N of asparagine or to the O of serine or threonine. Glycoproteins are usually located on the *extracellular* side of the bilayer.

C. Fluid Mosaic Model of Membrane Structure

Figure 9-1 summarizes current views regarding membrane structure. Amphipathic lipids are arranged in a bilayer, in which proteins are embedded. Major functional attributes of this structure are described below.

♦ *Fluidity*. The presence of *cis*-double bonds in naturally-occurring fatty acids is of critical importance. By introducing a bend in the hydrocarbon chain, a double bond prevents tight packing of chains, thus **lowering the melting point** of the lipid. Thus above a critical melting temperature or transition temperature (T_m) the fatty acid chains tend to be disordered, loosely packed, more mobile. This fluidity is favored by shorter chain lengths and by increased numbers of double bonds, whereas long saturated chains promote greater rigidity.

♦ *Role of cholesterol*. The effect of cholesterol is complex. Inserted into spaces between fatty acid chains, it increases the density of packing, stiffening the membrane and reducing fluidity. But by preventing exact alignment or crystallization of lipid molecules, it prevents congealing of the membrane at low temperatures, and so it increases fluidity under some conditions. The effect is to **broaden T_m**.

♦ *Mobility of components*. Individual lipid molecules are free to **diffuse laterally** within the plane of each leaflet. They are NOT free to cross ("flip-flop") to the opposite leaflet. The proteins, like icebergs in a sea, can also move laterally, but more slowly.

♦ *Asymmetry*. As stated above, the two leaflets of a membrane are different. Generally, **glycolipids** are located on the **extracellular** side, as are **carbohydrate residues** on proteins. Directional functions, such as ion pumps, are inserted with proper directionality.

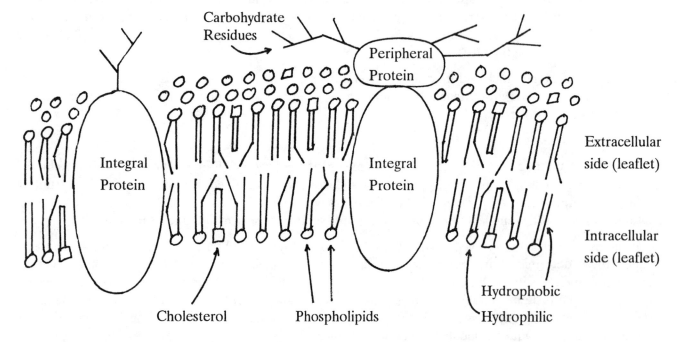

Figure 9-1. Fluid Mosaic Model of Biological Membranes.

III. FUNCTIONS

A. Barrier Function

With respect to diffusion, membranes are **impermeable to ions** and most polar molecules. Exceptions are H_2O and urea, which can cross a membrane freely, as do small gas molecules such as O_2, N_2, and CO_2. Membranes are readily **permeable to lipid-soluble substances** such as steroid hormones.

However, as described below, membranes contain a variety of highly selective transmembrane processes.

B. Transport Processes — Small Molecules

Some definitions (these concepts will be expanded and illustrated in subsequent sections):

♦ **Endergonic:** ΔG is positive (unfavorable); an input of energy is required.

♦ **Exergonic:** ΔG is negative (favorable); energy is released.

♦ **Passive diffusion:** movement of a substance exergonically down its electrochemical gradient in response to thermal motion of molecules, but limited by its solubility in the hydrophobic interior of the membrane (as measured by the permeability coefficient).

♦ **Facilitated diffusion:** movement down an electrochemical gradient with the aid of a transport protein which may be specific and saturable. The rate may be very many times greater than that of passive (or simple) diffusion. No extra energy is required; this process also is exergonic.

♦ **Active transport:** movement of a substance against its electrochemical gradient (endergonic, ΔG with respect to the substance is positive). A transport protein *and* input of energy are required. ΔG for the total system (substance transported and source of energy, such as splitting of ATP) *must* be negative; the combined processes are exergonic.

♦ **Uniport:** a system which moves only one type of molecule; movement may be bidirectional.

♦ **Cotransport:** a system which transports two solutes simultaneously.

♦ **Symport:** the cotransport system moves two types of molecule in the same direction.

♦ **Antiport:** the cotransport system moves two types of molecule in opposite directions.

Passive transport (simple diffusion). In simple diffusion, molecules move from a region of **higher** to one of **lower concentration**, without use of special transporting mechanisms. But the *rate* of diffusion may be severely limited by the solubility of the substance in a hydrophobic medium. Since the non-polar interior of a membrane is a hostile environment for a polar molecule to traverse, diffusion is very slow for many substances. This is illustrated by the extremely small permeability coefficients for glucose and chloride in simple artificial lipid bilayers (Table 9-1).

Facilitated diffusion systems. **Ion pores** or **channels** provide a polar path through the membrane, so that hydrophilic substances are insulated from contact with the non-polar interior of the membrane. Gramicidin A is an example: it is an antibiotic which forms a hydrophilic transmembrane channel through which alkali metal cations can pass.

Channels are gated, i.e., their opening and closing are controlled. **Ligand-gated** channels are regulated by a specific molecule that binds to a receptor; **voltage-gated** channels respond to a change in membrane potential. Ions move down their electrochemical gradients, and no input of energy is required.

Transport proteins (permeases) may be highly specific, such as the permease for glucose of erythrocyte membrane (Table 9-1). D-Glucose moves down its electrochemical gradient at an accelerated rate, and again, no input of energy is involved. L-Glucose hardly passes at all. This is an example of a **uniport**.

Other transport proteins may be **antiports**. An example is the anion exchange protein of erythrocyte membrane, which promotes the electrically neutral exchange of one ion (Cl^-) for another (HCO_3^-). The permeability coefficient for Cl^- is increased by a factor of 10,000,000 (Table 9-1).

Transported Species	Medium	Permeability Coefficient
Cl^-	Simple Lipid Bilayer	10^{-11}
Cl^-	Red Cell Membrane	10^{-4}
Glucose	Simple Lipid Bilayer	10^{-10}
Glucose	Red Cell Membrane	10^{-5}

Table 9-1. **Facilitated Diffusion** allows much faster penetration of red cell membrane by selected ions and small molecules.

Ionophores. Some transport substances have a hydrophobic exterior with a hydrophilic cavity that may be quite specific for particular ions. An example is valinomycin, an antibiotic which specifically complexes and transports K^+ ions. Formerly it was thought that this complex would actually cross the membrane, taking the ion from one side to the other like a ferry boat. But this carrier model, demonstrated in artificial bilayers, may not be applicable in natural membranes.

Characteristics of simple vs. *facilitated diffusion (Figure 9-2).* Facilitated transport, being a process in which a protein is used to lower an activation-energy barrier, is not unlike an enzyme-catalyzed reaction. The protein may exhibit high specificity for the target molecule, and the rate is greatly increased compared to that of the unassisted process, especially at low concentrations of substrate. At higher concentrations the carrier, like an enzyme, can become saturated, and the rate of transport levels off, as shown in the figure. The kinetics, like enzyme kinetics, may be described in Michaelis-Menten terms, with K_M, and may exhibit competitive inhibition. These considerations do not apply to simple diffusion.

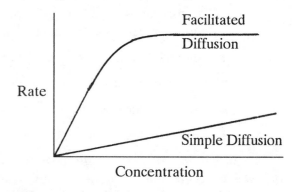

Figure 9-2. Effect of Concentration on Rate of Diffusion.

Active transport. Since a substance is being forced to move against its electrochemical gradient, a source of energy is required. The overall ΔG is then negative.

ATPases use energy directly from the hydrolysis of ATP. This is called **primary** active transport. An important example is the **Na^+ - K^+ pump.** This is a Mg^{++}- requiring ATPase which has membrane-spanning subunits that undergo conformational changes in a directional manner so that for each ATP used, 3 Na^+ are extruded from the cell and 2 K^+ are brought in. Because the movement of charges across the membrane is unequal, the pump is said to be **electrogenic**. It is inhibited by cardiotonic steroids such as ouabain and digitoxigenin.

Another important ATPase is the **calcium ion pump**, especially abundant in muscle tissue. Structurally and functionally related to the Na^+ - K^+ pump, it uses one ATP to transport two Ca^{++}. In muscle, it pumps Ca^{++} from cytosol into the sarcoplasmic reticulum. The $H^+ - K^+$ ATPase of stomach is also related to the sodium and calcium pumps.

The preceding are examples of **P-type ATPases**, so called because their mechanism of action involves a phosphorylated intermediate. Two other types occur: **V-type ATPases**, which transport H^+ and occur in membranes of lysosomes and of vesicles involved in endocytosis and exocytosis, and **F-type ATPases**, such as the F_o-F_1 ATPase of oxidative phosphorylation, which uses a proton gradient to *synthesize ATP* (Chapter 4).

Co-transport systems can be driven by the energy of the Na^+ gradient, and thus only indirectly by ATP hydrolysis. Absorption of glucose from small intestine, an example of **secondary** active transport, works in this way. A **Na^+ - glucose symport** protein is located in membranes of the apical surface of brush border cells, and obligatorily cotransports one glucose along with one Na^+. The Na^+ - K^+ ATPase is located only on the basolateral (serosal) membranes, and keeps the intracellular $[Na^+]$ very low, ensuring a concentration gradient between lumen and cell. Sodium ion, moving down its gradient, thus "drags" glucose with it into cells where the concentration of the latter may actually be high. Glucose then will pass into the bloodstream by facilitated diffusion.

Many other cells have similar uptake systems, especially for **amino acids**.

The sodium gradient is also linked to the transport of **calcium**, but in this case through an **antiport**, which pumps calcium endergonically out of cells, in exchange for 3 Na^+ entering exergonically down their gradient. This is particularly significant in heart muscle cells. A **Na^+ - H^+ antiport** is also used to regulate intracellular pH.

Transport by modification occurs when a substance such as a sugar is phosphorylated the instant it enters a cell. Though ΔG for initial entry may be unfavorable, phosphorylation effectively removes the product of the equilibrium (lowers the intracellular concentration of the free form), promoting entry of more sugar molecules since the overall ΔG is now negative (favorable) and the phosphorylated molecules are unable to escape.

C. Transport Processes — Large Molecules

Receptor-mediated endocytosis is a process well-illustrated by the uptake of LDL particles into extrahepatic cells (see also the chapter on Lipids). Small regions of the membrane become coated on the cytoplasmic side with a filamentous peripheral protein, **clathrin**, which participates in the formation of pits (concave to the outside of the cell) which are incipient vesicles. On the extracellular side, receptors congregate in the **coated pits**; for the uptake of LDL-cholesterol, these receptors are specific for apo B-100. Binding of LDL to receptors triggers internalization of the pits and bound LDL, which now become vesicles. Internalization is also triggered by Ca^{++}. The vesicles fuse with lysosomes, which release hydrolytic enzymes which degrade large molecules to smaller units for metabolism by the cell. Further endocytosis is regulated by a feedback mechanism which controls the number of receptors.

Membranes also engage in **exocytosis**, which in many ways is a reverse of endocytosis, and in **pinocytosis**, which results in entry of fluid and small solutes into the cell.

D. Transmembrane Signaling: Concepts

Signal transduction by a "second messenger" can be thought of as occurring in several distinct phases, as described in general terms below. This mechanism is applicable to the mode of action of a variety of agents, especially the **peptide hormones**. Specific illustrations will follow.

Events occur in the following sequence:

1. The original stimulating agent, glucagon for example, binds to a **receptor** which is specific for it and is located only on the **extracellular side** of the membrane. The hormone remains where it is

and *does not enter the cell.* But the receptor is a transmembrane protein, and binding of the hormone induces a conformational change which transmits a signal across the membrane to the cytoplasmic side.

2. The transmitted conformational signal then activates an intermediary protein, one of the "**G proteins**" (see below).

3. The activated intermediary protein then goes on to activate a special **enzyme**.

4. The newly-activated enzyme produces a new intracellular entity, known as the "**second messenger**," which is, in a sense, an intracellular hormone.

5. The new entity proceeds to induce various intracellular effects, such as **phosphorylation** of certain target proteins, which, in turn, **activate or inhibit** various metabolic processes.

6. Finally, there must be a way to reverse the procedure, **to turn the signal off.** A permanent "on" state, induced by some agents such as cholera toxin, leads to disease.

G-Proteins are the intermediary proteins that carry a signal from the intracellular side of the activated hormone receptor to the enzyme that catalyzes the formation of the second messenger. They bind guanyl nucleotides (hence the name), the complex with GDP being the inactive form, and consist of dissociable α- and other subunits. They comprise a family of related proteins, which are designated G_s if the are stimulatory to adenylate cyclase, or G_i if they inhibit adenylate cyclase. Other G-proteins work through different signaling systems.

On receipt of a hormonal signal, GTP replaces the bound GDP, causing the α-subunit to dissociate. The GTP-α-subunit is the active form that then activates (or inhibits) the enzyme that produces the second messenger. But the GTP-α-subunit has a "built-in" slow GTPase activity, which causes the bound GTP to revert to GDP, allowing re-association of subunits and inactivation of the signal. The process can be considered cyclic (Figure 9-3).

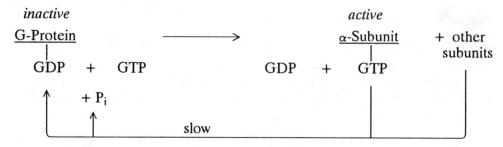

Figure 9-3. Dissociation - Reassociation Cycle of a G-Protein.

E. Transmembrane Signaling: Examples

The Adenylate Cyclase System is used with many peptide hormones. Some hormones that stimulate adenylate cyclase are ACTH, ADH, β-adrenergics, glucagon, PTH, etc. The list of agents that inhibit adenylate cyclase is shorter but includes acetylcholine, angiotensin II, somatostatin, and others. Notably, insulin does NOT involve this system.

Components of the system are (a) three proteins: the hormone **receptor** which is a transmembrane protein, a G_s (or G_i) **protein**, and **adenylate cyclase**; and (b) the second messenger itself, **cyclic AMP (cAMP,** shown in Figure 9-4, next page).

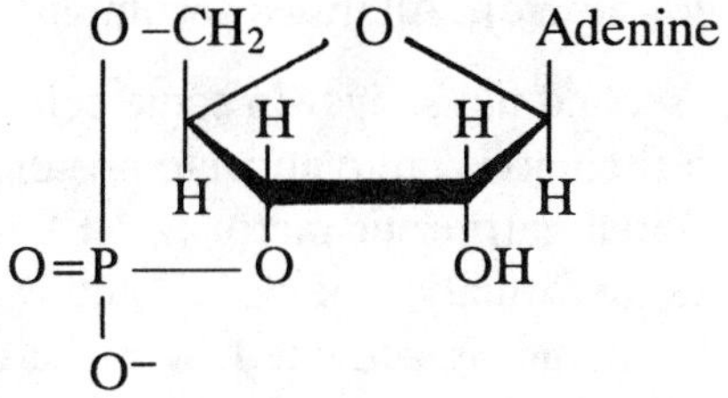

Operation of the cAMP system (Figure 9-5) begins when a hormone (e.g., glucagon) binds to the extracellular receptor. The signal is transmitted across the membrane to the G-protein on the cytoplasmic side, which becomes activated and exchanges GTP for bound GDP. Dissociation occurs, and the **GTP-α-subunit** then activates adenylate cyclase which promotes the reaction:

Figure 9-4. Cyclic AMP (cAMP)

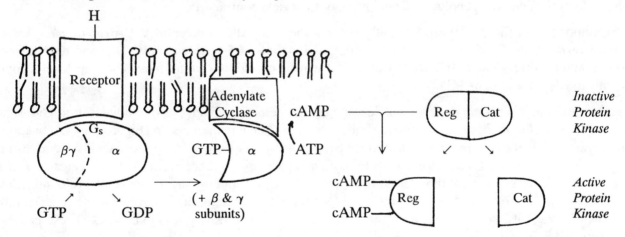

$$ATP \longrightarrow cAMP + PP_i$$

The cAMP then activates a **protein kinase A** (so-called because it is cAMP-dependent) such as *phosphorylase kinase kinase* in muscle. Typically these cAMP-dependent kinases consist of **regulatory subunits** bound to **catalytic subunits**, the whole being inactive. Binding of cAMP to regulatory subunits causes them to dissociate, allowing the catalytic subunits to become active. These, in turn, go on to phosphorylate proteins, either activating or inactivating these, depending on whether the phosphorylated target protein is the catalytically active or inactive form. Since a variety of proteins may become phosphorylated, metabolic effects from a single hormonal stimulus may be quite diverse.

Figure 9-5. The cAMP System.

Inactivation of the system may occur in various ways: (a) a **phosphodiesterase** is present that degrades cAMP to AMP; (b) the G-protein slowly inactivates itself through its built-in **GTPase activity**; (c) **phosphatases** dephosphorylate the protein phosphates. All this is assuming that the original hormone has diffused away from its extracellular receptor.

Some hormones, such as somatostatin, *lower* cAMP levels through an inhibitory G protein (G_i). If somatostatin is bound by its receptor, the effects of glucagon or epinephrine are reversed. In adipose tissue PGE_1, by inhibiting adenylate cyclase, blocks the activation of hormone-sensitive lipase, slowing the release of fatty acids from stored lipid.

Specificity of the cAMP system has two aspects. Specificity to the stimulatory hormone lies in the receptor for the hormone, on the extracellular surface of the membrane. Specificity with respect to the physiological response depends on the type of cell, and what target proteins are being phosphorylated and turned on or off inside the cell.

One way to remember what the effects will be when the cAMP system is stimulated is to consider cAMP as a "hunger signal." In this sense, cAMP activates those processes that produce energy, such as the *glycogen phosphorylase* cascade of liver, or lipolysis in adipose tissue. cAMP inhibits those processes that store energy, such as glycogen synthesis (via inactivation of *glycogen synthase*), or fatty acid synthesis (via inactivation of *acetyl CoA carboxylase*). All this is accomplished through phosphorylation of the appropriate enzymes.

Another second messenger in some cell types is cyclic GMP (cGMP), produced from GTP by guanylate cyclase, which then goes on to activate protein kinase G, which is the term used for a cGMP-dependent protein kinase. Atrial natriuretic factor (ANF), released by cells of the heart atrium, activates guanylate cyclase in kidney, causing formation of cGMP and excretion of Na^+ and water. Another type of guanylate cyclase, which contains heme, is activated by nitric oxide (NO). The action of NO (and of nitroglycerin which releases NO) to relax heart muscle is mediated by cGMP. cGMP is also involved in vision.

Some diseases involve a malfunction of cyclic nucleotide-mediated signaling. Cholera toxin causes transfer of the ADP-ribose moiety of NAD^+ to the α-subunit of the G_s protein of intestinal cells. This binding is irreversible and permanently "turns on" adenylate cyclase. Pertussis toxin, also through ADP-ribosylation, inactivates the inhibitory function of a G_i protein by destroying its GTPase activity, again leading to an irreversible activation of adenylate cyclase.

The Phosphoinositide Cascade is used with some other peptide hormone systems, especially those that are calcium-dependent. Among these are α_1-adrenergics, angiotensin II, and acetylcholine (muscarinic). Other effects include glycogenolysis in liver and smooth muscle contraction.

Components of the system are (a) four proteins: the extracellular **receptor**, a **G-protein**, *phospholipase C*, and *protein kinase C*; (b) two second messengers: **inositol-1,4,5-trisphosphate** (IP$_3$) and **diacylglycerol** (**DAG**); and (c) **calcium ion**. In this context, Ca^{++} may be considered a *third* messenger: its level responds not only to IP$_3$ but also to cAMP.

Operation of the system depends on cleavage of the membrane lipid, phosphatidylinositol-4,5-bisphosphate (PIP$_2$) by *phospholipase C* to produce both second messengers in one step (Figure 9-6). Activation of the enzyme is linked to the extracellular hormone receptor by a G-protein, as in the cAMP system. But the production of dual second messengers leads to more complex consequences: (a) DAG activates *Protein Kinase C*, which goes on to phosphorylate target proteins; (b) IP$_3$ opens calcium channels, releasing Ca^{++} from storage in the ER. Ca^{++}, via calmodulin, a Ca^{++}-binding protein, then activates another set of target proteins. There is also some crossover in that Ca^{++} can further activate protein kinase C. These processes are illustrated in Figure 9-7.

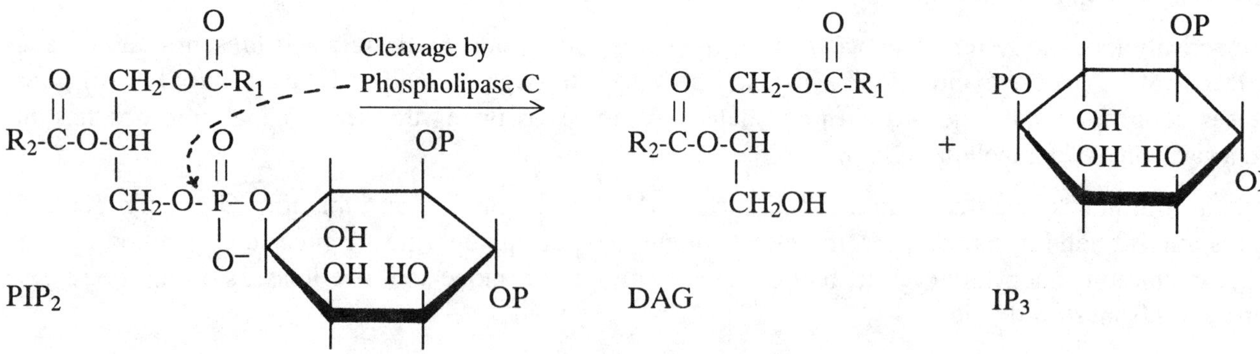

Figure 9-6. Conversion of Phosphatidylinositol-4,5-bisphosphate to DAG and IP$_3$.

In muscle, IP$_3$ causes release of stored Ca^{++}, which, in turn, triggers contraction. Ca^{++} also further activates phosphorylase kinase and stimulates the kinase that inactivates glycogen synthase. Thus contraction and glycogen metabolism are coordinated.

Turning the system off is accomplished by a phosphatase (*IP$_3$ase*) which converts IP$_3$ to IP$_2$. The latter is further metabolized to free inositol, which is then recycled. (Li^+ blocks removal of the last phosphate, interfering with re-use of inositol).

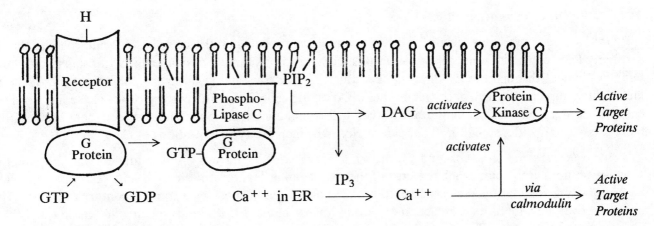

Figure 9-7. The Phosphoinositide System.

Insulin also binds to a transmembrane receptor, but its method of action is not well understood. It does not work through cAMP or phosphoinositides (though it may modulate the effects of these other systems), but the receptor itself is a protein kinase of a different sort. Activation by insulin results in stimulation of a tyrosine kinase activity on the cytoplasmic side of the receptor, which phosphorylates tyrosine residues in selected proteins (Figure 9-8). The target proteins include itself, which it autophosphorylates. How this translates into uptake of glucose and other nutrients is not clear. Internalization of the receptor occurs, which may be part of the action, or may simply be a way to control receptor concentration. Related hormone systems, e.g., those involving the receptor for epidermal growth factor (EGF) and that for the insulin-like growth factor I (IGF-I) also have tyrosine kinase activity, as do certain oncogene products (e.g., PDGF).

Insulin also affects nuclear processes, especially the production of specific mRNA's. For instance, the mRNA for PEPCK, a key enzyme of gluconeogenesis, is decreased by insulin. Many other mRNA's are known to be regulated by insulin, and this type of activity is becoming recognized as an important aspect of insulin's action.

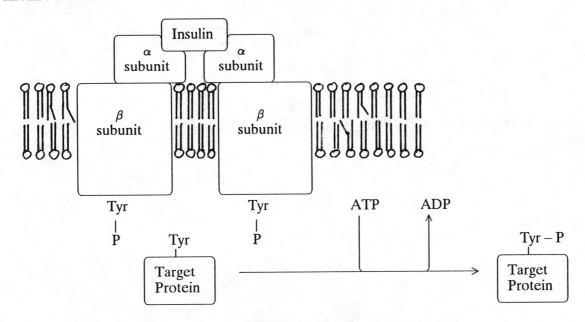

Figure 9-8. The Insulin Receptor System and Tyrosine Kinase Activity.

F. Other Membrane Phenomena

Gap junctions provide a means for direct communication between the cytoplasm of one cell and that of another. They consist of units called **connexons**, composed of subunits of **connexin**, a transmembrane protein. A channel through the middle is large enough to pass ions and small hydrophilic molecules (such as metabolites) directly through both membranes of adjoining cells, from cytosol to cytosol. These junctions occur in large numbers, and are important for cell-cell communication, nourishment of cells that are far from blood vessels, and differentiation. They can be closed by increased concentrations of Ca^{++} or H^+.

Intercellular communication and various *recognition phenomena* are often mediated through **gangliosides** and other carbohydrate-rich features of the extracellular leaflet of the membrane. Examples are **contact inhibition**, some **antigenic activities** (e.g., ABO blood-group substances), and **receptor functions**. The latter can even be a feature of the action of certain toxins: **cholera toxin** binds specifically to the ganglioside G_{M1}, with eventual irreversible activation of adenylate cyclase.

Synthesis of membranes occurs only on pre-existing membranes. Since they are asymmetric, it is essential that new membranes are produced so that structural features such as carbohydrate residues are on the extracellular side, and directional functions such as ion pumps are properly oriented.

Synthesis of membrane proteins begins on membrane-bound ribosomes (rough endoplasmic reticulum), where glycosylation also begins. Vesicles bud off and fuse with the Golgi apparatus where further processing, including additional glycosylation, takes place. Finally the vesicles fuse with plasma membrane, preserving their original asymmetry (Figure 9-9), in which the luminal side of the ER is equivalent to the outside of the cell.

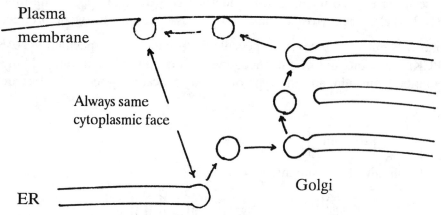

Figure 9-9. Synthesis of Membranes, Showing Preservation of Asymmetry.

IV. REVIEW QUESTIONS ON MEMBRANES

> **DIRECTIONS:** For each of the following multiple-choice questions (1 - 15), choose the ONE BEST answer.

1. All of the following contain glycerol EXCEPT:

A. lecithin
B. sphingomyelin
C. phosphatidyl ethanolamine
D. phosphatidyl inositol
E. cardiolipin.

Ans B because all the others are derivatives of phosphatidate, which contains glycerol.

2. Biological membranes do all of the following EXCEPT:

A. freely pass small ions such as H^+
B. freely pass gas molecules such as O_2
C. freely pass small lipophilic molecules such as $CHCl_3$
D. constitute a barrier to the passage of metabolites such as glucose-6-phosphate
E. create ionic gradients.

Ans A because hydrophilic particles cannot easily traverse the hydrophobic interior of membranes without a special carrier or transport mechanism.

3. Which of the following will NOT be found in membranes?

A. a triacylglycerol
B. a derivative of phosphatidate, such as lecithin
C. cholesterol
D. a protein with hydrophobic and hydrophilic regions
E. a lipid with carbohydrate residues attached.

Ans A because triacylglycerols are not amphipathic and so are not useful in membrane structure.

4. Transmembrane signaling by the insulin receptor is mediated by:

A. ligand-gated ion channels
B. a sodium symport system
C. interaction with a specific "G" protein
D. a receptor-associated protein tyrosine kinase activity
E. an intrinsic GTPase activity.

Ans D: The insulin system is not known to work through cyclic nucleotides or G proteins. The receptor does have tyrosine kinase activity.

5. The presence of cholesterol in a membrane

A. increases fluidity at both low and high temperatures
B. decreases fluidity at both low and high temperatures
C. increases fluidity at low temperatures and decreases it at high temperatures
D. decreases fluidity at low temperatures and increases it at high temperatures
E. has no effect on fluidity.

Ans C: Cholesterol prevents congealment of lipids at low temperatures (promotes fluidity), and stiffens the membrane at high temperatures. The effect is to keep fluidity at a more nearly constant value over a range of temperatures.

6. The side-chains of amino acids in the transmembrane domains of intrinsic proteins are likely to be rich in:

A. alanine and aspartate
B. lysine and leucine
C. glutamate and glutamine
D. valine and isoleucine
E. histidine and arginine.

Ans D: Transmembrane domains must be hydrophobic, so the R groups you would expect would be only those of alanine, leucine, isoleucine, etc

7. The stability of membranes stems from:

A. ionic interactions between polar head-groups of amphipathic lipids and water
B. ionic interactions between charged R-groups of proteins and water
C. steric hindrance among hydrophobic portions of amphipathic lipids
D. covalent bonds between lipid and protein
E. hydrophobic repulsion of lipid chains by water, and other non-covalent interactions.

Ans E: It is mainly the hydrophobic and non-covalent interactions which stabilize membranes, especially repulsion between hydrocarbon chains and water.

8. Inactivation of the phosphoinositide system may occur as a result of all of the following EXCEPT:

A. diffusion of the stimulating hormone away from its receptor
B. phosphorylation of the G protein
C. GTPase activity of the G protein
D. conversion of IP_3 to IP_2
E. dephosphorylation of phosphorylated proteins by a phosphatase.

Ans B is the exception. This is actually part of the process that activates the system.

9. Fluidity of membranes

A. is favored by longer fatty acid chains in the lipids
B. is favored by a greater degree of saturation in the lipids
C. is favored by the presence of double bonds in the lipids
D. is insufficient to allow lateral movement of proteins within the plane of the leaflet
E. allows crossing over of phospholipid molecules from one leaflet to the other.

Ans C because double bonds in the *cis* configuration prevent tight packing or crystallization of lipid chains.

10. Carbohydrate groups on glycoproteins destined for the plasma membrane might be found at which of the following locations?

A. at the cytoplasmic surface of the endoplasmic reticulum
B. at the inner (luminal) surface of membranes of the Golgi apparatus
C. at the intracellular surface of the plasma membrane
D. at the outer (cytoplasmic) surface of membranes of vesicles budded from the endoplasmic reticulum
E. at the outer (cytoplasmic) surface of membranes of the Golgi apparatus.

Ans B: Carbohydrate groups are attached at the inner surface of the ER and Golgi. When vesicles bud off and eventually fuse with the plasma membrane, this becomes the outside of the cell, where carbohydrate groups need to be.

11. In the cyclic AMP system for signal transduction, all of the following occur EXCEPT: If all occur, no exception, choose E.

A. The original stimulating hormone stays outside of the cell.
B. A conformational shift in a transmembrane protein causes a G-protein to dissociate and bind GTP.
C. Adenylate cyclase is activated, which promotes the conversion of ATP to cAMP.
D. Certain proteins are phosphorylated, which changes their activities.
E. All the above occur without exception.

Ans E: A - D all occur, and this is their correct sequence.

12. All of the following statements (A - D) are correct EXCEPT: If all are correct, no exception, choose E.

A. Phospholipids are the major class of membrane lipids
B. Many membranes also contain glycolipids and cholesterol
C. Phospholipids and glycolipids readily form bilayers
D. Lipid bilayers are highly permeable to ions and most polar molecules
E. All of the above are correct without exception.

Ans D is the exception. Membranes are effective barriers against most polar molecules.

13. Hormones that act primarily through the cAMP system include all of the following EXCEPT:

A. epinephrine
B. α_1-adrenergics
C. β-adrenergics
D. glucagon
E. ACTH.

Ans B: This uses the phosphoinositide system.

14. In a membrane, which of the following is most likely to be found in the interior of the bilayer, away from water?

A. carbon chains of fatty acids
B. glutamic acid-containing portions of intrinsic proteins
C. phosphate portions of amphipathic lipids
D. the portion of a hormone receptor that binds the hormone
E. the sugar moiety of a glycolipid.

Ans A: Carbon chains are hydrophobic and so are repelled by water. They must then be on the inside of the bilayer.

15. The major function of the sodium-potassium ATPase is to:

A. degrade intracellular ATP
B. generate ATP from electrochemical gradients of potassium and sodium ions
C. provide intracellular enzymes with adequate sodium
D. co-transport sodium and potassium down their electrochemical gradients
E. pump sodium out of cells and potassium into cells.

Ans E: The gradients of these ions across the plasma membrane are 10 - 20 times or more, and this requires much energy to produce and maintain. Na^+ is the major extracellular cation, K^+ the major intracellular cation.

> **DIRECTIONS:** For each set of questions, choose the ONE BEST answer from the lettered list above it. An answer may be used one or more times, or not at all.

Questions 16 - 24:

A. Simple diffusion
B. A facilitated transport system that is a uniport
C. A facilitated transport system that is an antiport
D. An active transport system that is a symport

E. Receptor-mediated endocytosis
F. An F-type ATPase
G. A P-type ATPase that is a uniport
H. A P-type ATPase that is an antiport

16. The Cl^- - HCO_3^- transporter of red cell membranes.

Ans C: These two ions flow down their gradients in opposite directions; no energy is needed.

17. Movement of O_2 across the inner mitochondrial membrane.

Ans A because neither a transporter nor expenditure of energy is required.

18. The ATP-generating system of mitochondrial oxidative phosphorylation.

Ans F by definition.

19. Uptake of LDL.

Ans E: This is a larger particle, which must be taken up by endocytosis.

20. Glucose permease of red cell membranes.

Ans B: Glucose moves down its gradient; no other species is involved.

21. The Na^+ - K^+ pump.

Ans H: This is a P-type ATPase by definition; antiport because the ions move in opposite directions.

22. Calcium ion pump of muscle.

Ans G: A P-type ATPase by definition; only one type of ion is involved, hence a uniport.

23. Valinomycin.

Ans B: No expenditure of energy is needed; only one type of ion is transported.

24. Absorption of glucose from intestinal lumen into brush border cells.

Ans D: Glucose moves into a compartment where its concentration is greater; this requires energy, supplied by Na^+ moving in the same direction, but down its concentration gradient.

Questions 25 - 35:

A. Extracellular receptor
B. G protein
C. Phospholipase C
D. Phosphatidylinositol-4,5-bisphosphate (PIP$_2$)
E. Inositol-1,4,5-trisphosphate (IP$_3$)

F. Diacylglycerol (DAG)
G. Protein kinase C
H. Calcium ion
I. Calmodulin
J. Phosphodiesterase

25. In a sense, a third messenger.

Ans H: The first two messengers being the original hormone and IP$_3$, in the phosphoinositide system.

26. A transmembrane protein.

Ans A: The receptor is an integral protein which needs to traverse the width of the membrane to transmit a signal.

27. Second messenger that opens calcium channels.

Ans E: IP$_3$ in the phosphoinositide system.

28. A calcium-binding protein.

Ans I: Calmodulin, after binding Ca^{++}, further modulates target proteins.

29. A GTPase.

Ans B: G proteins have intrinsic GTPase activity.

30. Phosphorylates target proteins.

Ans G: A protein kinase phosphorylates proteins. "C" stands for Ca^{++}-dependent.

31. Second messenger that activates protein kinase C.

Ans F: DAG is one of the two second messengers resulting from cleavage of PIP$_2$.

32. Enzyme that produces two second messengers in one step.

Ans C: Phospholipase C cleaves PIP$_2$ to produce DAG and IP$_3$.

33. Substrate whose cleavage produces two second messengers in one step.

Ans D: See question 32.

34. Binds acetylcholine.

Ans A: The receptor must bind the hormone before anything else can happen.

35. NOT part of the phosphoinositide cascade.

Ans J because phosphodiesterase is what cleaves (and inactivates) cAMP.

Questions 36 - 41:

    A. Interaction between a hormone and its extracellular receptor.   E. Adenylate cyclase.
    B. Transmission of a signal across the membrane.   F. cAMP
    C. Passage of the hormone across the membrane.   G. Phosphorylated proteins.
    D. G Protein.   H. Phosphodiesterase.

36. Intermediary between transmembrane protein and the enzyme that produces a second messenger.

Ans D: The G protein picks up the signal from the receptor, dissociates, binds GTP, then activates the enzyme.

37. Specificity as to whether a cell will respond to a particular hormonal stimulus.

Ans A: The receptor is what determines whether a hormone will have any effect at all.

38. One way to turn the system off.

Ans H: Phosphodiesterase hydrolyzes and inactivates the cyclic nucleotide.

39. NOT a part of the system by which glucagon regulates glycogen metabolism in liver.

Ans C because glucagon uses the cAMP system and does not need to enter the cell.

40. Specificity as to type of physiological response.

Ans G: Once the hormone is bound and the signal crosses the membrane, then the type of response depends on what target proteins are phosphorylated.

41. Second messenger.

Ans F: cAMP is the second messenger because the original extracellular hormone is the first messenger.

10. NUTRITION

Thomas Briggs

I. MAJOR NUTRIENTS

A. Energy Nutrition

The unit generally used is the **Kilocalorie** (Kcal, popularly but incorrectly also known as the Large Calorie, Cal): amount of energy needed to raise the temperature of 1 Kg water from 14.5° to 15.5°C. The **Megajoule** is gaining some acceptance: 1 MJ = 239 Kcal; 1 Kcal = 4.2 KJ. Ten MJ is roughly a day's supply of energy, at a moderate level of activity.

Caloric yields from the metabolism of fuels are:

Carbohydrate	4 Kcal/g	Fat	9 Kcal/g
Protein	4 Kcal/g	Ethanol	7 Kcal/g

Respiratory Quotient (RQ) is the volume of CO_2 produced / volume of O_2 consumed. This varies depending on the type of fuel being oxidized: 1.0 for carbohydrate, to 0.7 for fat.

Basal Metabolic Rate (BMR) is the rate of oxygen consumption, or equivalent heat production, of an awake individual, at rest, who has not eaten for at least 12 hours. To compensate for size differences, the BMR is usually expressed per unit surface area. A typical figure might be 35 Kcal/hr/m². It is higher in children, males, hyperthyroidism, etc. Resting energy expenditure (REE) is less precisely defined, but differs little from BMR.

Thermic Effect of Food (formerly known as specific dynamic action, SDA) is an increase in metabolic rate after eating, resulting from the energy cost of metabolism, particularly of protein, and may be about 6% on a normal mixed diet.

The daily requirement of energy varies greatly, depending on BMR and especially on muscular activity. For a sedentary 70 Kg man a typical value may be 2400 Kcal, but very strenuous activity could double this.

B. Fuels

1. *Carbohydrates*

The major dietary form is the polysaccharide, **starch**. Important oligosaccharides include **sucrose, lactose**. Monosaccharides occur increasingly in the US, due to the widespread use of high-**fructose** corn syrup (which may also contain glucose). Carbohydrates function mainly as fuels, with minor amounts incorporated into glycolipids and glycoproteins.

2. *Fats (lipids)*

Triacylglycerols (TAG, or TG) are the major dietary form. The dietarily essential lipids are the unsaturated fatty acids **linoleic** (18:2[9,12]) and **linolenic** (18:3[9,12,15]) **acids**. Recommended dietary allowances for these have not been established.

Although lipids have a major function as fuels and in the storage of energy, they also serve in the structure of membranes (phospholipids, cholesterol), in the formation of prostaglandins (poly-unsaturated fatty acids), as other hormones (steroids), as thermal insulators, and even for decoration (deposits of fatty tissue).

The amount and type of fat in the diet influence the levels of cholesterol in the body. A diet with a high ratio of polyunsaturated to saturated fatty acids, but low in total fat, tends to promote low levels of cholesterol in the blood. Omega – 3 (ω – 3, sometimes n – 3) fatty acids are polyunsaturated fatty acids with three carbon atoms beyond the last double bond in the chain, such as linoleic acid. Occurring especially in some fish oils, they have anticholesterogenic and antithrombogenic effects.

C. Protein

Though proteins consumed in excess of the daily requirement are simply used as fuels, the most significant dietary function is as a source of amino nitrogen for synthesis of body constituents.

Ten amino acids are considered dietarily essential because the human cannot synthesize the carbon skeleton (mnemonic: PVT TIM HALL):

Phenylalanine	Threonine	Histidine
Valine	Isoleucine	Arginine*
Tryptophan	Methionine	Leucine
		Lysine

*A dietary requirement has not been rigorously established for arginine in the <u>adult</u> human.

The quality of dietary proteins depends on (1) digestibility and (2) the content of essential amino acids. The biological value (BV) depends mainly on amino acid composition. A more useful index, **Net Protein Utilization** (NPU), considers both (1) and (2) above. The NPU of human milk is 95%; of cow's milk, 81%; of wheat protein, 49%; of corn protein, 36%. Cooking can increase the NPU, as can mixing proteins with complementary amino acid compositions. This is especially important in vegetarian diets.

The **nitrogen balance** is zero when intake just equals loss (as urea, digestive losses, etc.). For this to occur, all essential amino acids must be present in the diet in sufficient quantities. If even one is deficient, negative balance results. Growth, convalescence, and pregnancy are accompanied by positive nitrogen balance; illness, fever, starvation by negative balance.

The Recommended Dietary Allowance (RDA, see below) for protein consumed by adult males and females on a typical US diet is **0.8 g/Kg body weight/day**.

D. Other

1. *Fiber*

Dietary fiber is non-digested (especially plant) material such as cellulose, lignin, etc. Though not generally regarded as a dietary essential, fiber nevertheless has beneficial effects on digestion. Through its *bulking* action, it promotes good mechanical functioning of the digestive tract; by a *speeding* action, it reduces the transit time for intestinal contents and therefore the time available for bacteria to produce possible carcinogens; by *binding* bile salts it is thought to increase the turnover of the bile salt pool and thus to promote excretion of cholesterol through increased conversion of cholesterol to bile acids. The latter function is especially true of "soluble" fiber, found in fruits and legumes. Fiber may, however, decrease absorption of certain nutrients, such as iron and calcium.

2. *Water*

Requirement is highly variable. The most critical nutrient of all, it functions as a *solvent* for components of blood and tissues, and as a medium for excretion of wastes. Another important use is in *regulation of body temperature*.

E. Health-Related Issues

Obesity is a pervasive health problem related to energy intake because it predisposes to numerous conditions including cardiovascular disease, diabetes, and gallbladder disease.

Lactose intolerance is caused by loss, in adulthood, of the digestive enzyme β-galactosidase (lactase) needed to digest the lactose in milk and some other dairy products. Relatively common, it results in undigested lactose being converted to gas and acid by colonic bacteria, with intestinal disturbances such as bloating and diarrhea.

Malabsorption syndromes result from improper digestion and/or absorption of dietary lipid, brought about by, e.g., failure of bile flow, ileal resection or disease, or overgrowth of bacteria. Among other effects are losses of fat-soluble vitamins, and actual deficiencies if the condition persists long enough.

Atherosclerosis is often associated with dietary characteristics of two sorts: (a) the actual cholesterol in the food, from animal products, and (b) the cholesterogenicity of the diet, i.e., the amount and degree of unsaturation of the ingested fats. These are two separate issues which are frequently but not necessarily connected.

There is a correlation between the occurrence of **dental caries** and the amount of sucrose in the diet, because sucrose promotes the growth in the mouth of certain decay-causing bacteria.

On a world-wide basis, **malnutrition** is also a severe problem. There are two "pure" manifestations of malnutrition: *marasmus*, where total caloric intake is deficient, and *kwashiorkor*, where calories are sufficient but intake of protein is deficient. A real situation may be a combination of these.

For diseases related to deficiency of specific nutrients, see the appropriate sections to follow.

F. Dietary Recommendations

Recommended Dietary Allowance (RDA): issued for each nutrient by the Food and Nutrition Board of the National Academy of Sciences, and periodically revised in the light of new knowledge. Defined as the amount of a nutrient which, if consumed by every member of a population, will keep nearly everyone in good health, it can also have political overtones as some public health measures may be tied to it. Since people vary greatly in their dietary requirements, no one figure is applicable to all. The RDA is based on an average requirement (which may not be known with precision), two standard deviations are added, and often a safety factor besides, so that it is a generous excess for most people. Detailed tables are subdivided according to gender, age, etc. It should not be confused with an individual's minimum requirement.

RDA values for individual nutrients are given in the following sections that discuss each.

Caloric Balance: The typical diet in the US derives about 35 - 40% of its caloric value from fat. This is considered high; it is now recommended that no more than 30% of calories come from fat, and of these, a maximum of one-third from saturated fat. If about 10% of calories are derived from protein, then the remainder, at least 60%, should be provided by carbohydrate.

II. MICRONUTRIENTS

A. Vitamins, Fat-Soluble: isoprenoid derivatives with varying degrees of unsaturation.

1. *Vitamin A*

Structure: derived from a pro-vitamin, the carotenoid β-carotene, a yellow-orange plant pigment which is cleaved by an enzyme in intestinal mucosa. Vitamin A occurs in three forms with partial interconvertibility:

retinol (vitamin A alcohol) ⇌ **retinal** (vitamin A aldehyde) → **retinoic acid** (vitamin A acid)

The structure shown above is that of retinol. It is insoluble in water, so must be transported in blood bound to a retinol-binding protein (RBP).

Function: not known with precision except in vision, in which retinal is combined with the protein, opsin, to form the visual pigment, **rhodopsin**. The process of vision depends on the reversible conversion of **11-*cis*-retinal** to **all-*trans*-retinal**. Other functions of vitamin A (for which a dietary supply of any form will usually suffice) are in differentiation of epithelial cells, growth, reproduction, and in the immune system. It seems to be particularly important in fetal development. It may function in a way similar to that of steroid hormones acting, along with a receptor, as a transcription factor or in other ways to influence genetic expression.

A large number of **retinoids** have been synthesized, some of which are useful in dermatology (though some have been shown to be teratogens). β-Carotene and some retinoids are anti-oxidants and may also have anti-cancer activity.

Occurrence: **β-carotene** in green and yellow vegetables, carrots, pumpkin, cantaloupe melons. Vitamin A itself occurs only in animal products, especially butter (but often there by fortification), eggs, fish liver oil, liver (polar bear liver may contain toxic amounts).

Deficiency: night blindness is an early warning. This may advance to keratinization of epithelial tissues, especially of the eye, causing **xerophthalmia**, a form of blindness which is of major concern in some areas.

Requirement:[*] **0.8 - 1.0 mg/day** (as retinol), or six times that amount as β-carotene, since the conversion and/or absorption is variable and incomplete. Sustained high intake of vitamin A causes many toxic symptoms and abnormalities including birth defects. Carotenoids are not known to be toxic.

2. *Vitamin D*

Structure: cholecalciferol is derived from **7-dehydrocholesterol** through cleavage of the 9,10-bond by UV light, then rotation of ring A 180° around the 6,7-bond (Figure 10-1).

Function: Cholecalciferol is like a prohormone. The active form is generated by (1) insertion of a hydroxyl group at C-25 by the liver, and (2) 1α-hydroxylation by a kidney enzyme whose activity is enhanced by parathyroid hormone. The active **1α,25-dihydroxycholecalciferol (calcitriol)** acts in a manner similar to that

---

[*] In this chapter, requirements are the RDA for adults, as presented in the Tenth Edition, *Recommended Dietary Allowances*, National Academy Press, Washington, 1989.

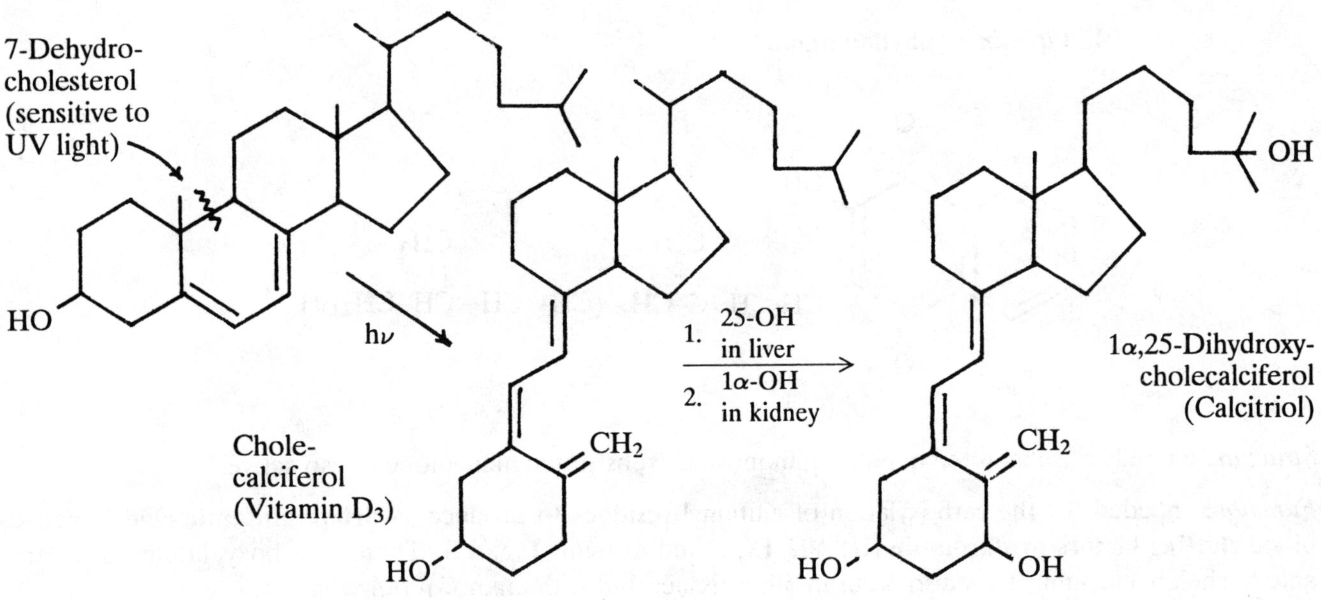

Figure 10-1. Formation and Metabolism of Vitamin D

of the steroid hormones, to cause synthesis of a calcium-binding protein by intestinal cells. It may also be active in bone and kidney. The effect is to raise the level of blood **calcium** and to promote mineralization of bone.

Occurrence in foods is not necessary if a person has a sufficient exposure to sunlight. Otherwise, dairy products, especially fortified milk, and fish liver oils are good sources.

Deficiency: **rickets**. Adult form is **osteomalacia**. Primary deficiency is rare in developed countries, but is sometimes found in individuals with kidney failure.

Requirement: **5 - 10 µg/day** (in the absence of sunlight). Excess can cause irreversible damage; the toxic level may be only 5 times the RDA.

3. *Vitamin E* (tocopherol)

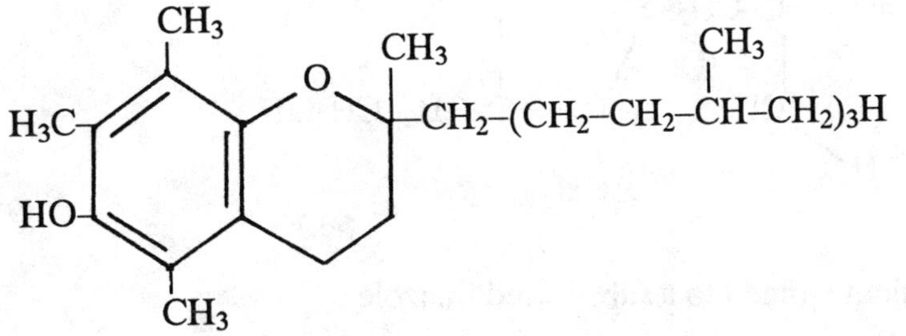

Structure: α, β-tocopherols, etc., depending on the length of the isoprenoid side-chain.

Functions as an **antioxidant**, probably by virtue of its ability to act as a trap for free radicals.

Occurs widely, especially in vegetable oils.

Deficiency: in animals, sterility and muscular dystrophy. In humans, anemia and neurological disorders may occur rarely, in association with malabsorption syndromes.

Requirement: **8 - 10 mg/day**, but may be greater with increased consumption of polyunsaturated oils. These, however, are also good sources of vitamin E.

4. *Vitamin K* (**phylloquinone**)

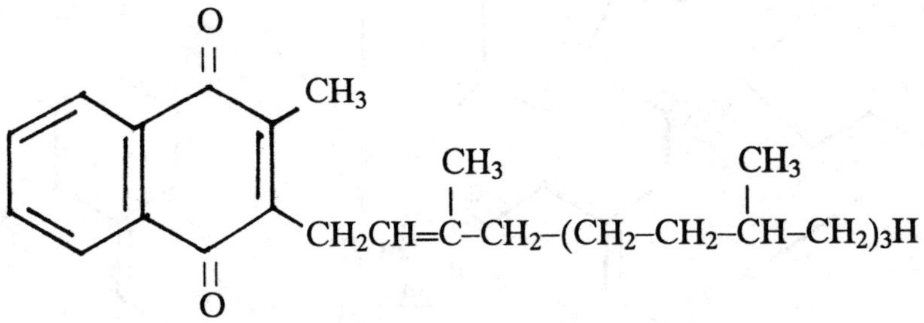

The structure shows a naphthoquinone ring with two C=O (O) groups, a CH_3 substituent, and a side chain:

$CH_2CH=C-CH_2-(CH_2-CH_2-CH-CH_2)_3H$ with CH_3 groups.

Structure: a family of substituted naphthoquinones. Unsubstituted menadione is also active.

Function: needed for the carboxylation of glutamyl residues to produce **γ-carboxyglutamic acid** in active blood **clotting factors prothrombin** (**II**), **VII, IX, X** and **proteins C** and **S**. These γ-carboxyglutamic acids are able to chelate calcium. They also occur in other tissues, but with unknown function.

Occurrence: ubiquitous in foods, especially green leafy vegetables, and also synthesized by intestinal bacteria.

Deficiency is rare but can occur in the newborn infant and in malabsorption syndromes, causing **hemorrhage** due to hypoprothrombinemia. **Dicoumarol** and related compounds (warfarin, coumadin) are antagonists of Vitamin K and are used to prevent thrombosis.

Requirement: **1 μg/Kg** body weight, or **60 - 80 μg/day**. Some bacterially-synthesized K can be absorbed, but not in sufficient amounts to replace the need for dietary K.

B. Water-Soluble Vitamins, Energy Releasing. So-called because most of their derivatives function as coenzymes in energy-producing metabolic pathways.

1. *Thiamin* (**Vitamin B₁**)

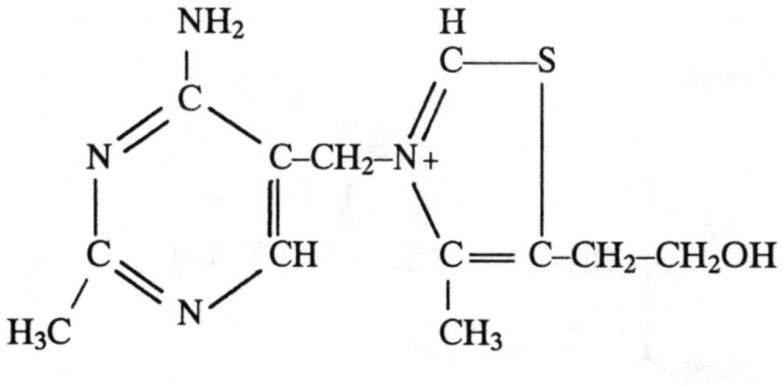

The structure shows a substituted pyrimidine ring (NH_2, N, N, H_3C, CH) joined via $C-CH_2-N^+$ to a substituted thiazole ring (H, C, S, CH_3, $C=C-CH_2-CH_2OH$).

Structure: a substituted pyrimidine joined to a substituted thiazole.

Cofactor form and function: **Thiamin pyrophosphate (TPP)**. Part of the *pyruvate dehydrogenase* complex, TPP functions in the **decarboxylation** of pyruvate to form acetyl CoA. It is also part of *α-ketoglutarate dehydrogenase* and of *transketolase*.

Occurrence: meat, liver, whole grains, legumes.

Deficiency: the classical disease is **beriberi**. Alcoholics may have a multiple B-vitamin deficiency known as **Wernicke's disease**.

Requirement: **1 - 1.5 mg/day**, sometimes expressed as a function of daily energy expenditure — 0.5 mg / 1000 Kcal.

2. *Riboflavin* (**Vitamin B$_2$**)

Structure: a heterotricyclic system joined to ribitol.

Cofactor form and function: **flavin mononucleotide** or **FMN** (riboflavin phosphate); **flavin adenine dinu-cleotide** or **FAD** (a combined nucleotide with AMP). These, as part of flavoproteins, act in **transfer of hydrogen and electrons** from NAD to CoQ, and from succinate, from acyl CoA in β-oxidation, etc. FAD is part of *pyruvate* and *α-ketoglutarate dehydrogenases*. Derivatives of riboflavin are also involved in functioning of B$_6$ and niacin.

Occurrence: milk, liver, meat, green vegetables. Cooking or exposure to light tends to destroy riboflavin.

Deficiency: lesions of the lips, skin, genitalia.

Requirement: **1.2 - 1.8 mg/day** or 0.6 mg/1000 Kcal.

3. *Nicotinamide, Nicotinic Acid* (**Niacin, Vitamin B$_3$**)

Structure: a substituted pyridine.

Cofactor form and function: **Nicotinamide adenine dinucleotide** or **NAD** (nicotinamide + adenine + 2 ribose + 2 phosphate), **NADP** (NAD + a third phosphate). These act as **carriers of hydrogen and electrons** in a multitude of dehydrogenase reactions. In general, NAD acts in catabolism while NADP acts in synthetic reactions. NADPH and O$_2$ are used by mixed function oxidases, particularly in metabolism of drugs and in various hydroxylations. Nicotinic acid has been useful as a cholesterol-lowering drug: it reduces production of VLDL, and so of LDL.

Occurrence: meat, liver, peanuts, legumes, whole grains. Nicotinate can be biosynthesized from tryptophan, but inefficiently (1 mg from 60 mg). Corn is a poor source of both niacin and tryptophan.

The *deficiency* disease is **pellagra**, characterized by diarrhea, dermatitis, dementia, and death ("4 D's"). Pellagra is of historical interest as it used to occur in the US over a geographic area where the population was particularly dependent on corn as a dietary staple.

Requirement: **15 - 20 mg/day**, or 6.6 mg/1000 Kcal.

4. *Pyridoxine* (Vitamin B_6)

Structure: a substituted pyridine. Also occurs as the aldehyde and amine forms.

Cofactor form and function: **pyridoxal phosphate** and **pyridoxamine phosphate**. These act as coenzymes of *amino acid transaminases*, some *decarboxylases*, *glycogen phosphorylase*, *ALA synthase*, etc. They reversibly form a Schiff base with amino groups.

Occurrence: fish, chicken, liver, whole grains.

Deficiency: convulsions, oxalate kidney stones. Deficiency can be caused by treatment with the drug isoniazid, which forms a rapidly-excreted hydrazone with pyridoxal.

Requirement: **1.5 - 2 mg/day**. Prolonged ingestion of megadoses may cause neurological effects.

5. *Pantothenic Acid*

Structure: pantoic acid + β-alanine.

Cofactor form and function: **Coenzyme A** (phosphoadenosine diphosphate attached to the pantoic acid moiety, + thioethanolamine attached to the β-alanine moiety). The terminal -SH combines with acyl groups to form a "high-energy" thioester; this is the way acyl compounds are activated before undergoing metabolism; example: acetyl CoA. The pantothenic acid and thioethanolamine are also part of *acyl carrier protein*, in which the -SH carries the growing chain during fatty acid synthesis.

Occurrence: ubiquitous in foods, especially animal tissues, whole grains, legumes. Some may be produced by intestinal microflora.

Deficiency: extremely rare.

Requirement: **4 - 7 mg/day** ("safe and adequate," no RDA has been set).

6. *Biotin*

Structure:

$$
\begin{array}{c}
O \\
\parallel \\
C \\
HN \qquad NH \\
| \qquad\quad | \\
HC \text{———} CH \\
| \qquad\qquad | \\
H_2C \qquad\quad CH\text{–}(CH_2)_4\text{–}COOH \\
\diagdown \quad \diagup \\
S
\end{array}
$$

Cofactor form and function: biotin acts in several **carboxylations**, being a part of *pyruvate carboxylase* and *acetyl CoA carboxylase*, among others. It is linked to its enzymes by an ϵ-amino group of lysine, forming a swinging arm which enables it to transfer a -COOH from one active site to another in an enzyme complex.

Occurrence: in many foods, and is made by intestinal bacteria in amounts possibly sufficient to satisfy needs.

Deficiency: very rare; can be induced by consumption of large amounts of raw egg whites which contain a glycoprotein, **avidin**, which binds biotin and renders it unabsorbable.

Requirement: **30 - 100 μg/day** (estimated safe and adequate). No RDA has been set, as knowledge is incomplete regarding requirement *vs.* relative amounts provided by diet and intestinal flora.

C. Water-Soluble Vitamins, Hematopoietic. So-called because blood-forming tissues are particularly dependent on them.

1. *Folic Acid* (**Folacin**)

Structure: pteroyl glutamic acid.

Cofactor form and function: **tetrahydrofolic acid (THFA)**, with several glutamic acids attached. Active in metabolism of **one-carbon** units, THFA carries a C-1 fragment, at various states of oxidation, between N^5 and N^{10} (Chapter 5). THFA is also essential in biosynthesis of **purines** and **thymidylate** (Chapter 11), in the reversible transformation of **serine** to glycine, and in other aspects of one-carbon metabolism. The transported form of the coenzyme is N^5-methyl THFA; regeneration to active THFA requires adenosyl cobalamin, one of the forms of Vitamin B_{12}.

Certain analogues of folic acid are useful as anticancer agents: **aminopterin**, **amethopterin**, and **methotrexate** (Chapter 11).

Occurrence: kidney, liver, dark green leafy vegetables (hence the name).

Deficiency: megaloblastic bone marrow and macrocytic anemia.

Requirement: **200 µg/day.** Folate and the drug phenytoin compete for absorption. There is some potential for toxicity from excessive intake.

2. *Cobalamin* (Vitamin B$_{12}$)

Structure: a **corrin** derivative, a complex tetrapyrrole related to the porphyrins but lacking one of the methenyl bridges, and containing **cobalt**.

Cofactor form and function: as the **deoxyadenosine** or **methyl** derivative. (Cyanocobalamin is a form often obtained from isolation procedures). B$_{12}$ functions in two areas: (1) regeneration of **active tetrahydrofolic acid** (during which a methyl group is transferred to homocysteine to form methionine); (2) *mutase* reactions, as in propionate metabolism when methylmalonyl CoA is converted to succinyl CoA.

Occurrence: made by certain microorganisms, and obtained in the diet only from foods of animal origin, especially meat and dairy products. Absent in plant foods.

Deficiency: all the symptoms of folate deficiency plus neurological abnormalities. In **pernicious anemia**, the deficiency is really one of **intrinsic factor**, a glycoprotein produced by the stomach, which is necessary for ileal absorption of B$_{12}$. **Methylmalonic aciduria** is one effect. Dietary deficiency may rarely occur in strict vegetarians.

Requirement: **2 µg/day.** But unlike other B-vitamins, B$_{12}$ can be stored; the liver can hold a several years' supply. It also undergoes efficient enterohepatic circulation, which greatly prolongs its lifetime in the body.

D. Other Water-Soluble Vitamin: *Ascorbic Acid* (Vitamin C)

Structure: related to that of the six-carbon sugars.

Cofactor form and function: cofactor (if any) not known. Vitamin C functions as a **reducing agent**, helping to keep Cu^+ and Fe^{++} of mono- and di-oxygenases, respectively, in their reduced states, and is also important in reducing dietary iron to the more easily absorbable ferrous form. It is essential in some **hydroxylations**, especially of **proline** and **lysine** in **collagen** synthesis, and in bile acid formation. It also probably has a function (unknown) in the metabolism of the adrenal cortex.

Occurrence: citrus fruits, tomatoes, other fruits and vegetables.

Deficiency: The classical disease is **scurvy**, which is particularly a disease of collagen formation: poor wound healing, fragile bones, loosening of teeth, hemorrhage. Though blood levels of ascorbic acid decline rapidly on a deficient diet, scurvy usually takes many weeks to develop. Some authorities maintain that a state of deficiency short of outright scurvy is a very common condition.

Requirement: **60 mg/day.** Since the turnover of vitamin C is increased in cigarette smokers, it is recommended that these should ingest 100 mg/day.

E. Minerals: Electrolytes

1. *Sodium*

The major **extracellular cation**, sodium functions in osmotic, water, and acid-base balance. It is ubiquitous in foods, especially those of animal origin and in preserved, prepared, and processed foods. A deficiency is uncommon, but can occur in an unacclimatized person through prolonged heavy sweating. An excess is widespread and insidious, and may cause **hypertension** in susceptible individuals. Estimated requirement: **0.5 g/day**, but not over 2.4 g/day (equivalent to 1.3 g and 6 g NaCl, respectively).

2. *Potassium*

The major **intracellular cation**, potassium also functions in osmotic, water, and acid-base balance. It occurs widely in (unprocessed) foods, especially those of plant origin. Deficiency may occur after prolonged vomiting, in diarrhea, diabetics, and in those on diuretics. Cardiac abnormalities, possibly fatal, may ensue. Excretion of sodium and potassium by the kidney is regulated by the **renin-angiotensin system**. Estimated requirement: **1.6 - 2.0 g/day**.

3. *Chloride*

The principal **anion** for Na^+ and K^+, it occurs everywhere and deficiency is not a problem. Estimated requirement: **1.8 g/day**.

F. Other Major Minerals

1. *Calcium*

The major inorganic element in the body, calcium is 1.5 - 2% of the total mass. Most is in the **skeleton**, but calcium has vital physiological roles in **muscle contraction**, **blood clotting**, as an intracellular **messenger**, and in many other ways. The level of blood calcium is regulated by **parathyroid hormone** (raises level), **calcitonin** (lowers level), **vitamin D** (promotes absorption in intestine), and by other hormones. Bone serves as a buffer or reservoir.

Bone density increases during the first $2\frac{1}{2}$ decades of life, during which time it is most important that adequate calcium intake be maintained. Later, losses of calcium predominate which may lead to osteomalacia, especially among those whose maximal bone density is low due to earlier inadequate intake. Post-menopausal women are also at particular risk.

Dairy products are good sources, as are cruciferous vegetables and some calcium-prepared foods. A primary dietary deficiency in adults is rare, but can occur secondary to vitamin D deficiency, and in women after multiple pregnancies and lactation. It sometimes is seen in a newborn on cows' milk (which has plenty of calcium but a low Ca/P ratio, which may make regulation difficult).

The adult requirement is **0.8 - 1.2 g/day** but adaptation or high bioavailability may make the true need much less. Post-menopausal women and the aged may have an increased requirement. It is not clear to what extent, if at all, calcium supplementation reverses loss of bone in the older population.

2. *Phosphorus*

An important **intracellular anion** (as phosphate), phosphorus is about 1% of the body's mass, occurring mostly as a constituent of **bone**. Chemically it has major functions as a component of **nucleic acids**, many **nucleotides**, and **phosphate esters** of sugars and other intermediates. A dietary deficiency is virtually unknown, but low levels may occur in the diabetic and in individuals receiving prolonged antacid treatment with aluminum hydroxide. Requirement: **0.8 - 1.2 g/day**.

3. *Magnesium*

Also an important intracellular cation. It occurs in **bone**, and also functions in many enzyme activities, especially those involving **ATP**. Most foods of vegetable origin are good sources, especially whole grains. Deficiency may occur in alcoholics. Requirement: **280 - 350 mg/day**.

4. *Iron*

Functioning in oxygen transport and storage, iron is part of the heme proteins *hemoglobin* and *myoglobin*. It is also part of the *cytochromes* of the electron transport chain. Other heme-containing enzymes are *catalase* and *peroxidase* which are involved in the metabolism of H_2O_2. The average adult has about 4 g, $2/3$ of which is in hemoglobin, and $1/4$ in storage, particularly as liver ferritin. The phases of iron metabolism are outlined below:

Absorption of dietary iron: highly variable. Only about 10% or less (1 - 2 mg/day) of non-heme iron may be absorbed, as the **reduced**, or **ferrous** form; heme iron is more efficiently absorbed, up to 40%.

Regulation is at the level of **absorption**. According to one proposal, in response to adequate body iron stores, **intestinal cells** produce a **ferritin trap** which prevents further passage of absorbed iron into the bloodstream before the cells slough off. When stores are low and there is a need for iron, ferritin is not produced by intestinal cells, and iron that is taken in is allowed to pass through into the circulation.

Oxidation: ferrous iron is oxidized to the ferric form by *ferroxidase (ceruloplasmin)*, a **copper**-containing enzyme.

Transport: iron is tightly bound by *transferrin*, two ferric atoms per molecule.

Storage: mostly in *ferritin* and *hemosiderin* of liver, some in bone marrow. Many ferric ions per molecule.

Excretion: virtually non-existent, as iron is efficiently conserved and re-used. The only way for the body to be significantly depleted of iron is by **bleeding** or **childbearing**. Iron overload may occur in some alcoholics, recipients of blood transfusions for hemolytic anemia, and overzealous practitioners of self-medication. Hemochromatosis is an inborn error in which too much dietary iron is absorbed, leading to excessive accumulation.

Occurrence: meat (heme iron is well-absorbed), egg yolk, legumes. Milk and spinach are poor sources. Simultaneous intake of **vitamin C** enhances the efficiency of absorption by helping to keep the iron in the reduced state. Substances that inhibit absorption include certain fibers, antacids, and polyphenols of tea.

Deficiency: iron-deficiency anemia is relatively common since dietary iron is inefficiently absorbed. It is most often seen in young children, early adolescence, young mothers, and in persons with chronic loss of blood.

Requirement: **10 mg/day** (**12 - 15** for teenagers; **15** for women of childbearing age).

G. Trace minerals with established RDA

1. *Iodide*

Iodide is required for **thyroid** function — T_4 and T_3. Found in sea foods, iodized salt. Requirement: **150 µg/day**. Deficiency is manifested as endemic **goiter**, in areas where food is locally-grown in soils poor in iodide.

2. *Selenium*

Selenium functions in *glutathione peroxidase*, which, as an **antioxidant**, is complementary to vitamin E. Requirement: **55 - 70 µg/day**. Toxic in the mg/day range.

3. *Zinc*

Zinc is important in the immune system and as a cofactor for many enzymes, e.g., **carbonic anhydrase, alcohol dehydrogenase, nucleic acid polymerases**. Deficiency is rare, but has been seen in alcoholics and in cases of renal disease. Absorption is inhibited by some other minerals and bran. Requirement: **12 - 15 mg/day.**

H. Other trace minerals with estimated safe and adequate dietary intakes

1. *Chromium:* Involved with the action of **insulin**. Estimated need: **50 - 200 μg/day.**

2. *Cobalt:* the only need is as a component of **vitamin B$_{12}$.**

3. *Copper*

Part of a number of enzymes, notably **cytochrome oxidase** and other oxidases including **ceruloplasmin**. Wilson's disease is an inborn error involving excessive deposition of copper in the liver. Estimated need: **1.5 - 3 mg/day.**

4. *Fluoride*

Probably not an essential element, but 1 ppm in the water supply is beneficial in helping to prevent **dental caries**. Estimated need: **1.5 - 4 mg/day.**

5. *Manganese*

Widely distributed in vegetable foods, manganese participates in a variety of enzymic activities, such as *pyruvate carboxylase*. Estimated need: **2 - 5 mg/day.**

6. *Molybdenum*

Required for *xanthine oxidase, aldehyde oxidase, sulfite oxidase*. Antagonistic to copper. Estimated need, easily furnished by diet: **75 - 250 μg/day**

I. Miscellaneous

1. *Carnitine*

Carnitine can be synthesized by the human, and it occurs widely in foods of animal origin. Deficiency due to a heritable defect in its synthesis has been described.

2. *Choline*

Widely present in foods, choline is a constituent of membranes as phosphatidyl choline (lecithin), and is required in the diet by some animals, as a source of methyl groups that can partially spare the methionine requirement. No dietary requirement has been shown in adult humans, though it may be essential in the diet of neonates.

3. *Inositol*

Important in membranes and in a second messenger system, inositol is provided by diet and synthesis. Deficiency has been produced in some animals, but has never been observed in humans.

Vitamin	Cofactor Form	Association or Function
A	11-cis Retinal	Rhodopsin in vision
	Other forms	Growth, differentiation, epithelial tissues
D	1α,25-Dihydroxycholecalciferol	Intestinal calcium-binding protein
E	none	Antioxidant
K	none	Carboxylation of clotting factors
Ascorbic acid (C)	none	Reducing agent, hydroxylations
Biotin	none	Carboxylations
Cobalamin (B$_{12}$)	Deoxyadenosyl or methyl derivative	Mutases, recovery of active THFA
Folic acid	Tetrahydro derivatives	Metabolism of 1-Carbon units
Niacin (B$_3$)	NAD	Dehydrogenases, catabolic
	NADP	Dehydrogenases, anabolic
Pantothenic acid	Coenzyme A	Activation of acyl groups
	Acyl Carrier protein	Growing chain in fatty acid synthesis
Pyridoxine (B$_6$)	Pyridoxal-P	Transaminases, decarboxylases
Riboflavin (B$_2$)	FMN, FAD	Dehydrogenases, catabolic
Thiamin (B$_1$)	Thiamin PP	Pyruvate, α-KG decarboxylases

Mineral	Association or Function
Calcium	Bone, coagulation, muscle contraction, second messenger
Chlorine	As chloride, anion for Na^+, K^+, etc.
Chromium	Potentiation of insulin
Cobalt	Component of vitamin B$_{12}$
Copper	Cytochrome oxidase, ferroxidase
Fluorine	Prevent dental caries
Iodine	Component of thyroid hormones
Iron	Hemoglobin, myoglobin, in cytochromes of e^- transport, catalase, peroxidases
Magnesium	Bone, kinase reactions
Manganese	Arginase, acetyl CoA carboxylase
Molybdenum	Xanthine oxidase, etc.
Phosphorus	As phosphate, in bone, DNA, RNA, nucleotides, organic phosphates
Potassium	Intracellular cation, osmotic, acid-base balance
Selenium	Glutathione peroxidase
Sodium	Extracellular cation, osmotic, acid-base balance
Zinc	Carbonic anhydrase, alcohol dehydrogenase, etc.

Table 10-1. Summary of nutrient - function associations

III. REVIEW QUESTIONS ON NUTRITION

> **DIRECTIONS:** For each of the following multiple-choice questions (1 - 24), choose the ONE BEST answer.

1. Quantitatively, the most prevalent dietary lipid is:

A. triacylglycerol
B. monoacylglycerol
C. cholesterol
D. oleic acid
E. cholesteryl oleate

Ans A: Most dietary lipid is plain fat, i.e., triglyceride or triacylglycerol.

2. A disaccharide that requires β-galactosidase for digestion is:

A. sucrose
B. maltose
C. fructose
D. lactose
E. galactose

Ans D: Lactose is a β-galactoside, and requires lactase, a β-galactosidase.

3. A person with a gallstone blocking the bile duct is at some risk of developing:

A. scurvy
B. night blindness
C. pellagra
D. Wernicke's disease
E. Cushing's syndrome

Ans B because blockage of biliary flow may cause a malabsorption syndrome, leading to deficiency of fat-soluble vitamins.

4. Which of the following has a mode of action like that of a steroid hormone?

A. thiamin
B. riboflavin
C. ascorbate
D. calcitriol
E. niacin

Ans D: Calcitriol, a sterol, induces the expression of selected genes. The others are all water-soluble vitamins.

5. "Hemorrhagic disease of the newborn" is usually the result of deficiency of:

A. vitamin B_{12}
B. folic acid
C. biotin
D. vitamin K
E. vitamin E

Ans D: K is needed for production of several blood-clotting factors.

Questions 6 - 8: You went to the grocery store to get a snack for yourself and your [spouse/room-mate/other]. You purchased a package labeled as follows:

portion size	2 ounces
Portions per package	3
Content per portion:	
Calories	200
Carbohydrate	20 g
Protein	5 g
Fat	11 g
Calcium	400 mg
Ascorbic acid	20 mg

But greed got the better of you and you ate the whole thing!

6. How many kilocalories did you consume?

A. 36
B. 66
C. 100
D. 200
E. 600

Ans E: 3 portions x 200 Kcal/portion = 600 Kcal.

7. What percent of calories came from fat?

A. 10
B. 15
C. 30
D. 50
E. 80

Ans D: In a portion, 11 g fat x 9 Kcal/g = 99 Kcal, which is half of 200 Kcal.

8. Your snack met nutritional guidelines or daily recommendations for which of the following? (Your weight is 65 Kg).

A. fat
B. calcium
C. vitamin C
D. protein
E. calcium and vitamin C

Ans E: Not protein (your RDA is 52 g); not fat (in excess of 30% calories from fat); you took in 60 mg vitamin C and 1200 mg calcium, both of which meet the respective RDA's.

9. The Recommended Dietary Allowance (RDA) is all of the following EXCEPT:

A. periodically revised
B. the amount of a nutrient needed to keep nearly everyone in good health
C. always greater than the average requirement for a particular nutrient
D. a valid measure of an individual's requirement for a nutrient
E. very hard to determine and may contain some approximations.

Ans D: The RDA is almost always well in excess of a particular individual's requirement.

10. Below is shown the structure of a lipid derived from the diet:

$$CH_3\text{-}CH_2\text{-}CH=CH\text{-}CH_2\text{-}CH=CH\text{-}CH_2\text{-}CH=CH\text{-}CH_2\text{-}CH_2\text{-}CH_2\text{-}CH_2\text{-}CH_2\text{-}CH_2\text{-}CH_2\text{-}COOH$$

Which of the following statements about this lipid is NOT true?

A. It is a dietarily essential fatty acid.
B. It is an n − 3 (ω − 3) fatty acid.
C. It will tend to lower blood cholesterol levels.
D. It will tend to retard formation of blood clots.
E. It is oleic acid.

Ans E is not correct because the substance is linolenic acid, an essential ω − 3 fatty acid, with anticholesterogenic and antithrombogenic effects.

11. Minerals act as prosthetic groups for metalloenzymes. Correct pairings include all of the following (A - D) EXCEPT: If all are correct, no exception, choose E.

A. zinc - carbonic anhydrase
B. selenium - glutathione peroxidase
C. copper - cytochrome oxidase
D. iron - cytochrome oxidase
E. all the above are correct, without exception.

Ans E.

12. A diet high in fiber is said to do all of the following (A - D) EXCEPT: If all occur, no exception, choose E.

A. Fiber binds to certain substances such as bile acids and retards their absorption.
B. Fiber promotes efficient functioning of the alimentary tract.
C. Fiber promotes absorption of certain nutrients such as calcium.
D. Fiber decreases the transit time of intestinal contents.
E. All of the above occur, without exception.

Ans C is the exception. Fiber actually decreases the efficiency of absorption of some nutrients such as minerals that may bind to it.

13. The recommended dietary allowance of protein for a 60-Kg person is about:

A. 24 g
B. 36 g
C. 48 g
D. 60 g
E. 72 g

Ans C: 0.8 g/Kg x 60 Kg = 48 g.

14. All of the following statements (A - D) about pantothenic acid are correct EXCEPT: If all are correct, no exception, choose E.

Pantothenic acid is:

A. a water-soluble vitamin
B. incorporated into the covalent structure of fatty acid synthase
C. a precursor for the synthesis of NAD$^+$
D. incorporated into the structure of coenzyme A
E. all the above are correct, without exception.

Ans C is the exception. A water-soluble vitamin, it is part of the structure of ACP and CoA.

15. All of the following statements (A - D) about the basal metabolic rate are correct EXCEPT: If all are correct, no exception, choose E.

A. It is often expressed as a function of the body surface area.
B. It is often expressed as a function of the amount of food consumed.
C. It is higher in children than in adults.
D. It can be calculated from the rate of oxygen consumption.
E. All of the above are correct without exception.

Ans B is the exception. BMR is measured during fasting. It varies by body size, corrected for by dividing by body surface area.

16. Malnutrition in regard to protein, in the presence of adequate calories, is called:

A. pernicious anemia
B. marasmus
C. kwashiorkor
D. pellagra
E. beriberi

Ans C by definition.

17. In the omega (ω) notation of fatty acids, the aspect that is emphasized is:

A. number of carbon atoms
B. number of double bonds
C. content of essential fatty acids in a dietary sample
D. position of the first double bond in relation to the first C atom in the chain
E. position of the last double bond in relation to the last C atom in the chain.

Ans E. In a particular family of ω fatty acids, the common feature is the number of carbon atoms beyond the double bond.

18. Which of the following groups contains ONLY substances that are nutritionally essential?

A. threonine, tyrosine
B. arginine, alanine
C. oleate, linoleate
D. pyridoxine, phenylalanine
E. sodium, serine

Ans D: In the other groups, non-essentials are serine, oleate, alanine, tyrosine (made from phenylalanine).

19. In iron metabolism:

A. the major plasma transport protein for iron is hemosiderin
B. most of the iron from the degradation of hemoglobin is excreted in the feces
C. nearly all iron in the diet is absorbed
D. ferritin is a major storage form of iron
E. most of the iron from the degradation of hemoglobin is excreted in the urine.

Ans D: Iron transport is by transferrin; iron is inefficiently absorbed and very little is excreted.

20. Which of the following is NOT a function or role of phosphate?

A. constituent of bone
B. part of some amphipathic lipids
C. part of the structure of nucleic acids
D. part of many metabolic intermediates
E. part of the structure of adenosine.

Ans E: Adenosine is a nucleo<u>side</u>, i.e., base-pentose, without phosphate.

21. Which of the following is required in the diet in SMALLEST amount?

A. calcium
B. iron
C. protein
D. sodium
E. vitamin C

Ans B: Aside from protein (a macronutrient), recommended daily intakes for the others are respectively: 0.8 - 1.2 g, 10 - 15 mg, 0.5 g, and 60 mg.

22. Obesity is a risk factor for all of the following (A - D) EXCEPT: If it is a risk factor for all, no exception, choose E.

A. atherosclerosis
B. diabetes
C. gallstones
D. xerophthalmia
E. all of the above, without exception.

Ans D is the exception. Xerophthalmia results from deficiency of vitamin A.

23. Amino acids are described as essential when:

A. they do not serve as substrates for transaminases
B. their carbon chains cannot be synthesized by the body
C. they are found only in animal protein
D. they are important intermediates in metabolic pathways
E. their carbon chains cannot be completely degraded by the body.

Ans B: The carbon skeletons are critical; amino groups can usually be added by transamination.

24. Which of the following would be most likely to have a negative nitrogen balance?

A. an experimental subject on a diet deficient in valine
B. an experimental subject on a diet deficient in proline
C. a person recovering from influenza
D. a growing child
E. a man beginning a weight-lifting regimen.

Ans A: Valine is dietarily essential. Proline deficiency is of no consequence; C, D, and E would be in positive N balance.

DIRECTIONS: For each set of questions, choose the ONE BEST answer from the lettered list above it. An answer may be used one or more times, or not at all.

Questions 25 - 32: For each condition or deficiency disease in the following set of questions, choose the associated nutrient lack from the list below:

A. Retinol
B. Niacin
C. Ascorbate
D. Cholecalciferol
E. Iodide
F. thiamin
G. riboflavin
H. folic acid

25. Xerophthalmia.

Ans A: Xerophthalmia is caused by severe A deficiency.

26. Defective collagen.

Ans C: Vitamin C is needed to complete the production of collagen.

27. Pellagra.

Ans B: Pellagra is the classical niacin deficiency disease.

28. Scurvy.

Ans C: Scurvy is the classical deficiency disease of vitamin C.

29. Rickets.

Ans D: Rickets is the classical deficiency disease of vitamin D.

30. Endemic goiter.

Ans E: Endemic goiter occurs in areas where iodide is lacking in the diet.

31. Inability to properly metabolize C-1 fragments.

Ans H: Most one-carbon units are carried as folate derivatives.

32. Wernicke's disease.

Ans F: Although multiple B-vitamin deficiency may be involved, lack of thiamin is the major cause.

Questions 33 - 42: Choose the vitamin from the list below that is associated with the enzyme or function in the following set of questions:

A. Niacin
B. Riboflavin
C. Biotin
D. Cobalamin
E. Pyridoxine

F. ascorbate
G. pantothenic acid
H. folic acid
I. retinol
J. cholecalciferol

33. ALA synthase.

Ans E: Pyridoxal phosphate is the coenzyme.

34. Succinate dehydrogenase.

Ans B: Requires FAD.

35. Lactate dehydrogenase.

Ans A: Requires NAD^+.

36. Acetyl CoA carboxylase.

Ans C: Carboxylases require biotin.

37. Methylmalonyl CoA mutase.

Ans D: Some mutases use B_{12}.

38. Carrier of activated acyl groups.

Ans G: Pantothenic acid is part of CoA and ACP.

39. Needs B_{12} to complete its metabolism.

Ans H: Regeneration of the pool of active folate derivatives from methyl folate requires participation of B_{12}.

40. Transaminations.

Ans E: Pyridoxal phosphate is needed for transfer of amino groups.

41. Needs intrinsic factor for absorption.

Ans D: B_{12} is absorbed very poorly in the absence of intrinsic factor, which makes an absorbable complex with it.

42. Some of its relatives and derivatives are useful in dermatology.

Ans I: Retinoids can be used to treat acne and other skin ailments.

Questions 43 - 46: For each condition or health problem in the following set of questions, choose the associated nutrient from the list below:

    A. protein (insufficiency)
    B. sucrose
    C. lactose
    D. cholesterol
    E. triglyceride

43. Intolerance to milk.   *Ans C* because many cannot produce lactase after reaching adulthood.

44. Atherosclerosis.   *Ans D:* Atherosclerosis involves deposits of cholesterol in arteries.

45. Kwashiorkor.   *Ans A:* This is the form of malnutrition caused by insufficient protein.

46. Dental caries.   *Ans B:* Sucrose (table sugar) fosters growth of cavity-causing microorganisms.

Questions 47 - 53: Choose the mineral from the list below that is associated with the reaction or function in the following set of questions:

    A. Magnesium   E. Sodium
    B. Molybdenum   F. Potassium
    C. Fluoride   G. Chloride
    D. Zinc   H. Iron

47. Glucose + ATP $\longrightarrow$ *Ans A:* This is a kinase reaction, which requires Mg^{++}.

    Glucose-6-phosphate + ADP

48. $CO_2 + H_2O \rightleftharpoons H_2CO_3$ *Ans D:* This is the carbonic anhydrase reaction, requiring Zn^{++}.

49. Xanthine $\longrightarrow$ Uric acid   *Ans B:* Xanthine oxidase uses molybdenum.

50. Resistance to dental caries.   *Ans C:* Fluoride, often added to water supplies, confers resistance to tooth decay.

51. Major intracellular cation.   *Ans F:* Potassium is intracellular; sodium is extracellular.

52. Its excessive intake may trigger hypertension in sensitive people.   *Ans E:* High dietary sodium is sometimes implicated in hypertension.

53. Its absorption is improved by vitamin C.   *Ans H:* Ascorbate helps keep iron is the more easily-absorbable ferrous form.

11. PURINES AND PYRIMIDINES

Leon Unger

Nucleic acids are comprised of nitrogenous bases (purines and pyrimidines), pentose sugars (ribose and deoxyribose) and phosphate groups. Specific sequences of purines and pyrimidines encode the genetic information of cells and organisms.

I. STRUCTURE AND NOMENCLATURE

A. Nitrogenous Bases

The purines, **adenine (A)** and **guanine (G)**, are present in both RNA and DNA. Catabolism of adenine and guanine produces the purines **inosine, hypoxanthine, xanthine** and **uric acid.**

The pyrimidines, **cytosine (C)** and **thymine (T)** are present in DNA. Cytosine and **uracil (U)** are contained in RNA (Figure 11-1).

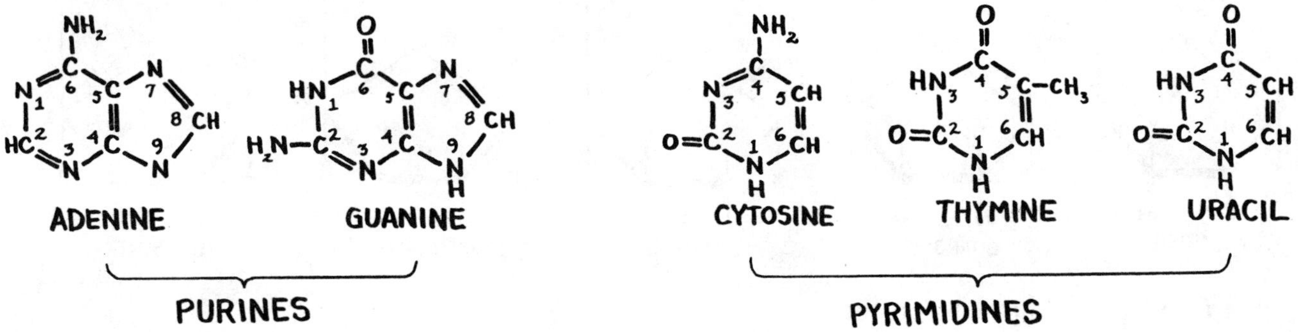

Figure 11-1. The Purine and Pyrimidine Bases.

B. Nucleosides

A **nucleoside** consists of a purine or pyrimidine base linked to a pentose sugar. In RNA, the pentose is **D-ribose,** therefore the nucleoside formed is termed a ribonucleoside (example: adenosine). The pentose is **2′-deoxy-D-ribose** in DNA, and the nucleoside is called a deoxyribonucleoside (example: deoxyadenosine). The atoms of the pentoses are designated by primed numbers to distinguish them from the numbers in the bases (Figure 11-2).

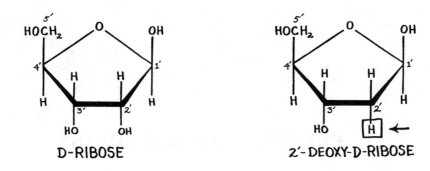

Figure 11-2. The Pentoses of Nucleic Acids.

In a nucleoside, carbon 1′ (C-1′) of the pentose is linked by a β-glycosidic bond to N-9 of the purine or N-1 of the pyrimidine.

<table>
<tr><td>N-9-purine</td><td>N-1-pyrimidine</td></tr>
<tr><td>|</td><td>|</td></tr>
<tr><td>pentose-C-1′</td><td>pentose-C-1′</td></tr>
</table>

The major ribonucleosides for the purines are **adenosine** and **guanosine** and for the pyrimidines, **cytidine** and **uridine**.

The major deoxyribonucleosides are **deoxyadenosine**, **deoxyguanosine**, **deoxycytidine** and (deoxy)-**thymidine** (Figure 11-3).

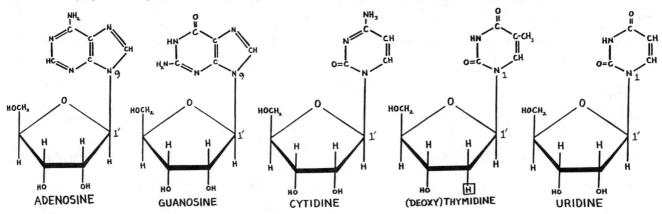

Figure 11-3. The Major Nucleosides.

C. Nucleotides

A nucleotide is a nucleoside in which a phosphate group has been esterified to a hydroxyl group on the pentose. The most common position to be esterified is at C-5′. Other positions for esterification include the hydroxyls at C-3′ (common) and C-2′ (less common).

II. SYNTHESIS OF NUCLEOSIDE DIPHOSPHATES AND TRIPHOSPHATES

The ribo- or deoxyribo-nucleoside 5′-monophosphates [NMP, (d)NMP] may be further phosphory-lated. Using ATP as a phosphate donor, *nucleoside monophosphate kinase* phosphorylates (d)NMP to (d)NDP. *Nucleoside diphosphate kinase* then converts (d)NDP to (d)NTP. These nucleotides are readily interconvertible.

$$(d)NMP + ATP \rightleftharpoons (d)NDP + ADP$$
$$(d)NDP + ATP \rightleftharpoons (d)NTP + ADP$$

III. PURINE METABOLISM

A. Synthesis of 5-Phosphoribosyl-1-Pyrophosphate (PRPP)

Synthesis: PRPP is synthesized from ribose-5-phosphate and ATP. The enzyme catalyzing this reaction is *ribose phosphate pyrophosphokinase (phosphoribosylpyrophosphate synthetase)*.

$$\text{Ribose-5-P} + ATP \longrightarrow PRPP + AMP$$

Functions: PRPP supplies ribose phosphate for the synthesis of <u>purine</u> ribonucleotides by both the *de novo* pathway and the salvage pathway. It also supplies ribose phosphate for the synthesis of <u>pyrimidine</u> ribonucleotides. PRPP is an intermediate in **histidine** and **tryptophan** biosynthesis.

B. Biosynthesis of Purine Nucleotides by the *De Novo* Pathway

The atoms for the purine ring are derived from **amino acids, tetrahydrofolate** and CO_2 and are assembled stepwise onto a pre-existing ribose phosphate molecule. The ribose phosphate comes from the **hexose monophosphate pathway**.

The first step specific to the synthesis of purine nucleotides is the formation of **5-phosphoribosylamine** from PRPP and glutamine. In this reaction, mediated by *glutamine-phosphoribosyl amidotransferase*, the pyrophosphate group on the C-1′ of PRPP is replaced by the amide group of glutamine whose N is destined to become N-9 of the purine. The irreversible formation of phosphoribosylamine commits PRPP to purine synthesis. The amidotransferase is inhibited by **azaserine**.

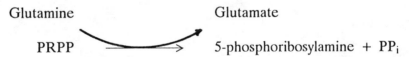

Glutamine   Glutamate

PRPP $\longrightarrow$ 5-phosphoribosylamine + PP$_i$

The next step involves the addition of the entire **glycine** molecule. The glycine atoms become C-4, C-5 and N-7 of the purine ring. This step requires the utilization of one ATP.

C-8 is provided by the formyl group of N^5,N^{10} **-methenyltetrahydrofolate**.

N-3 is added from the amide group of **glutamine**. This step utilizes an ATP.

The next step involves the cyclization of the molecule to form the five-membered **imidazole** ring.

C-6 is added by carboxylation with CO_2. **Biotin** is a cofactor in this process.

N-1 is provided from **aspartate**. Consumption of a molecule of ATP is required.

C-2 is donated by N^{10} **-formyltetrahydrofolic acid**.

The origins of the atoms in the purine ring are summarized in Figure 11-4. The amino acids which contribute C and N atoms are **glycine, aspartic acid** and **glutamine**. Glycine supplies C-4, C-5 and N-7. Aspartate provides N-1. Glutamine supplies N-3 and N-9 from its amide group.

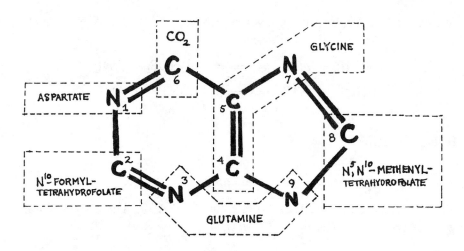

Figure 11-4. Origin of the Atoms in Purines.

Then the six-membered ring is closed to form the first complete purine, IMP (inosinate), or hypoxanthine ribose phosphate (Figure 11-5).

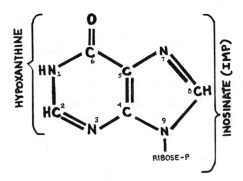

Figure 11-5. IMP (Inosinate).

IMP is then sequentially converted to AMP or GMP (Figure 11-6). These are further phosphorylated to ATP and GTP.

C. Control of *De Novo* Purine Nucleotide Biosynthesis

There are three sites at which feedback inhibition regulates *de novo* purine nucleotide biosynthesis (Figure 11-6):

(1) Excess AMP, GMP, and IMP inhibit the conversion of PRPP to phosphoribosylamine. These endproducts synergistically inhibit the amidotransferase, the first enzyme specific for purine biosynthesis.

(2) AMP inhibits the reaction converting IMP to **adenylosuccinate**, thereby preventing further AMP synthesis. This inhibition does not affect GMP formation.

(3) GMP inhibits the conversion of IMP to **xanthylate** (XMP), preventing its own formation without affecting AMP synthesis.

In addition, there is a cross-regulation between the two branches of purine synthesis: ATP is required to convert xanthylate to GMP, and GTP is needed in the formation of adenylosuccinate from IMP.

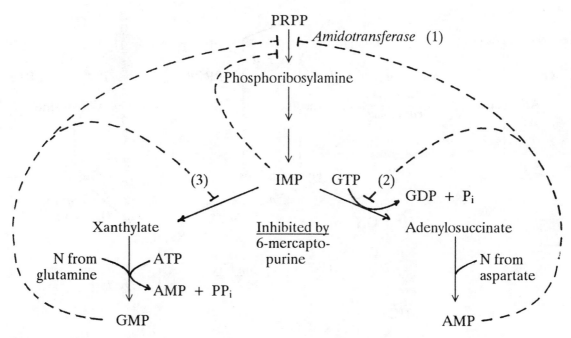

Figure 11-6. Synthesis of Adenylate and Guanylate, and Regulation by Nucleotides.

D. Catabolism of Purine Nucleotides and Their Regeneration by the Salvage Pathway

During digestion, nucleic acids are cleaved to purine and pyrimidine nucleotides by the action of pancreatic ribonucleases and deoxyribonucleases. These nucleotides are degraded to their corresponding nucleosides and then to the free bases. With respect to the purines, the bases may be <u>catabolized</u> to urate (in man, Figure 11-7) or <u>salvaged</u> to regenerate the parent nucleotides.

Catabolism. The phosphate of nucleotides AMP, IMP or GMP is hydrolytically removed by *purine 5'-nucleotidase* (1) to yield the nucleosides adenosine, inosine and guanosine, respectively. These nucleosides are catabolized to the free bases adenine, hypoxanthine and guanine by *purine nucleoside phosphorylase* (2), liberating ribose-1-phosphate. Instead of being converted to adenine, adenosine may be deaminated to inosine by *adenosine deaminase*.

If hypoxanthine is not processed through the salvage pathway and recycled, it is oxidized to **xanthine** and then to **uric acid**; these two final steps are both catalyzed by *xanthine oxidase*. Guanine may be deaminated to xanthine by *guanine deaminase*, then converted to uric acid. **Uric acid** is the final principal end-product of purine catabolism in man and is excreted in the urine (Figure 11-7).

Salvage. One salvage enzyme, *adenine phosphoribosyl transferase (APRT)* (3), catalyzes the transfer of ribosyl phosphate from PRPP to adenine to resynthesize AMP. A second salvage enzyme, *hypoxanthine-guanine phosphoribosyl transferase (HGPRT)* (4), mediates the transfer of ribose phosphate to hypoxanthine or guanine to regenerate IMP or GMP, respectively:

$$\text{Adenine} + \text{PRPP} \xrightarrow{\quad 3 \quad} \text{AMP} + \text{PP}_i$$

$$\text{Hypoxanthine} + \text{PRPP} \xrightarrow{\quad 4 \quad} \text{IMP} + \text{PP}_i$$

$$\text{Guanine} + \text{PRPP} \xrightarrow{\quad 4 \quad} \text{GMP} + \text{PP}_i$$

3 = *Adenine Phosphoribosyl Transferase (APRT)*

4 = *Hypoxanthine-Guanine Phosphoribosyl Transferase (HGPRT)*

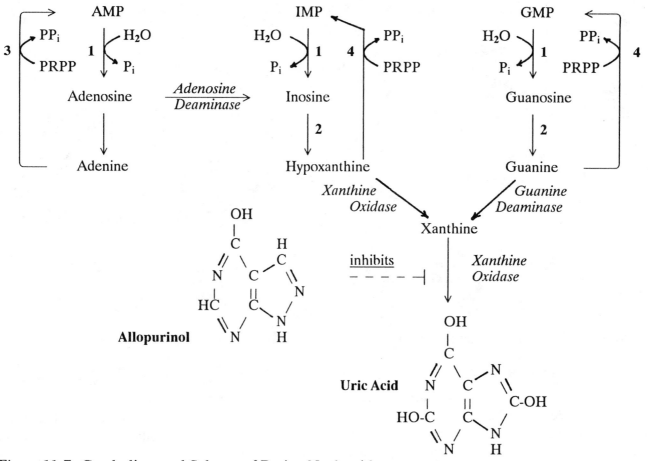

Figure 11-7. Catabolism and Salvage of Purine Nucleotides.
Numbers in boldface identify enzymes described in text.

E. Disorders of Purine Metabolism

Gout is the result of the excessive production of uric acid and its subsequent deposition, as poorly soluble sodium urate crystals, in the joints and kidneys. The overproduction of urate results from the catabolism of abnormally high levels of purines synthesized via the *de novo* pathway. The high concentration of purines may be due to an increase in PRPP either because of: (a) decreased utilization of PRPP by the salvage pathway caused by a partial deficiency of *HGPRT* activity, (b) enhanced production of PRPP because of overactivity of *phosphoribosyl pyrophosphate synthetase*, or (c) enhanced production of **phosphoribosylamine (PR-amine)** due to overactivity of *amidophosphoribosyl transferase* which is the rate-limiting enzyme in purine synthesis.

Gout is treated with **allopurinol**, an analog of hypoxanthine (Figure 11-7). Allopurinol inhibits *xanthine oxidase* and, therefore, prevents the conversion of hypoxanthine to xanthine and then xanthine to urate. This results in a reduction of serum urate concentration and increases the concentrations of the more soluble hypoxanthine and xanthine. In addition, the remaining active HGPRT uses allopurinol as a substrate, thereby diminishing the supply of PRPP available for *de novo* purine synthesis and further reducing production of urate.

The *Lesch-Nyhan Syndrome* results from a nearly complete absence of *HGPRT* because of a genetic defect. Consequently, there is an absence or marked decrease of hypoxanthine and guanine salvage, an elevated PRPP concentration and an increased purine nucleotide synthesis by the *de novo* pathway. The ensuing catabolism of the over-produced purines leads to increased urate production. Victims of this disease show <u>gout</u>, <u>mental retardation</u>, <u>self-mutilation</u>, <u>hostility</u> and a <u>lack of muscular coordination</u>. Allopurinol is not effective in alleviating the symptoms of this disease. The syndrome is inherited as a <u>sex-linked recessive</u> trait in males.

Immunodeficiency Diseases

A deficiency of *adenosine deaminase* is associated with a severe combined deficiency of both T-cells and B-cells and deoxyadenosinuria (see below, page 201).

A *purine nucleoside phosphorylase* deficiency leads to a severe T-cell deficiency but normal B-cell function. Additional characteristics of this disorder include inosinuria, guanosinuria and hypouricemia. Both of these disorders are inherited as <u>autosomal recessive</u> traits.

IV. PYRIMIDINE METABOLISM: BIOSYNTHESIS

Whereas the growing purine ring is assembled on ribose-P, the pyrimidine ring is made first and then joined to ribose-P to form a pyrimidine nucleotide. (The steps below are identified in Figure 11-8).

Step 1: The regulatory step in the synthesis of pyrimidines in mammals is the formation of **carbamoyl phosphate** by cytoplasmic *carbamoyl phosphate synthetase II* (*CPS II*). This enzyme uses the γ-amide nitrogen of glutamine as a nitrogen source and differs in this respect from the mitochondrial *carbamoyl phosphate synthetase I* of the Urea Cycle which uses ammonia as the nitrogen source (see Chapter 5). In mammals, *CPS II* is inhibited by a product of the pathway, UTP, and activated by ATP and PRPP.

Step 2: The committed step in the synthesis of the pyrimidine ring in bacteria is the formation of **N-carbamoylaspartate** from carbamoyl phosphate and aspartate. The carbamoyl phosphate atoms become C-2 and N-3. Aspartate donates C-4, C-5, C-6 and N-1. The carbamoylation of aspartate is catalyzed by *aspartate transcarbamoylase (ATC-ase)*. In bacteria, ATC-ase is the rate limiting enzyme; it is feedback-inhibited allosterically by CTP, the end-product of the pathway.

Steps 3 and 4: Ring closure and oxidation yields the pyrimidine, **orotic acid**.

Step 5: Ribose-P, derived from PRPP, is attached to orotate to yield the nucleotide, **orotidylate (orotidine-5′-P)**.

Step 6: Orotidylate is decarboxylated to form **UMP**. All other pyrimidine nucleotides are synthesized from UMP.

Steps 7 and 8: UMP is phosphorylated sequentially to **UDP** and **UTP**.

Step 9: UTP is aminated by glutamine to form **CTP**.

V. DEOXYRIBONUCLEOTIDES ARE FORMED BY REDUCTION OF RIBONUCLEOSIDE DIPHOSPHATES BY RIBONUCLEOSIDE DIPHOSPHATE REDUCTASE (RIBONUCLEOTIDE REDUCTASE).

A. Ribonucleotide Reductase

Deoxyribonucleotides are formed from ribonucleotides. The ribose of the ribonucleoside diphosphate is reduced at the 2′ carbon atom to form the 2′-deoxyribonucleoside diphosphate.

$$NDP \longrightarrow dNDP$$

$$\text{For example, UDP} \longrightarrow dUDP$$

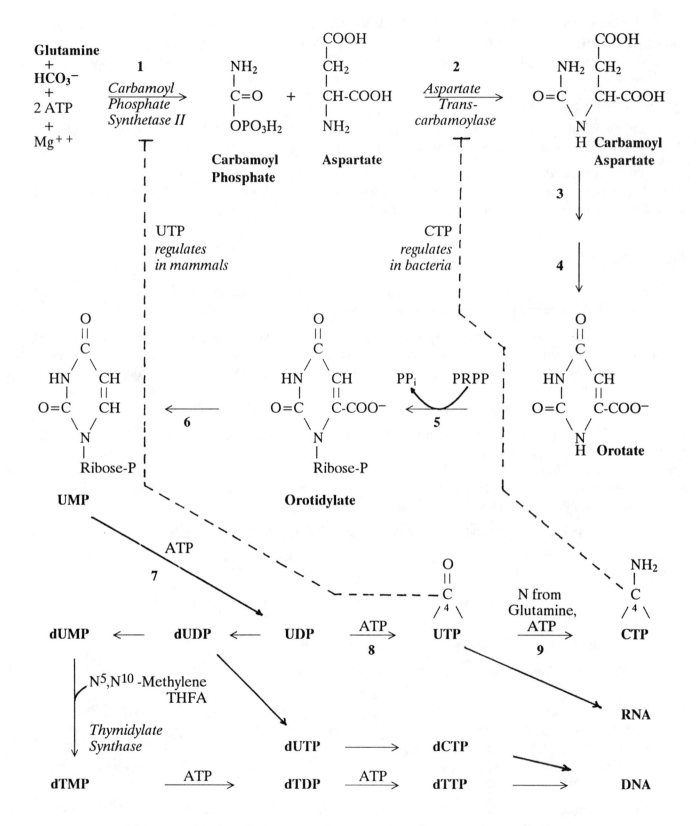

Figure 11-8. Biosynthesis of Pyrimidine Nucleotides, and Sites of Feedback Control. Numbers in boldface refer to steps described in text.

B. Regulation of Ribonucleotide Reductase Activity and Specificity

Both the activity and substrate specificity of ribonucleotide reductase are regulated in a complex fashion at two allosteric sites (sites A and S) which are discrete from the substrate-binding catalytic site (site C).

♦ *Activity Site (site A):* This regulatory site is an on/off switch for overall enzyme activity.

Enzyme on: binding of ATP to site A activates the enzyme to catalyze the reduction of the four rNDP's to their corresponding dNDP's. The resulting increase in the supply of dNTP's promotes DNA synthesis.

Enzyme off: binding of *deoxy*ATP inactivates the enzyme and prevents the reduction of rNDP's. The resulting decrease in the supply of the essential dNTP's interferes with DNA synthesis. The accumulation of high concentrations of the toxic dATP in T- and B-lymphocytes of patients deficient in *adenosine deaminase* explains their Severe Combined Immunodeficiency Disease (SCID).

♦ *Substrate Specificity Site (site S):* Allosteric site S regulates the substrate specificity of the catalytic site (site C).

Either ATP, dTTP, dGTP, or dATP can bind to site S. Which of these nucleotides occupies site S determines which rNDP is used as a substrate at catalytic site C. In this way, the right balance of dATP : dGTP : dCTP : dTTP required for DNA synthesis is produced.

DNA differs from RNA in that it contains **thymine** instead of uracil. Thymine is uracil methylated at the C-5 position. **Deoxythymidylate (dTMP)** is produced by methylation of deoxyuridylate (dUMP) by thymidylate synthase with N^5,N^{10}-methylene tetrahydrofolate serving as methyl donor (Figure 11-8).

VI. SOME ANTICANCER DRUGS ACT BY BLOCKING DEOXYTHYMIDYLATE SYNTHESIS

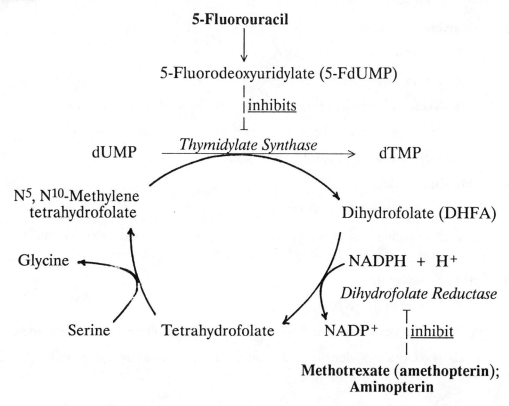

Figure 11-9. Synthesis of Deoxythymidylate, Showing Inhibition by Some Anti-Cancer Drugs.

Several drugs are used to prevent the conversion of dUMP to dTMP which is required for DNA synthesis. 5-Fluorouracil (5-FU) is converted to 5-Fluorodeoxyuridylate (5-FdUMP) by the salvage pathways in the cell. 5-FdUMP irreversibly inhibits (suicide inhibitor) *thymidylate synthase* (Figure 11-9) and is useful in treating solid tumors. The folic acid antagonists, **aminopterin** and **amethopterin (methotrexate)**, inhibit dihydrofolate reductase, preventing the regeneration of tetrahydrofolate and inhibiting dTMP synthesis. Methotrexate is used to treat leukemia and choriocarcinoma.

VII. FUNCTIONS OF NUCLEOTIDES

The nucleoside diphosphates and triphosphates are the "active" forms of the nucleotides. Some functions of the nucleotides are:

A. Adenine nucleotides

1. Immediate energy source for most enzymatic reactions requiring energy expenditure. The form in which most biological energy is stored.

2. Precursor to RNA and DNA.

3. Precursor to cAMP.

4. Precursor to important coenzymes like NAD^+, FAD and CoA.

5. Regulator of many enzymes by phosphorylation of serine or threonine or tyrosine residues. In some instances, an adenosyl group is transferred to the enzyme.

6. Regulator of many enzymes by allosteric mechanisms; effector may be in the form of ATP, ADP or AMP.

B. Guanosine nucleotides

1. Precursor to RNA and DNA.

2. Precursor to cGMP.

3. Nucleotide carrier for mannose and fucose in glycoprotein biosynthesis.

4. Involved in several key steps in peptide bond formation in protein biosynthesis.

5. Key nucleotide in biochemistry of vision.

C. Uridine nucleotides

1. Precursor to RNA and DNA.

2. Nucleotide carrier for glucose, galactose, N-acetylglucosamine and glucuronic acids in polysaccharide, glycoprotein and glycosaminoglycan biosyntheses.

D. Cytidine nucleotides

1. Precursor to RNA and DNA

2. Nucleotide carrier for diacylglycerol in phosphoglyceride synthesis and choline salvage.

3. Nucleotide carrier for neuraminic acid (sialic acids) in glycoprotein synthesis.

VIII. REVIEW QUESTIONS ON PURINES AND PYRIMIDINES

DIRECTIONS: For each of the following multiple-choice questions (1 - 14), choose the ONE BEST answer.

---

1. The four nitrogen atoms of the purine ring are derived from:

A. aspartate, glutamine, and glycine
B. glutamine, ammonia, and aspartate
C. glycine and aspartate
D. ammonia, glycine, and glutamate
E. urea and ammonia

Ans A: Aspartate supplies N-1, glycine provides N-7, and glutamine donates its amide nitrogen for N-3 and N-9.

2. Thymidylate synthase catalyzes the conversion of:

A. dUDP → dUMP
B. dUMP → dTMP
C. dTMP → dTDP
D. dTDP → dTTP
E. dCTP → dUTP

Ans B: Thymidylate synthase catalyzes the methylation of dUMP at the C-5 position to yield dTMP. N^5,N^{10}-Methylene tetrahydrofolate serves as methyl donor.

3. By which of the following means are bases linked to the pentoses in RNA and DNA molecules?

A. They are linked together by N-glycosidic bonds in the beta-configuration
B. They are linked together by $3',5'$ phosphodiester bridges
C. They are linked together by $2',5'$ phosphodiester bridges
D. They are linked together by alpha-glycosidic bonds
E. They are held together by electrostatic attraction

Ans A: There is a beta-glycosidic linkage from the N-9 of the purine or the N-1 of the pyrimidine to the $C-1'$ of the pentoses.

4. Thymine is:

A. 2,4-dioxy-pyrimidine
B. 6-aminopurine
C. 2-amino-6-oxy-purine
D. 2,4-dioxy-5-methyl-pyrimidine (5-methyl uracil)
E. 2-oxy-4-amino-pyrimidine

Ans D: Thymine is uracil (2,4-dioxypyrimidine) methylated at C-5.

5. Each of the following compounds is used in the biosynthesis of **BOTH** purine and pyrimidine nucleotides EXCEPT:

A. glutamine
B. aspartic acid
C. 5-phosphoribosyl-1-pyrophosphate
D. carbamoyl phosphate
E. tetrahydrofolic acid derivative

Ans D: Carbamoyl phosphate is formed by the first reaction of pyrimidine biosynthesis, catalyzed by carbamoyl phosphate synthetase II. Carbamoyl phosphate is not involved in the synthesis of purine nucleotides.

6. Ribonucleotide reductase:

A. catalyzes the formation of a 3',5'-phosphodiester bridge between successive pentose molecules
B. catalyzes the formation of a beta-glycosidic linkage between a base and a pentose
C. catalyzes the reduction of the 2C'-OH of rNDP's to yield the corresponding dNDP
D. is inhibited by ATP
E. catalyzes the reduction of dihydrofolate to tetrahydrofolate

Ans C: Ribonucleotide reductase catalyzes the reduction of the 2C'-OH of the rNDP to the 2'-deoxy derivative. The 2'-OH is replaced by a hydride (H:⁻) ion.

7. Lesch-Nyhan syndrome is associated with a severe deficiency of an enzyme of purine metabolism. The deficient enzyme is:

A. adenosine deaminase
B. adenine phosphoribosyl transferase
C. purine 5'-nucleotidase
D. hypoxanthine-guanine phosphoribosyl transferase (HGPRT)
E. purine nucleoside phosphorylase

Ans D: Lesch-Nyhan syndrome is a sex-linked recessive disorder caused by a severe deficiency of the salvage enzyme, HGPRT.

8. The regulated step in pyrimidine biosynthesis in *E. coli* involves which of the following substrates?

A. IMP and aspartic acid
B. orotic acid and 5-phosphoribosyl-1-pyrophosphate (PRPP)
C. aspartic acid and ribose-5-phosphate
D. aspartic acid and carbamoyl phosphate
E. PRPP and phosphoribosylamine

Ans D: In prokaryotes, aspartate transcarbamoylase is the rate-limiting step. The allosteric enzyme catalyzes the conversion of carbamoyl phosphate and aspartate to carbamoyl aspartate and is inhibited by CTP.

9. An inherited immunodeficiency disease in which there is a severe T-cell defect, but normal B-cell function, is associated with a deficiency of an enzyme of purine catabolism. The deficient enzyme is:

A. aspartic transcarbamoylase
B. xanthine oxidase
C. hypoxanthine-guanine phosphoribosyl transferase
D. purine nucleoside phosphorylase
E. purine 5'-nucleotidase

Ans D: A deficiency of purine nucleoside phosphorylase, a catabolic enzyme which deribosylates nucleoside monophosphates, underlies the loss of T-cell function with retention of B-cell function. This is an autosomal recessive immunodeficiency disease.

10. The committed step in purine synthesis is the step in which:

A. phosphoribosylamine is synthesized from PRPP and glutamic acid
B. glutamine is incorporated intact into the molecule on the sugar phosphate unit
C. N-formylglycinamide ribonucleotide is formed from glycinamide ribonucleotide with a methyl group transferred from N⁵,N¹⁰-methylene THFA
D. phosphoribosylamine is formed enzymatically from PRPP and glutamine
E. bicarbonate is utilized to carboxylate aminoimidazole ribo nucleotide

Ans D: The enzyme, glutamine-phosphoribosyl pyrophosphate amidotransferase, catalyzes replacement of the pyrophosphate at C-1 of PRPP with the amide of glutamine to form phosphoribosyl amine, in the committed step of *de novo* purine nucleotide biosynthesis. The amidotransferase is inhibited by AMP, GMP and IMP.

11. AMP is synthesized in a two step reaction sequence involving

A. inosinic acid, NAD^+ and glutamine
B. inosinic acid, ATP and glutamine
C. inosinic acid, GTP and aspartate
D. hypoxanthine and ribose-1-phosphate
E. none of these

Ans C: IMP, the parent purine, is first condensed with aspartate and converted to adenylosuccinate in a reaction which requires GTP. In the second step, adenylosuccinate is converted to AMP.

12. The oxidation of xanthine to urate is catalyzed by:

A. aspartic transcarbamoylase
B. xanthine oxidase
C. hypoxanthine-guanine phosphoribosyl transferase
D. purine nucleoside phosphorylase
E. purine 5′-nucleotidase

Ans B: Xanthine oxidase catalyzes two reactions: the oxidation of hypoxanthine to xanthine and the subsequent oxidation of xanthine to urate. Urate, the final product of purine degradation in humans, is excreted.

13. In the mammalian purine catabolic pathway, adenosine is:

A. oxidized directly to xanthine
B. converted to guanine and then deaminated to xanthine
C. deaminated to inosine
D. phosphorylated to give IMP
E. oxidized directly to urate

Ans C: In mammalian degradation of purine nucleotides, AMP is deaminated to inosine by adenosine deaminase.

14. Important controls of purine nucleotide *de novo* synthesis include:

A. availability of PRPP
B. inhibition by GMP, AMP, and IMP of glutamine-PRPP amido-transferase
C. requirement of GTP for AMP synthesis
D. requirement of ATP for GMP synthesis
E. all of the above.

Ans E: Conversion of PRPP to PR-amine by amidotransferase is the committed and rate-limiting step in *de novo* purine synthesis. The enzyme is inhibited by AMP, GMP and IMP. The reaction rate is also controlled by availability of PRPP, one of the substrates. In addition, conversion of IMP to AMP requires GTP as energy source; synthesis of GMP from IMP requires ATP. IMP is, then, diverted toward synthesis of the purine nucleotide (AMP or GMP) present in lower concentration.

DIRECTIONS: For each set of questions, choose the ONE BEST answer from the lettered list above it. An answer may be used one or more times, or not at all.

Questions 15 - 17:

    A. Severe combined (T-cell and B-cell) immunodeficiency disease
    B. Lesch-Nyhan syndrome
    C. T-cell immunodeficiency disease
    D. Beta-thalassemia
    E. I-disease

15. Hypoxanthine-guanine phosphoribosyl transferase (HGPRT) deficiency

Ans B: Deficiency of HGPRT impairs the utilization of PRPP to salvage hypoxanthine and guanine and to regenerate the corresponding nucleotides. Accumulation of PRPP leads to *de novo* overproduction and subsequent increased catabolism of excess purines resulting in increased urate formation.

16. Purine nucleoside phosphorylase deficiency

Ans C: Deficiency of purine nucleoside phosphorylase leads to impaired T-cell function but does not affect B-cell function.

17. Adenosine deaminase (ADA) deficiency

Ans A: Inability to deaminate deoxyadenosine to deoxyinosine results in the accumulation of dATP which inhibits ribonucleotide reductase. The resulting deficiency of the other dNTP's interferes with DNA synthesis and impairs lymphocyte development and function.

12. NUCLEIC ACIDS: STRUCTURE AND SYNTHESIS

Jay Hanas

I. DNA

A. Structure

Deoxyribonucleic acid (DNA) and ribonucleic acid (RNA) are responsible for the transmission of genetic information and are intimately involved in cellular metabolism, growth, and differentiation. The central dogma of Molecular Biology states that genetic information is transferred from DNA to RNA to protein.

$$DNA \longrightarrow RNA \longrightarrow Protein$$

The understanding of the structure, synthesis, and function of DNA and RNA is of fundamental importance in biology and medicine. DNA is a polymer of deoxynucleoside monophosphates (deoxynucleotides). The 5′ phosphate group of one deoxynucleotide is joined to the 3′ OH group of another, forming a phosphodiester linkage. A polymer of nucleotides (polynucleotide) results when many nucleotides are joined in linear fashion. In deoxyribonucleic acid (DNA, lacking a 2′ OH group on the nucleotide sugar), the nucleotide bases (adenine, guanine, cytosine, and thymine) of one strand form hydrogen bonds with the nucleotide bases of the other strand (Figure 12 - 1).

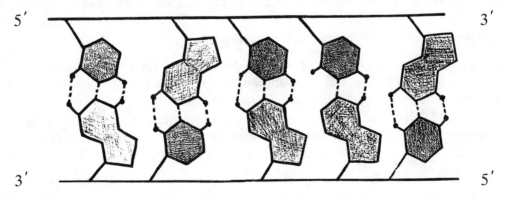

Figure 12 - 1. Pairing of Bases in Double-stranded DNA.

There are precise rules governing which bases form hydrogen bonds with one another: adenine forms two hydrogen bonds with thymine; guanine forms three hydrogen bonds with cytosine. When DNA forms a double stranded structure, the polynucleotides are anti-parallel, i.e. one strand is in the 5′- 3′ orientation and the other strand is in the opposite 3′- 5′ orientation. DNA can also exist in a single-stranded form.

DNA strands can contain millions of nucleotides joined to each other by phosphodiester linkages. Genetic information is stored as unique sequences of the four nucleotide bases. The **genome** of an organism is all of the DNA content expressed as base pairs. The human genome is much larger than those of bacteria and unicellular organisms (by several orders of magnitude) but smaller than that of some plants and amphibians. The significance of the differences in genome sizes is not known, but in general more complex organisms have larger genomes. Usually, more complex organisms also have large amounts of repetitive DNA which does not appear to code for functional proteins. The function of this repetitive DNA is unknown.

Double-stranded DNA is the repository of genetic information and in animal cells is located in the nucleus and mitochondria. The DNA in the nucleus is very large in size (millions of base pairs) whereas the DNA in animal mitochondria is small by comparison and exists in a circular form (the ends are covalently joined). Mitochondrial DNA codes for some proteins involved in electron transport processes and for other molecules necessary for the functions of these organelles.

Double-stranded DNA forms a double helix structure in which the two strands are intertwined as two right-handed helices (the helices in DNA turn clockwise as one looks up the center of the molecule). One complete turn of the double helix requires 10.4 base pairs in the "B" form of DNA. The two strands of a duplex DNA molecule unwind (breaking of hydrogen bonds) in a process called **denaturation**. Denaturation can be brought about by high temperature, alkali, or specific enzymes. The T_m (melting point) of DNA is the temperature at which 50% of the bases become unpaired and is proportional to the GC content of the molecule. The individual strands can re-anneal in a process called **renaturation** if the DNA is cooled, or the pH brought to neutrality.

The ability of complementary DNA sequences to re-anneal or **hybridize** is the basis for identifying specific DNA sequences of clinical interest. For example, the presence of a particular virus in blood can be detected with a radiolabeled, single-stranded DNA **probe** which hybridizes only to the viral genome and not to lymphocyte DNA.

The individual nucleotide bases in double-stranded nucleic acids absorb less UV light than in single-stranded or free forms, a phenomenon called **hypochromicity**. When DNA is denatured, UV absorbance of the sample increases (hyperchromicity). Single-stranded nucleic acids can form intrastrand base-pairs giving partial double stranded character (hairpin structures).

The genomes of bacteria, some viruses, mitochondria and plasmid DNA are covalently closed, circular DNA molecules. Circular, double-stranded DNA is topologically restrained (as is DNA in which both ends are anchored by some means) and in many cases is **supercoiled**.

Supercoiling is governed by enzymes termed **topoisomerases**. There are basically two types of DNA topoisomerases, type I or type II. Type II enzymes require ATP as an energy source whereas type I enzymes require no exogenous energy source. Topo I in bacteria relaxes negatively supercoiled DNA whereas topo II in bacteria (also called DNA gyrase) promotes negative supercoiling. *In vivo*, the circular genome of bacteria and plasmids exists in a negative supercoiled state. Negative supercoiling makes DNA melting during replication or transcription energetically favorable. DNA gyrase is inhibited by **nalidixic acid** and **novobiocin**.

Eukaryotic organisms contain type I and II topoisomerases but not a DNA gyrase, the enzyme which promotes negative supercoiling. Negative supercoiling in eukaryotic organisms can be generated through the formation and removal of nucleosomes in which the DNA is wrapped around a nucleosome about one and three-quarters times.

B. DNA Replication

DNA replication results in the exact duplication of the double-stranded DNA molecule. DNA replication takes place by a semiconservative pathway in which the two strands of duplex DNA are separated and each single strand serves as a template for the synthesis of the second, complementary strand (Figure 12 - 2).

DNA replication in bacteria is better understood than in mammalian cells. The bacterium *E. coli* contains three distinct DNA polymerases (the enzyme which catalyzes the polymerization of deoxynucleotides on the template strand), called DNA pol I, II, and III. Mammalian cells also contain three DNA polymerases, two nuclear (one for repair) and a distinct mitochondrial enzyme. DNA polymerases require a free 3′ OH group in order to react with an incoming deoxynucleotide 5′ triphosphate. Thus, the template strand needs a hybridized primer to initiate DNA synthesis. The primer is usually a small RNA strand (RNA polymerase does not need a primer to begin synthesis). DNA pol I in bacteria is the most abundant DNA polymerase and is used for repair synthesis of DNA.

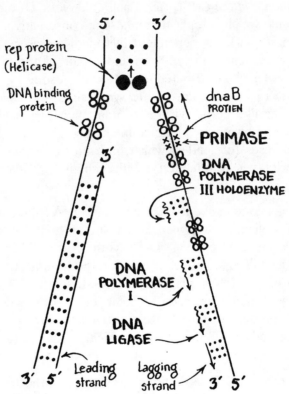

Figure 12 - 2. DNA Replication in Prokaryotes.

DNA synthesis is continuous on one strand (**leading strand**) and discontinuous on other (**lagging strand**). This means one strand of the parental DNA duplex is copied at the replication fork by DNA polymerase III in a continuous fashion whereas the other strand needs a "back-stitching" process to synthesize the other strand. Primers are of continual necessity for this back-stitching process (Figure 12 - 2). The discontinuous nature of DNA synthesis results in the production of newly synthesized DNA pieces called **Okazaki fragments**. DNA ligase joins the Okazaki fragments once the RNA primer has been degraded and the gaps filled in. DNA polymerase I also can correct errors in synthesis because it has a 3′ to 5′ exonuclease activity which is capable of cleaving off nucleotides which were mis-incorporated.

DNA replication is bi-directional, i.e. two replication forks (where the duplex DNA is denatured into two single strands), initiated at the same point, move in opposite directions on the same duplex strand. In bacteria there is usually only one origin of replication but in mammalian cells, there are thousands.

DNA replication requires the unwinding of the duplex DNA template to yield the single-stranded templates. Energy is needed to unwind or denature duplex DNA. Cells contain enzymes called **helicases** which can unwind the duplex DNA molecule utilizing ATP as an energy source. This unwinding process associated with DNA replication would eventually come to a halt because of topological constraints due to the build up of "knots" ahead of the replication fork. Topoisomerases relax these knots (by nicking and rejoining the DNA) and thus permit the replication fork to proceed.

A number of medically important animal viruses, e.g. HTLV (human T-cell lymphotropic virus) utilize an enzyme called **reverse transcriptase** to convert their single-stranded RNA genome into a duplex DNA genome which integrates into nuclear DNA. This is another example of DNA replication although in this case the template molecule is a single-stranded RNA molecule. HTLV viruses cause some forms of human leukemia. HIV (human immunodeficiency virus, formerly known as HTLV-III) is the causative agent in AIDS (acquired immune deficiency syndrome).

Eukaryotic DNA Replication and the Cell Cycle. Each eukaryotic chromosome is comprised of a linear, double-stranded DNA molecule which is replicated in a semi-conservative, bidirectional manner. Like bacteria, eukaryotic cells have multiple DNA-dependent DNA polymerases. At least two distinct DNA polymerases (alpha and delta) are involved in synthesis and repair of the chromosomal DNA found in the nucleus (six billion base pairs of DNA distributed over 46 chromosomes is the diploid human **genome**) and one distinct DNA polymerase (gamma) is devoted to synthesis of the small, circular DNA found in mitochondria. DNA animal viruses like herpes viruses contain their own DNA-dependent DNA polymerase which can be selectively inhibited by **acyclovir**, a guanine nucleotide analog which causes chain termination. Replication of animal retroviruses (e.g., HIV) which have an RNA genome and RNA-dependent DNA polymerase (reverse transcriptase) is inhibited by the chain terminators azidothymidine (**AZT**) and dideoxynucleosides.

Because mammalian nuclei contain about 1000 times the DNA found in bacteria, each chromosome has a large number of regions of replication (**replicons**, each with an origin of DNA synthesis) whereas the bacterial chromosome usually has just one replicon. DNA synthesis is initiated from a RNA primer synthesized by a eukaryotic primase. Okazaki fragments are formed on the trailing strand and the RNA primer is latter degraded. Unlike bacteria which have circular chromosomes, eukaryotic chromosomes contain DNA ends (**telomeres**). **Telomerases** are unique enzymes found in eukaryotes which synthesize and maintain the telomeric ends. Telomerase contains an integral RNA subunit similar to **ribozymes** which are enzymes that require an RNA molecule for activity. Telomerases are highly active in tumor cells and relatively absent in senescent cells.

Unlike bacteria, normal eukaryotic cells do not continuously synthesize DNA and divide but rather possess a **cell cycle** which is temporally subdivided into gaps and phases. The cell cycle begins with **G1** (gap) which is preparatory for the **S phase** in which DNA synthesis occurs resulting in a nucleus with a **4N** (tetraploid) complement of DNA. The transition from G1 to S is initiated by the phosphorylation activity of S-phase promoting factor (SPF) which is composed of specific **cyclin** proteins and a cyclin-dependent protein kinase (CdK). The tumor suppressor p53 inhibits SPF and DNA synthesis by promoting the synthesis of a protein that inhibits CdK. Histone proteins are synthesized at a high rate during S phase and are used in nucleosome formation on the newly synthesize DNA. Upon completion of DNA synthesis, cell enters **G2** which is preparatory for the **M** (mitotic) phase. Mitosis is characterized by chromosome condensation and pairing, dissolution of the nuclear membrane, and actual cell division (cytokinesis). Entry from G2 into the M phase is initiated by the phosphorylation activity of the M-phase promoting factor (MPF) which is composed of specific cyclins and cell division cycle (cdc) kinase. Phosphorylation of histones and nuclear lamin proteins by cdc kinase causes chromosomal condensation and nuclear membrane dissolution respectively.

C. Deoxyribonucleoproteins

In most cases, DNA in animal cells does not exist free but occurs tightly bound to specific proteins. This is because the functional units are actually protein-nucleic acid complexes. In addition, the protein components protect the nucleic acids from degradation by nucleases.

The proteins are basic proteins called **histones** which contain large amounts of **lysine** and **arginine**. There are five histones, designated H1, H2A, H2B, H3, and H4. Histones are highly conserved proteins with H3 and H4 being the most highly conserved of all proteins. They interact with each other in a very ordered way first to form dimers of H2A, H2B, H3, and H4, which then combine with each other to produce an **octomer**. About 145 base pairs of duplex DNA are wrapped around an octomer (slightly less than two turns) to form a structure called a **nucleosome** (Figure 12 - 3). H1 is located where DNA enters and leaves the nucleosome. About 25 - 100 bp (base pairs) of spacer or linker DNA span the region between adjacent nucleosomes. A single DNA molecule can wrap itself around many nucleosomes to form a **polynucleosome**.

Chromatin and Chromosomes. In animal cells, polynucleosomes can be packaged along with many non-histone proteins into a nucleoprotein complex called **chromatin**. If chromatin is washed extensively with concentrated salt solutions, all the proteins come off except the core histones, H2A, H2B, H3, H4. In the electron microscope the remaining structure looks like "beads" of nucleosomes on a "string" of DNA. Each "bead on a string" has a diameter of about 10 nm. When chromatin is less extensively washed with salt, the observed structure has a diameter of about 30 nm and appears to be composed of a helical array of nucleosomes (Figure 12 - 3).

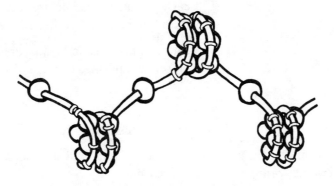

Figure 12 - 3. Interactions between DNA and Histones to Form Nucleosomes.

Chromosomes are highly condensed structures consisting of the 30-nm chromatin fibers discussed above. The chromatin exists as **heterochromatin**, which is highly condensed, and **euchromatin**, which is not so highly condensed. Human somatic cells (non-germ line) contain 22 pairs of non-identical chromosomes and 2 sex chromosomes (XX, female, and XY, male). Each eukaryotic chromosome contains one very large double-stranded DNA molecule consisting of millions of base pairs. During mitosis, all chromosomes are highly condensed and are visible in the light microscope. The ends of chromosomes are referred to as **telomeres**. The more central regions of mitotic chromosomes which become attached to spindle fibers during mitosis are called **centromeres** and contain large amounts of repetitive DNA.

Many pathological conditions result in chromosome alterations that are easily seen in mitotic chromosomes with the light microscope. These include chromosomal rearrangements, deletions, and amplifications as observed in tumors, as well as abnormal chromosome numbers and structure (trisomy of chromosome 21 in Down's syndrome and X chromosome alterations in the fragile X syndrome associated with mental retardation).

When DNA in the nucleus of an animal cell is replicated, the two strands are separated and new complementary strands are synthesized on each separated strand. The nucleosome structure must also be replicated as well. All the "old" nucleosomes remain associated with one of the separated strands and new nucleosomes (composed of newly synthesized histones from the cytoplasm) form on the other strand.

D. DNA Mutations

A gene mutation occurs when the unique sequence of deoxyribonucleotides in a gene is altered in any way. A mutation in the DNA sequence that codes for a protein can alter its amino acid sequence.

The consequences of DNA mutations on cellular metabolism can range from extremely harmful to beneficial. Mutations can occur spontaneously due enzymological errors in DNA replication, recombination, and cell division. DNA mutations can also result from environmental factors like ionizing radiation, ultraviolet light from the sun and other sources, chemical mutagens, and viruses and viral genes.

Mutagen	Type
5-bromouracil	causes transitions, replaces thymine in DNA and, after isomerization, pairs with G instead of A
aminopurine	causes transitions
hydroxylamine	causes transitions, alters C so it pairs with A
nitrous acid	causes transitions, deaminates cytosine to uracil so that it pairs with A rather than G
sulfur mustards nitrogen mustards ethyl ethanesulfonate methyl methanesulfonate	cause transitions and transversions by alkylating N7 of G, remove purines (apurination)
low pH	causes transitions, tranversions and apurination
intercalating dyes (bind between base pairs): proflavin, 5-aminoacridine, acridine orange, ethidium bromide, UV light	cause insertions and deletions resulting in frameshift mutations
ultraviolet (UV) light	dimerizes adjacent pyrimidines (thymine dimers) leading to deletions
other ionizing radiation (X-rays, radioactivity, cosmic rays)	cause severe damage to DNA by breaking covalent bonds

Table 12 - 1. Mode of Action of Mutagenic Agents

Mutations can be of a number of different kinds. **Point mutations** involve the alteration of a single nucleotide or nucleotide pair which can be of several types. A **transition** occurs when a purine is substituted for another purine or a pyrimidine is substituted for another pyrimidine. A **transversion** occurs when a purine is substituted for a pyrimidine or vice versa. A point **deletion** is the loss of a nucleotide or nucleotide pair and a point **insertion** is the gain of a nucleotide or nucleotide pair. **Gross mutations** involve the deletion or insertion of more than one nucleotide pair, up to very large segments of the genome. A **translocation** is the movement of a chromosomal segment to a non-homologous chromosome. **Inversions** are inverted pieces of DNA within one chromosome. Mutations which lead to defective proteins and metabolic alterations are usually deleterious to the cell. Some mutations can be harmless, leading to an amino acid change in a protein which has no effect on structure and function. Mutations can be beneficial if the mutated gene product confers a selective advantage on the organism. Table 12 - 1 is a partial list of environmental mutagens and the types of mutations they cause.

The mutation rate in organisms caused by ultraviolet and ionizing radiation should be much higher than is actually observed. This relatively low mutation rate is due in large part to the ability of organisms to repair their own DNA. **Thymine dimers** (T-T dimers), which are induced by UV irradiation, can be repaired by photoreactivation, involving the light-activation of specific enzymes (occurs in bacteria). T-T dimers are also removed by excision repair, in which an endonuclease (a nuclease which cleaves the phosphodiester bond between adjacent nucleotides) clips the phosphodiester backbone near the dimer, an exonuclease (a nuclease which cleaves phosphodiester bonds successively from an end) removes the dimer and surrounding nucleotides, and the gap is filled in by the action of DNA polymerase I and DNA ligase (Figure 12 - 4). T-T dimers can also be repaired by recombination events. In *xeroderma pigmentosum*, a human disease, there is a defect in the excision-repair enzymes causing these individuals to be susceptible to skin cancers.

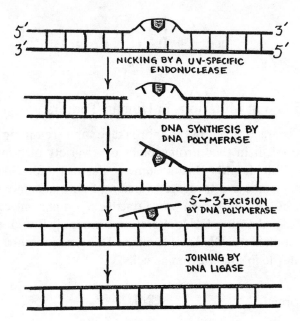

Figure 12 - 4. DNA Repair: Excision of Thymine Dimers

E. DNA Mutations and Cancer

Cancer is characterized by uncontrolled cell division and metastasis (transfer) of previously normal tissue cells. Gene mutations (either inherited, induced by carcinogens, radiation, or by viruses) are responsible for the carcinogenesis process which in most cases is multistep (mutations in many different genes are involved). Mutations in **proto-oncogenes** (genes which code mainly for proteins involved in stimulating normal cell division) and **tumor suppressor genes** (which code mainly for proteins whose normal function is to block cell division) are responsible for many human **carcinomas** (cancers derived from epithelial cells). For example, lung carcinoma can be induced by chemical carcinogens found in cigarette smoke and tar (e.g., benzpyrene) resulting in point mutations in both the *ras* proto-oncogene and the **p53** tumor suppressor gene. The *ras* proto-oncogene codes for a **G-protein** (attached to the inner surface of the plasma membrane by a covalently attached farnesyl group) which, when bound to GTP, promotes cell division via a **signal transduction cascade**. This cascade of protein phosphorylation-dephosphorylation is initiated by **growth factor** binding to a **receptor (tyrosine kinase)** at the cell surface and progresses through the cytoplasm into the nucleus where activation of **transcription factors** and proteins involved in stimulating DNA synthesis and cell division occurs. Carcinogen-induced point mutations in the *ras* proto-oncogene (converting it to an **oncogene**) occur mainly at codons 12 and 61 which are in the GTP binding site of the protein. The resulting amino acid changes inhibit the GTPase activity of the protein resulting in constitutive GTP binding and stimulation of cell division. 50 - 60% of all human carcinomas have point mutations in *ras* and are especially prevalent in lung, colon, and pancreatic tumors. Animal studies have revealed that drugs which block the farnesylation of the *ras* protein inhibit tumor development.

DNA point mutations in the tumor suppressor gene p53 are present in 70-80% of human carcinomas and, unlike the *ras* gene, are found over a wide range in the 393 amino acid sequence (Figure 12 - 5, arrows). The normal function of p53, which is a transcription factor, is to inhibit cell entry into the S (DNA synthesis) phase of the cell cycle and also to promote **apoptosis** (programmed cell death) which tumor cells lack. Point mutations in p53 therefore lead to continual DNA synthesis and cell immortality. Cells contain **mismatch repair** enzymes for correcting point mutation mistakes in double-stranded DNA. In **hereditary non-polyposis colon carcinoma** (HNPCC) and possibly some other tumors, these mismatch repair enzymes are themselves mutated resulting in a "mutator" phenotype.

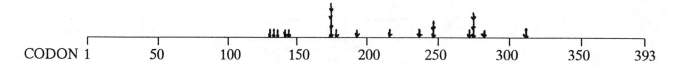

CODON 1   50   100   150   200   250   300   350   393

Figure 12 - 5. Location of p53 Point Mutations in Human Tumors.

DNA translocations are characterized by the breakage and rejoining of double-stranded DNA molecules at different sites and occur in the chromosomes of a variety of human tumors. The reciprocal DNA translocation forming the **Philadelphia chromosome** is diagnostic for **chronic myelogenous leukemia** (CML). The Philadelphia chromosome is a truncated version of chromosome 22 (q⁻) and is formed by the reciprocal translocation of the *abl* tyrosine kinase proto-oncogene on chromosome 9 with the bcr region of chromosome 22 (see Figure 12 - 6). Another DNA alteration commonly found in tumor cells is chromosome **aneuploidy** which is caused by anomalous DNA replication. This leads to chromosome numbers above the normal **diploid** (2N) complement (46 in somatic human cells).

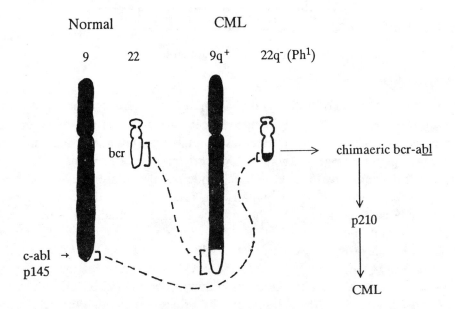

Figure 12 - 6. Philadelphia chromosome results from translocation of the *abl* gene on chromosome 9 to the *bcr* region of chromosome 22.

II. RNA

A. Structure

Ribonucleic acid (RNA) contains ribose in place of deoxyribose as the nucleotide sugar. The pyrimidine is uracil instead of thymine. RNA is usually single-stranded and contains rarely more than a few thousand nucleotides joined in phosphodiester linkages. Compared to DNA, cellular RNA is much more varied in size, structure, and function. The various kinds of RNA in the animal cell include **heterogeneous nuclear RNA** (hnRNA), **messenger RNA** (mRNA), **small nuclear RNA** (snRNA), **ribosomal RNA** (rRNA), **transfer RNA** (tRNA) and other minor RNA species. The majority of the RNA in cells is rRNA (80%), followed by tRNA (15%), mRNA (2%), and other (3%). RNA is readily hydrolyzed in alkaline solutions because of the adjacent $2'$, $3'$ OH groups on the ribose, whereas DNA (containing only the $3'$ OH group) is denatured but not hydrolyzed by alkali. RNA can contain unusual nucleotide bases like pseudouracil, ribothymidine, methyl adenine, and dihydrouracil. DNA modification in animal cells is usually limited to methylation of cytosine (5 position on base) when $5'$ to a guanosine (C^5 $^{methyl}pG$).

RNA is synthesized in the cell nucleus and also in mitochondria. hnRNA, which is unstable, is found in the nucleus and can be the largest of the RNA species (10 - 20,000 bases). It is the precursor of mRNA and is processed rapidly into mRNA in the nucleus. Messenger RNA (mRNA) contains the genetic information of DNA (unique sequence) in the form of the genetic code and is found primarily in the cytoplasm.

B. RNA Synthesis: Bacterial Transcription

RNA synthesis is the process in which a single-stranded RNA is synthesized from ribonucleotide triphosphate precursors by RNA polymerase utilizing a single-stranded DNA template. The template strand of the duplex DNA molecule (bottom strand, $3'$ - $5'$ orientation) is copied into a $5'$ - $3'$ RNA transcript (transcription as well as DNA polymerization is always in the $5'$ - $3'$ direction). *In bacteria there is one RNA polymerase which transcribes all the different genes in the bacteria.* The complete enzyme (called the holoenzyme) contains 5 protein subunits: beta, beta$'$, 2 alphas, and one sigma. RNA transcripts in bacteria are not as extensively modified and processed as are transcripts in eukaryotic cells (capping, polyadenylation, and splicing.

Messenger RNAs in bacteria can be polycistronic, i.e. coding for more than one protein. In many cases, transcription of the bacterial polycistronic mRNA is under control of a promoter and operator region at which coordinated regulatory events occur. This situation is referred to as an operon. The mRNAs in eukaryotic or higher organisms tend to be monocistronic, i.e. code for only one protein. In both bacterial and higher systems, there exist accessory proteins that interact with promoter elements and RNA polymerase and help (or hinder in the case of repressor proteins) the enzyme initiate transcription. These proteins are called transcription factors.

In bacteria, transcription factors include catabolite activator protein (CAP) and the lactose operon repressor. Although the *E. coli* genome contains only one site of initiation of DNA synthesis, there exist hundreds of RNA synthesis initiation sites, each containing specific DNA sequences called **promoters** which direct RNA polymerase to the proper initiation site. In bacteria, this promoter is characterized by a AT-rich region around nucleotide position –10 from the start of transcription (at +1) and also important sequences at nucleotide position –35. Unlike DNA polymerase, RNA polymerase does not require a primer to initiate transcription. The first nucleotide transcribed is a NTP, a purine. Transcription initiation in bacteria is inhibited by the antibiotic **rifampicin**.

Transcription factors can regulate the initiation of RNA synthesis in bacterial cells. For example, the lactose repressor (which is encoded in the lactose operon) is a protein which binds to a DNA sequence called an "operator" which lies between the transcription promoter and the transcription initiation site for the beta-galactosidase gene (and two other downstream genes) which produces an enzyme which degrades lactose. If lactose is not present in the cell, then the repressor binds to the operator and blocks RNA polymerase from initiating transcription. If lactose is present in the cell, then the sugar binds to repressor, alters its conformation so that it no longer binds to the operator. RNA polymerase is now able to transcribe the galactosidase gene. This results in the production of beta galactosidase that degrades lactose, allowing the cell to use the breakdown products for metabolism. Regulation of this type is an example of **negative control**.

However, if glucose is also present, the cell prefers to metabolize glucose and the lactose operon remains turned off even in the presence of lactose. This is accomplished by a phenomenon termed **catabolite repression**. A catabolite of glucose metabolism reduces the amount of cyclic AMP (cAMP) present in the cell. Cyclic AMP is referred to as a second messenger. It is synthesized by an enzyme called adenyl cyclase which condenses the $5'$ phosphate group with the $3'$OH group of AMP. cAMP is degraded by a phosphodiesterase. In the absence of glucose metabolism, cAMP levels are high and the nucleotide binds and activates the catabolite activator protein, CAP. CAP when bound to cAMP is a positive transcription factor for the lactose operon. That is, it binds upstream of the lactose promoter and facilitates binding and transcription by RNA polymerase when lactose is present (so the lactose repressor is not blocking RNA polymerase binding). Therefore if only lactose is present, the lactose operon is turned on transcriptionally but if glucose is also present, the operon will remain turned off because CAP will not activate transcription under these conditions. This is an example of **positive control**.

Transcription termination in bacteria proceeds primarily via two mechanisms. In some genes, the newly synthesized RNA at the end of a gene undergoes intrastrand base-pairing forming a hairpin structure. This hairpin is usually followed by a string of U's. This structure triggers release of the RNA polymerase from the DNA template and release of the RNA chain. Other bacterial genes require a protein factor termed **rho** for efficient transcription termination.

In some amino acid biosynthetic operons a form of transcription termination occurs under certain conditions and is referred to as **attenuation**. In the histidine biosynthetic operon, there exists a leader RNA sequence positioned before the protein coding regions of the operon that contains 7 contiguous codons for histidine. In the absence of histidine, the operon is transcribed by RNA polymerase and ribosomes bind the nascent RNA chain and begin translation in a process called coupled transcription-translation. Because histidine is absent (and also charged histidyl-tRNA), the ribosome stalls at the 7 histidine codons and the leader RNA assumes a folded structure that allows RNA polymerase to continue transcription of the operon. However, in the presence of excess histidine, the ribosome translates through the leader mRNA, altering its structure so as to produce a hairpin recognized as a termination signal by RNA polymerase moving just ahead of the ribosome. The result is the transcription is terminated under conditions where new histidine synthesis is not needed.

C. RNA Synthesis: Mammalian Transcription

RNA synthesis in higher organisms is more complicated than in bacteria because of the large number and variety of eukaryotic genes and because of the requirement for their temporal and spatial transcription during development and cell differentiation. Of the 100,000 unique genes found in vertebrate cells, only about 10% at most are actively transcribed in any cell type. Some genes transcribed in different cell types are the same and other genes are uniquely transcribed in that cell type. This phenomenon of differential gene transcription is responsible in large part for cell differentiation and development. Overall transcription repression of eukaryotic genes appears to be mediated through higher order structures in chromatin and RNA polymerase accessibility.

Eukaryotic cell nuclei contain three distinct DNA-dependent RNA polymerases (I, II, and III); mitochondria contain their own, distinct RNA polymerase. Nuclear RNA polymerases are significantly larger than bacterial enzymes and have 10 or more protein subunits; some of the subunits are shared between the three polymerases. **RNA polymerase I** transcribes the major ribosomal RNA genes which code for the 45S ribosomal RNA precursor. This precursor is processed into the 18S, 28S, and 5.8S RNAs found in the mature ribosome. Synthesis of these RNAs and assembly into the 40S and 60S ribosomal subunits occurs in the **nucleolus** of the eukaryotic nucleus. Tumor cells have large, multiple nucleoli and synthesize ribosomes at a high rate needed to sustain rapid cell division. **RNA polymerase II** transcribes all the messenger RNA-coding genes and **RNA polymerase III** transcribes transfer RNA, 5S ribosomal RNA, and other small RNAs including the 7S RNA found in the **signal recognition particle** (SRP). SRP is involved in translocation of secretory proteins into the endoplasmic reticulum. These three polymerases exhibit different sensitivities to the mushroom-derived toxin, **alpha amanitin**. Transcription by RNA pol I is resistant to this toxin; RNA pol II transcription is very sensitive and RNA pol III transcription is moderately sensitive to this toxin.

RNA polymerases I, II, and III can not bind specifically to transcription promoter sequences and initiate transcription by themselves. If this were the case, cellular chaos would result. But rather, gene transcription in eukaryotes is regulated by protein transcription factors whose role is to form specific complexes (protein-DNA, protein-protein) on transcription promoters, upstream activating sequences, and enhancers. The respective RNA polymerases then bind to these specific protein-DNA complexes and initiate RNA synthesis. Importantly, the cell's repertoire of transcription factors ultimately determines its phenotype. The **core promoter** for mRNA genes transcribed by RNA polymerase II usually contains a **TATA** cis-acting element (thymine and adenine nucleotide stretches) on the non-template strand ($5'$ to $3'$) about 35 nucleotides upstream from the start of transcription on the template strand ($3'$ to $5'$). About 70 nucleotides upstream from this −35 region, many pol II genes contain **CAAT** and/or **GC-rich** nucleotide regions. Even further upstream (and sometimes downstream of the start site of transcription) sequence stretches referred to as **enhancer elements** are found which, when present, significantly increase basal levels of transcription. Transcription factor **TFIID** which contains the **TATA-binding protein** (TBP) and TBP-associated factors (**TAFs**) binds to the −35 region of pol II genes. Subsequent complexes with transcription factors **TFIIB and TFIIF** and other factors are formed on the core promoter region followed by RNA polymerase II binding and transcription initiation. ATP (in addition to its substrate need along with the other NTPs) is required for pol II initiation and a modified "**cap**" nucleotide (7-methyl GTP) is condensed to the $5'$ triphosphate of the first nucleotide of the nascent RNA chain shortly after initiation. RNA polymerase II mRNA transcripts (hnRNA or heteronuclear) are cleaved just downstream of an AU-rich region and **poly A polymerase** adds a string of As (20 - 200) to the $3'$ end of the processed transcript. This poly A tail is useful in the isolation of eukaryotic mRNA by oligo dT affinity chromatography. The initial pol II transcript is further processed by splicing together of coding sequences. In general, mRNA molecules are less stable than ribosomal RNA and tRNA.

Additional RNA polymerase II transcription factors interact with upstream activating sequences (e.g., GC box) and enhancers. Their function is to provide needed stability to the overall transcription initiation complex necessary for initiation of mRNA synthesis. Transcription factors **Sp1** and CAAT/enhancer binding protein (**CBP**) bind to the GC-rich and CAAT boxes respectively. Unique enhancers and enhancer binding proteins are numerous; one such protein is the estrogen receptor (**ER**) which binds to the estrogen response element (**ERE**) enhancer in certain genes of estrogen-responsive tissue. **Breast tumors** usually progress from ER positive to ER negative as tumorigenesis advances; the estrogen receptor therefore is diagnostic for tumor stage. Since one of the genes regulated by the ER is the progesterone receptor (**PR**) gene, advanced breast tumors are usually PR negative as well.

Transcription factors responsible for regulating initiation of RNA synthesis by RNA polymerase III are **TFIIIA, TFIIIB, and TFIIIC**. 5S ribosomal RNA synthesis requires all three factors whereas tRNA synthesis requires only factors TFIIIB and TFIIIC. 5S and tRNA genes have internal promoters within the coding regions to which these factors bind. TFIIIB is a multisubunit complex containing and requiring TATA-binding protein even though 5S and tRNA genes do not contain a TATA element. The termination signal for pol III transcription is a string of Ts on the non-template DNA strand. Synthesis of the major ribosomal RNA genes by RNA polymerase I is regulated by transcription factors **UBF** (upstream binding factor) and **SL1**. The pol I promoter to which these factors bind encompasses the start site of transcription and extends upstream 70 - 80 nucleotides. The SL1 factor is a multisubunit complex containing TBP even though the pol I promoter has no TATA element. TBP is therefore an essential factor for all three eukaryotic RNA polymerases.

A multitude of eukaryotic transcription factors exist and many can be classified according to distinct structural motifs. TFIIIA contains multiple **zinc finger** domains in which the metal is bound by two cysteine and two histidine amino acids forming a stable structure. This Cys_2His_2 DNA binding motif is found in hundreds of vertebrate transcription factors including **Sp1** and the tumor suppressor, **WT1**. WT1 is mutated in **Wilms' tumor**, a pediatric kidney tumor. Steroid hormone receptors like **ER** and **PR** contain two zinc fingers in which the metal is coordinated by four cysteine residues. They also have hormone binding regions and domains involved in protein-protein interactions. Other zinc-binding transcription factors include **p53** and **GL1**, a Cys_2His_2 finger protein present in large amounts in glioblastoma brain tumors. The **leucine zipper** is another structural motif found in many eukaryotic transcription factors and is present in the CAAT box binding protein. This motif allows protein-protein interactions through hydrophobic leucine repeat elements on separate polypeptide chains. The **MYC** oncogene protein, which is overexpressed in **Burkitt's lymphoma** and some other tumors, is a transcription factor which contains the **helix-loop-helix** motif. This motif promotes protein-protein interactions and differs from the **helix-turn-helix** DNA binding motif of bacterial transcription factors. The activity of many eukaryotic transcription factors is regulated by modifications like phosphorylation and are important mediators in signal transduction pathways.

III. REVIEW QUESTIONS ON NUCLEIC ACIDS

DIRECTIONS: For each of the following multiple-choice questions (1 - 43), choose the ONE BEST answer.

---

1. DNA ligase joins segments of DNA:

A. forming a β-glycosidic linkage between the preceding base and the next sugar
B. catalyzing the formation of a phosphodiester bond between a free $5'$-nucleotide phosphate group and a $2'$-sugar hydroxyl group
C. inserting an RNA primer between adjacent segments
D. catalyzing the formation of a phosphodiester bond between a free $3'$-hydroxyl at the end of one DNA chain and a $5'$-phosphate at the end of the other DNA chain
E. sealing gaps between RNA primers and growing DNA strands.

Ans D: DNA ligase reforms the phosphodiester bond between adjacent $5'$ phosphate and $3'$ OH groups in double-stranded DNA.

2. Nucleosomes are:

A. found primarily in prokaryotic cells and not in mammalian cells
B. nucleoprotein particles of DNA wrapped around a core of basic proteins containing a high proportion of lysine and arginine
C. nucleoprotein particles of RNA wrapped around a core of eight histone proteins
D. particles in which the histones are bound to the nucleic acid by a phosphodiester bond
E. nucleoprotein particles in which the amino acids are negatively charged at neutral pH.

Ans B: Nucleosomes are protein-DNA complexes in which the DNA in eukaryotic chromosomes is wrapped about twice around the surface of a histone octomer.

3. It was established that DNA is replicated semi-conservatively by demonstrating that:

A. only part of the DNA is replicated at any particular time
B. during replication none of the genetic information is conserved
C. after replication each daughter molecule contains one parental strand and one new strand
D. after replication both parental strands were in one daughter cell and both new strands were in the other daughter cell
E. The mechanism of DNA replication involved $5' \rightarrow 3'$ synthesis of one parental strand and $3' \rightarrow 5'$ synthesis of the other strand.

Ans C: Semi-conservative replication is when a new strand is synthesized on an old strand.

4. RNA is transcribed in the (referring to the RNA):

A. $5'$ to $3'$ direction
B. $3'$ to $5'$ direction
C. amino-terminal to carboxy-terminal direction
D. carboxy-terminal to amino-terminal direction.
E. both the $5'$ to $3'$ and $3'$ to $5'$ direction.

Ans A: Like DNA, RNA is also polymerized in the $5'$ to $3'$ direction.

5. The sequence of a short duplex DNA is

$$5'\,GAACCTAC\,3'$$
$$3'\,CTTGGATG\,5'.$$

What is the corresponding mRNA sequence transcribed from this region?

A. GAACCTAC
B. CUUGGAUG
C. CAUCCAAG
D. GAACCUAC
E. GUAGGUUC

Ans D: By convention, the bottom DNA strand is designated the template strand for transcription of the newly synthesized RNA strand in the $5'$ to $3'$ direction with ribo U substituted for deoxyribo T in DNA.

6. When two DNA molecules are compared according to buoyant density and their "melting points" (T_m), the one with the greatest A - T content will have:

A. the higher density and higher T_m
B. the lower density and lower T_m
C. the higher density and lower T_m
D. the lower density and higher T_m
E. no detectable difference.

Ans B: High A-T content of DNA leads to lower density (T has a methyl group) and lower melting temperature because only two hydrogen bonds are involved in base pairing as opposed to three for G-C pairs.

7. By which of the following means are bases linked to the pentoses in RNA and DNA molecules?

A. They are linked together by β-glycosidic bonds.
B. They are linked together by $3',5'$ phosphodiester bridges.
C. They are linked together by $2',5'$ phosphodiester bridges.
D. They are linked together by α-glycosidic bonds.
E. They are held together by electrostatic attraction.

Ans A: The covalent bond between the sugar and the base in DNA and RNA is a β-glycosidic linkage.

8. DNA polymerase I is extremely accurate in its ability to make a complementary copy of the template. The key activity that allows the polymerase to be so accurate is the:

A. ligase activity
B. adenosine deaminase activity
C. gyrase activity
D. $3' \to 5'$ exonuclease activity
E. $5' \to 3'$ exonuclease activity.

Ans D: The $3'$-$5'$ exonuclease activity allows most DNA polymerases to back track and remove a misincorporated nucleotide and polymerize the correct one.

9. Energy-dependent unwinding of duplex DNA can be accomplished by:

A. helicases
B. restriction enzymes
C. telomerases
D. nucleases
E. ligases.

Ans A: Helicases are enzymes that use ATP to melt duplex DNA

10. What mammalian genes does RNA polymerase II transcribe?

A. tRNA
B. mRNA
C. ribosomal RNA
D. 5S RNA
E. SRP RNA.

Ans B: Pol II transcribes mRNA genes, pol I transcribes ribosomal RNA genes, and pol III transcribes tRNA, 5S RNA, and SRP RNA genes.

11. RNA polymerases require as substrates:

A. ATP, GTP, TTP, CTP
B. ATP, GTP, UTP, CTP
C. dATP, dGTP, dTTP, dCTP
D. an RNA template
E. ribosomes.

Ans B: RNA polymerases require ribonucleotides for substrates and DNA polymerases require deoxyribonucleotides.

12. Mutations in mismatch DNA repair enzymes are most likely inherited in what disease?

A. Wilms' tumor
B. Burkitt's lymphoma
C. breast tumor
D. non-polyposis colon carcinoma
E. chronic myelogenous leukemia.

Ans D: Mutations leading to hereditary non-polyposis colon carcinoma (HNPCC) have been localized to DNA mismatch repair genes (helicase, endonuclease).

13. Carcinogenic point mutations in the mRNA codons for RAS protein cause:

A. increased GTPase activity
B. decreased GTPase activity
C. altered DNA binding
D. cell apoptosis
E. protein degradation.

Ans B: RAS is a G-protein and mutations in codons 12 and 61 in many human tumors cause increased binding of GTP to the protein with less hydrolysis to GDP resulting in increase signal transduction and cell division.

14. The nucleosome core contains:

A. 1 copy each of histone H2A, H2B, H3, and H4.
B. 2 copies each of histone H2A, H2B, H3, and H4
C. DNA covalently linked to histones
D. histone H1
E. RNA polymerase.

Ans B: The nucleosome core contains an octomer of histones, two each of the four histones.

15. In polynucleotides the individual nucleosides are linked together with:

A. hydrogen bonds
B. ionic bonds
C. glycosidic bonds
D. phosphodiester bonds
E. phosphomonoester bonds.

Ans D: Phosphodiester bonds link adjacent nucleotides in DNA and RNA.

16. An inhibitor of bacterial transcription is:

A. nalidixic acid
B. tetracycline
C. rifampicin
D. acyclovir
E. azidothymidine.

Ans C: Rifampicin inhibits initiation of RNA synthesis by RNA polymerase holoenzyme in bacteria.

17. Which nucleic acid is most rapidly degraded in mammalian cells?

A. rRNA
B. hnRNA
C. tRNA
D. rDNA
E. 5S RNA.

Ans B: hnRNA is the precursor to mRNA in mammalian cells and is rapidly processes and is less stable than the other nucleic acids.

18. Denatured human lymphocyte DNA will not hybridize with human:

A. lymphocyte rRNA
B. kidney tRNA
C. denatured mitochondrial DNA
D. denatured liver DNA
E. brain mRNA.

Ans C: Lymphocyte DNA refers to nuclear DNA so little hybridization would be seen with mitochondrial DNA which is unique.

19. Tissue cells incubated with [^{3}H] thymidine will most rapidly accumulate the radioactive isotope in:

A. mitochondria
B. ribosomes
C. hnRNA
D. rough ER
E. nuclei.

Ans E: Thymidine is a precursor to dTTP and would be incorporated mostly into nuclear DNA during S phase.

20. SPF regulates the eukaryotic cell cycle by stimulating the transition from:

A. G1 into S
B. G2 into S
C. M into S
D. G2 into M
E. M into G0.

Ans A: SPF is S-phase promoting factor which stimulated DNA synthesis by promoting the transition from G1 to S.

21. In the Watson-Crick model for DNA, the molecule is a(an):

A. single stranded helix
B. α-helical structure
C. right handed, double-stranded helix of the same polarity
D. left-handed, double-stranded helix of opposite polarity
E. right-handed, double-stranded helix of opposite polarity.

Ans E: Double-stranded DNA is a right-handed double helix in which the strands are anti-parallel (run in 5′ to 3′ direction opposite to each other).

22. The substrate which provides the 3' OH group needed for extension of newly synthesized DNA and the enzyme that synthesizes this substrate is:

A. RNA, reverse transcriptase
B. RNA, RNA polymerase I
C. DNA, DNA polymerase III
D. RNA, primase
E. DNA, primase.

Ans D: The substrate is an RNA molecule which is synthesized by a unique RNA polymerase called a primase.

23. Replication is a sequential process. In the following list of the proteins involved in replication, which one is NOT in proper sequence?

A. helicase
B. RNA polymerase
C. DNA ligase
D. DNA polymerase III
E. DNA polymerase I.

Ans C: DNA ligase action would be last in this series when it covalently closes the remaining nick or gap in the phosphodiester backbone.

24. A codon mutation resulting from a deletion of a nucleotide base results in a:

A. translocation
B. inversion
C. frameshift change
D. transversion
E. transition.

Ans C: Deletion of a single nucleotide in the protein-coding region of a gene would result in a frameshift in translating the mRNA and eventual polypeptide chain termination because a new stop codon would be generated.

25. Which statement about the estrogen receptor (ER) transcription factor is FALSE?

A. the protein has a leucine zipper motif
B. Advanced breast tumors are usually ER negative
C. the protein has a zinc finger motif
D. the factor binds to the ERE enhancer
E. the protein activates transcription of the progesterone receptor gene.

Ans A: The estrogen receptor has a zinc finger motif.

26. In the presence of the second messenger cAMP and lactose in the bacterium *E. coli*, the catabolite activator protein (CAP) will:

A. bind the operator of the lactose operon and shut off transcription of the operon
B. bind to the upstream activator sequence of the lactose operon and stimulate operon transcription
C. bind to the upstream activator sequence of the lactose operon and repress operon transcription
D. bind lactose and dislodge the lactose repressor
E. bind sigma factor of RNA polymerase and repress transcription.

Ans B: cAMP binds the CAP protein which allows the protein to bind the upstream activator sequence of the lactose operon. Lactose binds the lactose repressor removing it from the operator allowing CAP to stimulate RNA polymerase binding to the promoter and initiating transcription.

27. DNA and RNA molecules in solution absorb ultraviolet light strongly at 260 nm. This absorption is due to:

A. pyrimidines
B. purines
C. both pyrimidines and purines
D. phosphate
E. ribose and deoxyribose.

Ans C: The electrons of both purine and pyrimidine bases absorb UV light of wavelength 260 nanometers. This wavelength can induce DNA mutations.

28. At the growing fork of replicating DNA, small DNA fragments are found and are referred to as:

A. Watson fragments
B. Cairn fragments
C. Meselson and Stahl fragments
D. Okazaki fragments
E. Avery fragments.

Ans D: Okazaki is the last name of the scientist who discovered the discontinuous nature of DNA synthesis on the lagging DNA strand.

29. Regulation of histidine and other amino acid biosynthetic operons by a coupled transcription-translation mechanism is called:

A. translocation
B. recombination
C. attenuation
D. transposition
E. synergism.

Ans C: Attenuation is the transcription termination event caused by ribosome read-through of a message containing multiple amino acid codons for amino acids present in the bacterial cell.

30. The complementary strands of double-stranded DNA are:

A. held together by ionic bonds
B. pair A with U
C. held together by specific hydrogen bonds
D. not denatured by high temperature
E. denatured by topoisomerases.

Ans C: base pairing by hydrogen bonding is responsible for the specific interactions holding together complementary DNA strands.

31. Base pairing rules state that in DNA:

A. A = U
B. G = T
C. A = C
D. A = T
E. C = U

Ans D: T is in DNA and U is in RNA.

32. The 3′ end of a nucleic acid strand contains:

A. an amino group
B a phosphate group
C. a hydroxyl group attached to the base
D. a triphosphate group
E. a hydroxyl group attached to a ribose or deoxyribose sugar.

Ans E: Nucleic acids have 3′ hydroxyl groups attached to the sugar.

33. All the following statements about the transcription factor p53 are correct EXCEPT:

A. this protein is a tumor suppressor
B. this factor promotes cell apoptosis
C. this protein is commonly mutated in human carcinomas
D. this protein is rarely mutated in human carcinomas
E. this protein promotes transcription of an inhibitor of SPF.

Ans D: p53 point mutations are the most commonly observed mutations in human tumors.

34. Histones are rich in the amino acid(s):

A. cysteine
B. phenylalanine
C. tryptophan
D. leucine and valine
E. lysine and arginine.

Ans E: Histones, which are constituents of nucleosomes, are rich in the basic amino acids (positively charged), lysine and arginine.

35. p53 blocks DNA synthesis by:

A. blocking the cell cycle from G0 to G1
B. binding to DNA polymerase
C. binding to RNA polymerase II
D. activating transcription of a cyclin-dependent kinase inhibitor
E. inhibiting protein synthesis.

Ans D: p53 turns on the gene for WAF1 (wild-type p53 activating fragment) which binds to CdK and inhibits its kinase activity.

36. The DNA consensus sequences TATA and CAAT are:

A. initiation signals for DNA synthesis
B. initiation signals for RNA synthesis
C. histone binding sites
D. DNA methylation sites
E. p53 binding sites.

Ans B: These are promoter and upstream activator sequences for RNA polymerase II in eukaryotic cells.

37. DNA mutations which convert proto-oncogenes to oncogenes can be caused by all of the following EXCEPT:

A. radiation
B. viruses
C. benzpyrene
D. plasmids
E. DNA polymerase.

Ans D: Bacterial plasmids, which are self-replicating circular DNA molecules, transfer antibiotic resistance genes from one bacterium to another and have not been shown to be involved in carcinogenesis.

38. The DNA translocation generating the Philadelphia chromosome involves which proto-oncogene and is diagnostic for which disease?

A. ras G-protein, colon carcinoma
B. ras G-protein, chronic myelogenous leukemia
C. abl tyrosine kinase, colon carcinoma
D. abl tyrosine kinase, chronic myelogenous leukemia
E. p53 transcription factor, chronic myelogenous leukemia.

Ans D: The Philadelphia chromosome ($22q^-$) results from the reciprocal translocation of the abl proto-oncogene from chromosome 9 to chromosome 22 and the bcr region of chromosome 22 to 9.

39. The AIDS virus (HIV) contains an RNA genome. Which enzyme is responsible for converting this genome into DNA in T lymphocytes?

A. ligase
B. EcoRI
C. DNA polymerase
D. reverse transcriptase
E. RNA polymerase.

Ans D: Reverse transcriptase is a RNA-dependent DNA polymerase and copies retrovirus RNA genome into complementary DNA (cDNA). This enzyme action is inhibited by the chain terminator, azidothymidine (AZT).

40. DNA contains:

A. pyrimidine ribose monophosphates
B. purine deoxyribose diphosphates
C. purine deoxyribose monophosphates
D. purine deoxyhexose monophosphates
E. 5′- 2′ phosphodiester bonds.

Ans C: Although nucleoside triphosphates are the substrates for nucleic acid synthesis, 5′ monophosphate is attached to the 3′ end of the growing nucleotide chain with hydrolysis of pyrophosphate (pp).

41. Genetic information in retroviruses flows from:

A. DNA to RNA to protein
B. RNA to DNA to RNA to protein
C. DNA to RNA to DNA to protein
D. RNA to DNA to protein
E. protein to RNA to DNA.

Ans B: Retroviruses like HIV contain reverse transcriptase which first copies their single-stranded RNA genome in the cytoplasm into double-stranded complementary DNA. This cDNA is then inserted into a nuclear chromosome where it is transcribed into RNA that returns to the cytoplasm to form new viruses.

42. A promoter:

A. contains conserved sequences usually upstream from the start of a gene to which RNA polymerase binds
B. is a binding site for DNA polymerase I
C. is a nick in a DNA duplex
D. is the site at which DNA replication begins
E. is required by DNA polymerase III.

Ans A: A transcription promoter is a unique DNA sequence (usually conserved through evolution) which directs RNA polymerase to the proper transcription initiation site. In bacteria, the enzyme can bind the promoter directly in some cases; in higher organisms, transcription factors must first bind to promoter sequences to stimulate RNA polymerase binding.

43. Which transcription factor is commonly used for RNA polymerase I, II, and III transcription in mammalian cells?

A. SL1
B. TBP
C. TFIIIA
D. UBF
E. TFIID.

Ans B: TATA-binding protein (TBP) is present in SL1 for pol I transcription, TFIID for pol II transcription, and TFIIIB for pol III transcription. SL1, TFIID, TFIIIB have additional, distinct proteins which make them function only for their respective polymerases.

13. PROTEIN BIOSYNTHESIS
Jay Hanas and Albert M. Chandler

I. INTRODUCTION

A protein consists of 20 different amino acids linked head to tail by peptide bonds. The amino acids are arranged in a specific sequence in each particular protein. In order to accomplish this feat each cell must have: (1) cellular machinery by which the proteins are assembled; (2) informational mechanisms to designate the amino acid sequences; (3) enzymes and other factors to carry out the detailed assembly processes.

II. CELLULAR MACHINERY

Protein biosynthesis involves three major species of RNA: **ribosomal RNA (rRNA)**, **messenger RNA (mRNA)** and **transfer RNA (tRNA)**. In addition, in eukaryotic organisms, two other types, **heterogeneous nuclear RNA (hnRNA)** and **small nuclear RNA (snRNA)**, are also involved.

A. Ribosomal RNA, Ribosomes and Polysomes

If the post-mitochondrial supernatant from a mammalian cell source is centrifuged at 105,000 x g for 1 hour, a pellet is formed which is termed the **microsomal** fraction. This consists of the membranes of the **endoplasmic reticulum**. Membranes to which ribosomes are attached by their larger subunits are called **rough endoplasmic reticulum (RER)**. Membranes lacking these particles are termed **smooth endoplasmic reticulum (SER)**. In addition, free ribosomes will be found both as individual particles and attached together in clusters by a thread of messenger RNA. These clusters are called **polyribosomes** or **polysomes**. *Protein synthesis always takes place on polysomes.* As a general rule, proteins destined to be exported from the cell or destined to become part of cell membranes are synthesized on membrane-bound polysomes. Proteins destined to stay within the cell, i.e., "housekeeping proteins", are synthesized on free polysomes. The ribosomes and polysomes of bacteria are usually found free, seldom membrane-bound.

♦ Structure of Ribosomes

Bacterial. Individual ribosomes, when examined with an electron microscope, will be seen as spherical particles about 180 - 220 Å in diameter and consisting of two subunits. An example of the structure and composition of *E. coli* ribosomes is shown in Figure 13 - 1. The whole ribosome sediments with a value of 70S and can be broken down into a 50S subunit and a 30S subunit. "S" refers to the sedimentation constant and is indicative of the size of the RNA. Transfer RNA, for example, sediments during centrifugation as a 4S entity. The 50S subunit yields a characteristic 23S and a 5S rRNA while the 30S subunit yields a 16S rRNA. Both subunits yield a characteristic set of specific ribosomal proteins, 21 from the 30S and 31 from the 50S.

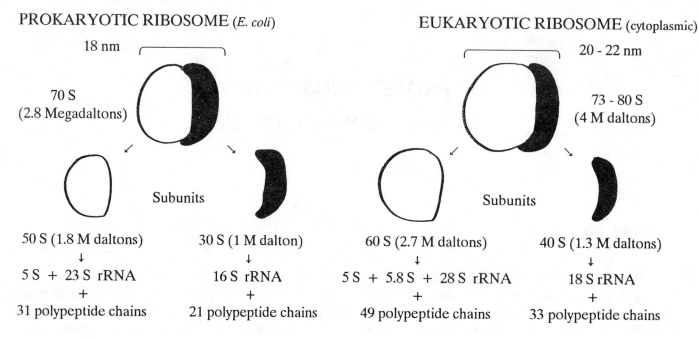

PROKARYOTIC RIBOSOME (*E. coli*)

18 nm

70 S
(2.8 Megadaltons)

Subunits

50 S (1.8 M daltons)
↓
5 S + 23 S rRNA
+
31 polypeptide chains

30 S (1 M dalton)
↓
16 S rRNA
+
21 polypeptide chains

EUKARYOTIC RIBOSOME (cytoplasmic)

20 - 22 nm

73 - 80 S
(4 M daltons)

Subunits

60 S (2.7 M daltons)
↓
5 S + 5.8 S + 28 S rRNA
+
49 polypeptide chains

40 S (1.3 M daltons)
↓
18 S rRNA
+
33 polypeptide chains

Figure 13 - 1. Composition of Prokaryotic and Eukaryotic Ribosomes.

Mammalian. The ribosomal RNAs in eukaryotic organisms have a sedimentation constant of 80S made up of 60S and 40S subunits. The 60S subunit contains three major rRNA species, the 5S, 5.8S, and 28S RNAs and 49 proteins, while the 40S subunit contains one 18S rRNA and 33 proteins. The 5S rRNA contains 120 nucleotides, the 5.8S rRNA contains 160 nucleotides, the 18S rRNA contains about 1900 nucleotides and the 28S rRNA contains about 5000 nucleotides (Figure 13 - 1).

♦ <u>Synthesis</u> (Figure 13 - 3, page 230)

Synthesis of eukaryotic ribosomes occurs in the nucleus and involves a special organelle, the **nucleolus**. Synthesis begins with production in the nucleolus of a large **45S precursor** by the enzyme **RNA polymerase I**. After a series of alkylations and cleavages this single precursor gives rise to the 28S, 18S, and 5.8S species of rRNA. (The clear areas in Figure 13 - 3 are portions of the original 45S molecule that are degraded). 5S rRNA is made elsewhere in the nucleus. Ribosomal RNA also contains methylated bases and sugars.

Mitochondria and chloroplasts of eukaryotic cells contain their own genetic systems and are capable of autonomous protein synthesis. They also have ribosomes that appear to be intermediate between cytoplasmic ribosomes of eukaryotes and those of bacteria but differ from both in size and RNA composition. Mitochondrial ribosomes are smaller than cytoplasmic ribosomes, having a sedimentation constant of 55S. *In biological behavior they closely resemble bacterial ribosomes.*

B. Messenger RNA

The diagram shown in Figure 13 - 2 is a schematic representation of the general linear structure of a typical eukaryotic mRNA molecule.

A "**cap**" is essential for the initiation of translation and acts to protect the molecule from degradation. The cap consists of a GTP molecule methylated in the 7 position of the guanine moiety and attached through the third phosphate residue $5' - 5'$ to the end of the mRNA. If the cap is removed or if it is not methylated

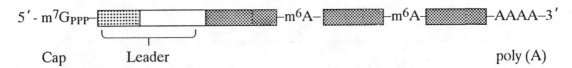

$5' - m^7G_{PPP}$—▨—□—▨—m^6A—▨—m^6A—▨—AAAA–$3'$

Cap   Leader   poly (A)

Figure 13 - 2. Structure of a Typical Eukaryotic Messenger RNA.

the mRNA will either not be translated at all or will be translated with markedly reduced efficiency. The "leader" is recognized by binding sites on the ribosome and by certain initiation factors. The function of the internal methylations is unknown. The **poly(A) "tail"** confers stability on the molecule and is suspected to be involved in the transport of newly synthesized mRNA from the nucleus to the cytoplasm. The poly(A) tail can range from 100 - 200 nucleotides long. There is no poly(T) region on the DNA template coding for the tail. The tail is synthesized from ATP molecules which are added sequentially by *poly(A) polymerase,* an enzyme that recognizes a unique sequence (**AAUAAA**) on the primary transcript. The polymerase can be inhibited by **cordycepin** ($3'$ - deoxyribose).

Bacterial mRNAs differ from eukaryotic mRNAs in that they have neither cap structures nor poly(A) tails. As a consequence bacterial mRNAs have very short half lives, (2 - 3 minutes), whereas mammalian mRNAs can have comparatively long half lives (hours or days).

Synthesis of mRNA in Eukaryotic cells. In prokaryotic cells the coding regions of the DNA usually form one long continuous sequence. In eukaryotic cells, however, the coding regions are usually "split", that is, coding sequences (**exons**) are separated from one another by non-coding sequences (**introns**) (Figure 13 - 4). The functions of these introns are unknown. When the gene is transcribed, the primary RNA transcript contains sequences from both exons and introns. The introns must be removed and the exons "spliced" together to form functional mRNA. This splicing process is carried out by ribonucleoprotein particles called **small nuclear ribonucleoprotein particles (snRNPs)** which come together to form complex structures called **spliceosomes.** The RNA components of these particles are small, stable RNAs designated U1, U2, . . . U8, ranging in size from 150 - 300 bases. Several of these RNAs appear to be involved in the processes converting hnRNA to mRNA.

The introns are removed as "lariat" structures, in which the $5'$ end of the intron is covalently attached to an internal adenosine residue to form a loop with a tail. All introns have **GU** at their $5'$ ends and **AG** at their $3'$ ends.

Patients with the autoimmune disease **systemic lupus erythematosus (SLE)** and other autoimmune diseases produce autoantibodies against these snRNPs as well as against other cellular nucleoproteins. Certain mutations in human globin genes will not allow the proper splicing of primary transcripts and can be the cause of **thalassemias.**

C. Transfer RNA

There are 50 - 60 different, but related, species. They are all similar in structure, but do have differences in base composition and structure, in amino acid specificity and in codon recognition ability. They contain 10 - 20% unusual "minor" bases. These minor bases are formed after the polynucleotide chain has been transcribed. Each reaction requires a specific enzyme.

Transfer RNAs are 70 - 90 nucleotides in length; M.W. about 25 - 30,000 and all can be folded into a "cloverleaf" secondary structure. Three-dimensionally they are "L" shaped. The structure of tRNA is very well understood due to X-ray crystallography as illustrated in Figure 13 - 5. All end in -**CCA** at the $3'$ end. The amino acid is bound to the ribose unit of the terminal adenosine. The "amino acid arm" contains 7 base pairs, the "anticodon arm" contains 5 base pairs and the anticodon loop always contains 7 nucleotides. Thus, all tRNAs are of the same length. All contain a TUC loop and a dihydrouracil loop which are close together

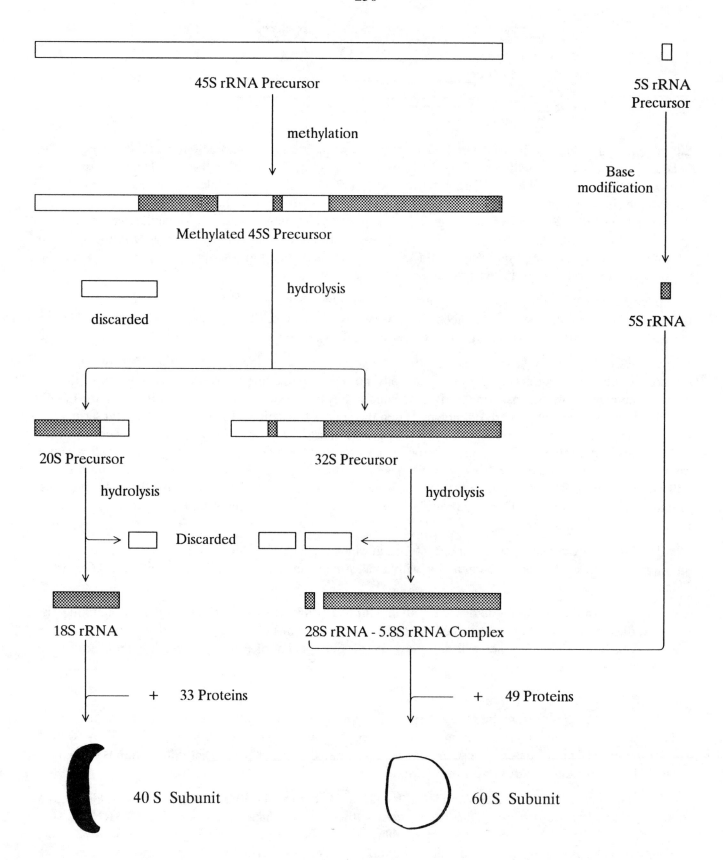

Figure 13 - 3. Assembly of Eukaryotic Ribosomes.

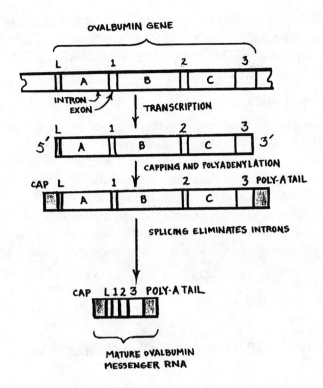

Figure 13 - 4. Structure of the Ovalbumin Gene.

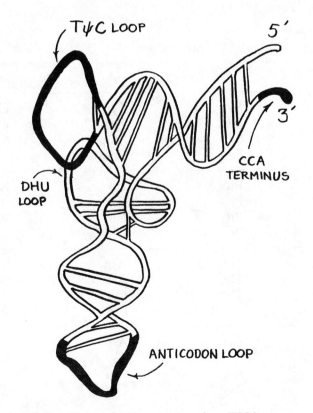

Figure 13 - 5. Three-dimensional Structure of a Representative tRNA.

in the 3-D structure of the molecule but far apart in the 2-D structure. These loops contain modified bases. These probably associate with specific binding sites on the ribosome. The **anticodon** (codon recognition site) usually has on one side a pyrimidine (U) and on the other an alkylated purine. The codon and anticodon always pair up in an antiparallel manner.

D. Amino Acid Activation

In order to be coupled into a growing polypeptide chain an amino acid must first be "activated", that is, coupled to a specific tRNA. This coupling requires specific *amino acid activating enzymes* also known as *aminoacyl-tRNA synthetases*. There is at least one amino acid activating enzyme for each amino acid. These enzymes have double specificity; that is, they have two specific binding sites, one for a specific amino acid and the other for a specific tRNA. They catalyze reactions (1 and (2) below. Although these are reversible *in vitro*, because of the presence of very active inorganic pyrophosphatases in living cells, the overall reaction favors the formation of aminoacyl-tRNAs. The amino acid is coupled via its carboxyl group in ester linkage with the **3′OH** of the ribose moiety of the terminal adenosine moiety of the tRNA.

(1) $Enz_1 + ATP + AA_1 \longrightarrow Enz_1\text{-}AA_1\text{-}AMP + PP_i$

(2) $Enz_1\text{-}AA_1\text{-}AMP + tRNA_1 \longrightarrow AA_1\text{-}tRNA_1 + Enz_1 + AMP$

Sum: $AA_1 + tRNA_1 + ATP \longrightarrow AA_1\text{-}tRNA_1 + AMP + PP_i$

III.   INFORMATION TRANSFER

A. The Genetic Code (Table 13 - 1)

1st pos. 5′ end	2nd position				3rd pos. 3′ end
	U	C	A	G	
U	Phe	Ser	Tyr	Cys	U
	Phe	Ser	Tyr	Cys	C
	Leu	Ser	**Stop**	**Stop**	A
	Leu	Ser	**Stop**	Trp	G
C	Leu	Pro	His	Arg	U
	Leu	Pro	His	Arg	C
	Leu	Pro	Gln	Arg	A
	Leu	Pro	Gln	Arg	G
A	Ile	Thr	Asn	Ser	U
	Ile	Thr	Asn	Ser	C
	Ile	Thr	Lys	Arg	A
	Met	Thr	Lys	Arg	G
G	Val	Ala	Asp	Gly	U
	Val	Ala	Asp	Gly	C
	Val	Ala	Glu	Gly	A
	Val	Ala	Glu	Gly	G

Table 13 - 1. The Genetic Code.

The genetic code is the relationship between the sequence of bases in DNA (or its RNA transcripts) and the sequence of amino acids in a protein. Deciphering the genetic code in the 1960's was a monumental achievement, ushering in the modern era of biology. Although genetic information is stored in DNA in unique sequences of the four deoxynucleotides and transcribed into an identical RNA sequence of four ribonucleotides (deoxythymidylate in DNA replaced by ribouridylate in RNA), this information must then be translated into a unique protein amino acid sequence. There are 20 distinct amino acids commonly found in proteins, linear arrays of amino acids joined by amide linkages between the alpha amino group of one amino acid and the alpha carboxyl group of the next. Given four bases (A, U, G, or C), a singlet or doublet genetic code would only give 4 or 8 combinations (4^1 or 4^2), not enough to code for 20 amino acids. A triplet code (4^3) would code for 64 unique possibilities, more than enough for 20 amino acids. By using synthetic ribonucleotide polymers in crude *in vitro* protein synthesizing systems the genetic code was indeed found to be **triplet** in nature and **codons** (three adjacent ribonucleotides) were assigned to all 20 amino acids.

B. General Features of the Code

The code is **degenerate**; that means that in many cases, several different codons code for the same amino acid. During protein synthesis the anticodons of charged tRNAs base pair with the codons in mRNA. In fact some charged tRNAs can bind to different codons due to a phenomenon called "**wobble**" in which non-Watson-Crick base pairing can occur between the third position of some codons and the first position of a tRNA anticodon. The code is **non-overlapping**; one codon never codes for more than one amino acid. The code **lacks punctuation**; mRNA is translated by ribosomes in one continuous process, three nucleotides at a time. If a mutation occurs in the mRNA such that one nucleotide has either been deleted or added, the reading phase is changed accordingly (a "**frameshift**" mutation). Once changed, the phase will not change unless another compensating mutation occurs downstream. Three triplets in the genetic code were found not to code for amino acids (viz. **UAA, UAG**, and **UGA**). These codons were later found to signal termination of protein synthesis and are referred to as **termination, stop** or **nonsense** codons. The codon, **AUG**, codes for **methionine** and is also used as the **initiation** codon.

With the exception of mitochondria, the code is universal. That is, all forms of life as well as viruses have the same codon/amino acid assignments. Mitochondria have some minor changes in the codon/amino acid assignments indicating a different evolutionary history for these organelles.

IV.   PEPTIDE BOND FORMATION

A. Prokaryotes

Protein synthesis can be divided into three steps, **initiation, elongation**, and **termination**. All prokaryotic proteins begin with N-formyl methionine (fMet). The initiation step of protein synthesis places the initiator codon of the mRNA (usually AUG, sometimes GUG in prokaryotes) and the fMet-tRNA at the proper site on the ribosome for the start of translation. The 5' leader sequence of mRNA in bacteria has a sequence which is complementary to the 3' end of the 16S RNA in the 30S ribosomal subunit (**Shine-Delgarno sequence**). Base pairing occurs between these complementary regions and assists in the proper placement of the mRNA on the 30S subunit.

♦ Chain Initiation

The role of N-formylmethionine tRNA$_f$. The methionine-accepting tRNAs consist of two distinct species, termed **tRNA$_f^{Met}$** and **tRNA$_m^{Met}$**. These can undergo the following series of reactions in prokaryotic cells:

$$ \text{ATP} \qquad\qquad \text{AMP} + \text{PP}_i $$

$$
(1) \qquad
\begin{array}{l}
\text{tRNA}_f^{Met} \\[4pt]
+ \text{ methionine} \\[4pt]
\text{tRNA}_m^{Met}
\end{array}
\xrightarrow[\text{enzyme}]{\text{activating}}
\begin{array}{l}
\text{Met–tRNA}_f^{Met} \\[4pt] \\[4pt]
\text{Met–tRNA}_m^{Met}
\end{array}
$$

$$
(2) \qquad
\begin{array}{l}
\text{Met–tRNA}_f^{Met} \\[10pt]
\text{Met–tRNA}_m^{Met}
\end{array}
\xrightarrow[\text{transformylase}]{N^{10}\text{-}formyltetrahydrofolate}
\begin{array}{l}
\text{fMet–tRNA}_f^{Met} \text{ (formylated)} \\[10pt]
\text{Met–tRNA}_m^{Met} \text{ (not formylated)}
\end{array}
$$

Both Met-tRNAs recognize AUG as a codon, but fMet-tRNA$_f$ is used only when AUG is in the chain initiation position, and Met-tRNA$_m$ only for chain elongation. The formyl group is attached to the α-amino group of the methionine residue. fMet-tRNA$_f$ is used to initiate protein synthesis in all prokaryotes (bacteria, viruses) and in mitochondria and chloroplasts of higher organisms.

All ribosomes have two aminoacyl-tRNA binding sites on their surfaces. One of the sites is called the **acceptor site (A-site)** and the other the **peptidyl site (P-site)**. These binding sites extend over both the large and small ribosomal subunits.

Initiation Factors. There are three protein factors involved in initiation. These are called **IF-1, IF-2** and **IF-3**.

Initiation Steps (Figure 13 - 6). Binding of mRNA to the initiation factor-30S subunit complex releases IF-3. Next, the 50S subunit binds to the 30S subunit, mRNA, fMet-tRNA, IF-1, IF-2, GTP complex. IF-1 and IF-2 are released at this stage upon GTP hydrolysis. The fMet-tRNA is positioned correctly (anticodon bound to initiator AUG in mRNA) in the P (peptidyl tRNA) site of the 70S ribosome, poised for the elongation phase of protein synthesis.

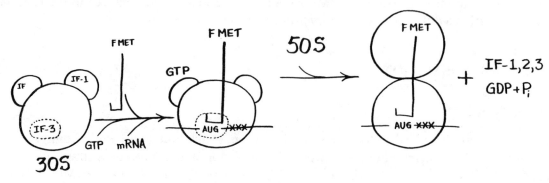

Figure 13 - 6. Initiation Phase.

♦ <u>Chain Elongation</u>

Elongation Factors. There are three elongation factors termed **EF-Tu, EF-Ts, EF-G**. In addition, there is a catalytic activity on the 50S subunit called *peptidyl transferase* which actually forms the peptide bond. This activity is activated only when the 50S subunit is combined with the 30S subunit on the polysome.

Elongation Steps (Figure 13 - 7). Aminoacyl tRNA binding depends upon the matching of its anticodon with the triplet codon of the mRNA positioned in the A site. Charged tRNA is complexed with elongation factor Tu and GTP. Once the charged tRNA is bound to the ribosome, the GTP is hydrolyzed to GDP and

Tu is released. Tu is recycled with GTP by factor Ts. Peptide bond formation between the amino group of the aminoacyl tRNA (amino group condensed with the tRNA 3′OH) in the A site and the carboxyl group of the peptidyl tRNA in the P site is catalyzed by peptidyl transferase, an enzyme located on the 50S subunit. Once peptide bond formation occurs, the tRNA in the A-site now containing the growing polypeptide chain is transferred to the P-site along with concomitant release of the deacylated tRNA in the P-site. This reaction is promoted by elongation factor G and GTP. GTP hydrolysis results in the release of G factor. The A site is now vacant and ready for another elongation round. In this manner, the polypeptide chain grows from N-terminal to C-terminal.

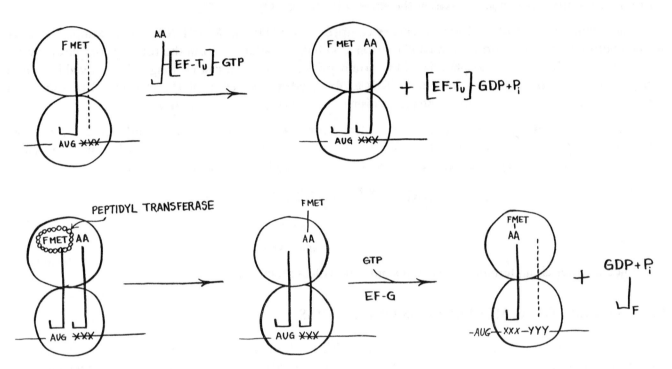

Figure 13 - 7. Elongation Phase.

Recycling of EF-Tu occurs as follows:

$$[\text{EF-Tu}]\text{-GDP} + \text{GTP} + \text{aa-tRNA} \xrightarrow{\text{EF-Ts}} [\text{aa-tRNA}]\text{-}[\text{EF-Tu}]\text{-GTP} + \text{GDP}$$

Correction of errors. The error frequency of protein synthesis (incorporation of an amino acid from an aminoacyl tRNA that was "misread" by the ribosome) is extremely low. This is due to proofreading of the peptidyl-tRNA bound to the P-site in the ribosome. The ribosome checks the match of the codon-anticodon interaction in the P-site and, if it is not correct, ejects the peptidyl tRNA from the ribosome in a GTP-dependent process. The antibiotic streptomycin interferes with this proofreading and causes misreading of the genetic code.

♦ Chain Termination

After many elongation cycles the ribosome eventually encounters one or more chain termination codons (**UAA, UAG** or **UGA**) on the mRNA. There are three release factors, **RF-1, RF-2** and **RF-3**, involved in chain termination in prokaryotes. These factors can bind to one or more termination codons in such a manner that the specificity of the 50S-bound peptidyltransferase is altered so that the peptide chain is transferred to water instead of to another α-amino group. The protein molecule is released at this stage. GTP hydrolysis results in the release of the termination factors.

VI.   SOME ANTIBIOTICS ACTING AS INHIBITORS OF PROTEIN SYNTHESIS

Inhibitors of Transcription	Mode of Action
Actinomycin D	Binds to DNA in GC-rich regions leading to inhibition of transcription, especially ribosomal RNA.
Á-Amanitin	Inhibits eukaryotic RNA polymerases I, II, and III to varying degrees.
Cordycepin	Inhibits chain elongation during RNA transcription and poly(A) tail formation.
Rifampicin	Inhibits bacterial RNA synthesis by binding to RNA polymerase.

Inhibitors of Translation (Act at ribosomal level)

Bacterial

Tetracyclines	Prevent initial binding at entry site. Maximum inhibition of aa-tRNA binding is 50%, thus it appears to be specific for one site only.
Chloramphenicol	Acts at a step after binding and during peptide bond formation. It binds to the 50S subunit and stops the formation of a peptide bond. It also inhibits the puromycin-induced premature release of peptides.
Streptomycin and di-hydrostreptomycin	Bind to 30S subunit of the ribosome causing distortion of the structural relationships between the ribosomes, mRNA, and aa-tRNA leading to misreading of the code. The products are non-functional proteins.
Erythromycin	Inhibits the translocation step during protein synthesis in bacteria.

Mammalian

Cycloheximide, eme-tine and dihydro-emetine)	Inhibit protein synthesis in L-cells, reticulocytes, liver slices, yeast but do not affect bacteria. Exact site of action unknown, but seems to be near site of peptide bond formation. The rate of translation is slowed down markedly and the release of completed peptide chains is inhibited. Do not act on prokaryotic ribosomes.

Miscellaneous antibiotics

Diphtheria toxin	Covalently modifies elongation factor EF-2 in mammalian cells, leading to inhibition of protein synthesis.
Nalidixic acid and Novobiocin	Inhibit topoisomerases leading to inhibition of DNA synthesis in bacteria.
Puromycin	Mimics the aminoacyl region of aminoacyl tRNA and causes premature release of polypeptide chains during protein synthesis in bacterial and eukaryotic cells.
Tunicamycin	Inhibits N-linked glycosylation events in animal cells.

Tu is released. Tu is recycled with GTP by factor Ts. Peptide bond formation between the amino group of the aminoacyl tRNA (amino group condensed with the tRNA 3′OH) in the A site and the carboxyl group of the peptidyl tRNA in the P site is catalyzed by peptidyl transferase, an enzyme located on the 50S subunit. Once peptide bond formation occurs, the tRNA in the A-site now containing the growing polypeptide chain is transferred to the P-site along with concomitant release of the deacylated tRNA in the P-site. This reaction is promoted by elongation factor G and GTP. GTP hydrolysis results in the release of G factor. The A site is now vacant and ready for another elongation round. In this manner, the polypeptide chain grows from N-terminal to C-terminal.

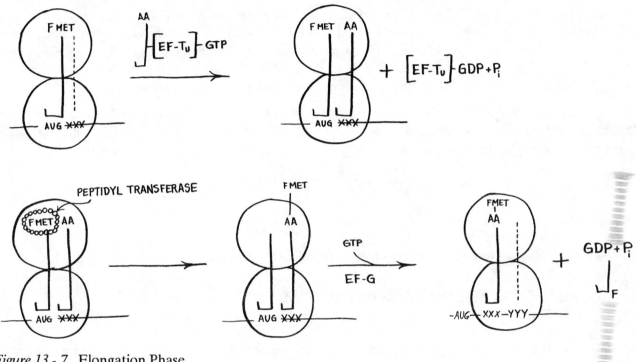

Figure 13 - 7. Elongation Phase.

Recycling of EF-Tu occurs as follows:

$$[\text{EF-Tu}]\text{-GDP} + \text{GTP} + \text{aa-tRNA} \xrightarrow{\text{EF-Ts}} [\text{aa-tRNA}]\text{-}[\text{EF-Tu}]\text{-GTP} + \text{GDP}$$

Correction of errors. The error frequency of protein synthesis (incorporation of an amino acid from an aminoacyl tRNA that was "misread" by the ribosome) is extremely low. This is due to proofreading of the peptidyl-tRNA bound to the P-site in the ribosome. The ribosome checks the match of the codon-anticodon interaction in the P-site and, if it is not correct, ejects the peptidyl tRNA from the ribosome in a GTP-dependent process. The antibiotic streptomycin interferes with this proofreading and causes misreading of the genetic code.

♦ Chain Termination

After many elongation cycles the ribosome eventually encounters one or more chain termination codons (**UAA, UAG** or **UGA**) on the mRNA. There are three release factors, **RF-1**, **RF-2** and **RF-3**, involved in chain termination in prokaryotes. These factors can bind to one or more termination codons in such a manner that the specificity of the 50S-bound peptidyltransferase is altered so that the peptide chain is transferred to water instead of to another α-amino group. The protein molecule is released at this stage. GTP hydrolysis results in the release of the termination factors.

B. Eukaryotes

The process of peptide bond formation on mammalian ribosomes is very similar to that for bacteria. Differences do exist, however. Eukaryotic mRNAs do not have Shine-Delgarno sequences. Chain initiation requires Met-tRNA$_i$ and the methionine is NOT formylated. There are at least thirteen initiation factors instead of three and in addition to GTP, ATP is required. The initiation factors in eukaryotes all have the prefix eIF-. Although all the initiation factors are important, two should be pointed out. The first is **eIF-4B**, the **cap recognition factor**. This factor, with the assistance of other proteins binds the methylated cap of the mammalian mRNA to the surface of the ribosome which then "scans" the message in the 3′ direction until it encounters the first AUG which serves as the chain initiation codon.

The second is **eIF-2** which forms a **ternary** complex with GTP and Met-tRNA$_i$. This complex must be formed before it can associate with the 40S ribosomal subunit. Failure to form the ternary complex prevents chain initiation. If eIF-2 is phosphorylated by certain protein kinases, it can no longer form a viable ternary complex and protein synthesis is inhibited. This is observed in the action of **Heme Controlled Repressor (HCR)** in reticulocytes and as part of the mechanism of action of certain **interferons**.

Eukaryotic chain elongation requires two factors termed **eEF-1** (Transferase I) and **eEF-2** (Transferase II). eEF-1 is equivalent to bacterial EF-Tu and EF-Ts; eEF-2 is equivalent to EF-G. eEF-2 but NOT EF-G is irreversibly inactivated by **diphtheria toxin** according to the following scheme:

$$eEF\text{-}2 \ + \ NAD \ \xrightarrow{\ toxin\ } \ \begin{array}{c} eEF\text{-}2 \ + \ Niacin \\ | \\ ADP \\ (inactive) \end{array}$$

In eukaryotes chain termination requires only one release factor, **RF**.

V. POSTRIBOSOMAL MODIFICATION OF PROTEINS

Very few proteins when isolated from cells have their N-termini starting with methionine and none with N-formyl-methionine. This is because of the presence of a series of specific aminopeptidases which cleave off these end groups. This process can occur after the protein is released from the ribosome or can even occur while the polypeptide chain is still growing on the ribosome.

A. The Signal Hypothesis

Proteins destined for export from the cell or to become membrane components usually have a specific hydrophobic portion, called a "signal" sequence, located at the amino terminus of the protein (about the first 20 amino acids). This acts as a "signal" for insertion of the growing polypeptide chain through the lipid bilayer of the endoplasmic reticulum. According to the "signal hypothesis", when a translating ribosome has synthesized enough of the protein so that the first 20 amino acids emerge from the large subunit, this signal is recognized by the **signal recognition particle (SRP)** The scheme shown in Figure 13 - 8 is a highly simplified version of **Blobel's signal hypothesis**.

The SRP is a ribonucleoprotein containing a 7S RNA and 6 - 8 specific proteins. The SRP binds to the signal sequence and also to the translating ribosome which results in a halt of protein synthesis. The SRP-ribosome complex is then recognized by a **docking protein** located on the surface of the rough endoplasmic reticulum. This docking protein binds the SRP-ribosome complex to the membrane resulting in the release of SRP and the resumption of translation. Translation causes the amino terminal signal sequence to be inserted through the lipid bilayer into the lumen of the rough ER. The protein is transported inside the rough ER by the translation process. During this process, the signal sequence is usually cleaved off.

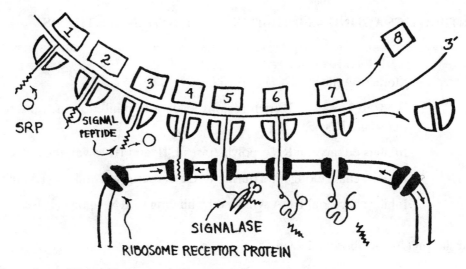

Figure 13 - 8. Secretion of Proteins through the Membranes of the RER.

B. Glycosylation of Proteins

The large majority of proteins synthesized in eukaryotic cells undergo glycosylation after the polypeptide backbone has been synthesized. The oligosaccharide side chains of these glycoproteins are formed in two stages. In the RER, a "mannose core", consisting of two N-acetylglucosamine, nine mannose and three glucose residues, is attached to the nascent glycoprotein after the polypeptide chain has been inserted into the lumen of the RER. The mannose core is assembled in the cytosol on a isoprenoid lipid called **dolichol phosphate**. The core is transferred from the dolichol phosphate, through the membrane, to the amide group of an asparagine residue. This process can be inhibited in animal cells by **tunicamycin** which blocks the formation of dolichol-PP-oligosaccharides.

While in the lumen of the RER the mannose core begins to be processed, i.e., certain glucose and mannose residues are removed. The glycopeptide is then encased in vesicles pinched off from the RER and transported to the *cis*-side of the **Golgi apparatus** where further processing occurs. As the glycopeptide traverses the *cis*, medial and *trans* structures of the Golgi, extensive remodeling of oligosaccharide side chains occurs and more peripheral sugars are attached until the glycoprotein is in finished form.

The completed glycoproteins are then enclosed in vesicles which are transported from the *trans* face of the Golgi apparatus with the aid of microtubules to the inner surface of the plasma membrane. The vesicles fuse with the plasma membrane and release their contents into the intercellular fluid.

Glycoproteins destined to become **membrane** components have regions consisting of hydrophobic amino acids inserted into the lipid bilayer of the plasma membrane, and serving as anchors to the membrane. Glycoproteins destined to become **lysosomal** enzymes have certain mannose residues phosphorylated in the 6 position, a signal that they are to be transported to pre-lysosomal vesicles.. The phosphorylation is a two-step process occurring in membranes of the *cis* Golgi. A transferase first attaches N-acetyl glucosaminyl phosphate to the mannose and then N-acetyl glucosamine is removed, leaving phosphate attached to the mannose residue. A genetic deficiency in the transferase is the cause of **I-disease**.

The role of oligosaccharide side chains appears to be topological, that is, they can act as signals for binding to receptors or to other proteins. They may also control protein folding and the lifetime of the glycoproteins. In diseases such as cancer, marked variations in the composition of the carbohydrate portions of membrane glycoproteins and glycolipids have been observed. These alterations may result in damage to cell-cell recognition signals leading to loss of contact inhibition and increased cellular mobility.

VI.   SOME ANTIBIOTICS ACTING AS INHIBITORS OF PROTEIN SYNTHESIS

Inhibitors of Transcription	Mode of Action
Actinomycin D	Binds to DNA in GC-rich regions leading to inhibition of transcription, especially ribosomal RNA.
Á-Amanitin	Inhibits eukaryotic RNA polymerases I, II, and III to varying degrees.
Cordycepin	Inhibits chain elongation during RNA transcription and poly(A) tail formation.
Rifampicin	Inhibits bacterial RNA synthesis by binding to RNA polymerase.

Inhibitors of Translation (Act at ribosomal level)

Bacterial

Tetracyclines	Prevent initial binding at entry site. Maximum inhibition of aa-tRNA binding is 50%, thus it appears to be specific for one site only.
Chloramphenicol	Acts at a step after binding and during peptide bond formation. It binds to the 50S subunit and stops the formation of a peptide bond. It also inhibits the puromycin-induced premature release of peptides.
Streptomycin and **dihydrostreptomycin**	Bind to 30S subunit of the ribosome causing distortion of the structural relationships between the ribosomes, mRNA, and aa-tRNA leading to misreading of the code. The products are non-functional proteins.
Erythromycin	Inhibits the translocation step during protein synthesis in bacteria.

Mammalian

Cycloheximide, **emetine** and **dihydroemetine**)	Inhibit protein synthesis in L-cells, reticulocytes, liver slices, yeast but do not affect bacteria. Exact site of action unknown, but seems to be near site of peptide bond formation. The rate of translation is slowed down markedly and the release of completed peptide chains is inhibited. Do not act on prokaryotic ribosomes.

Miscellaneous antibiotics

Diphtheria toxin	Covalently modifies elongation factor EF-2 in mammalian cells, leading to inhibition of protein synthesis.
Nalidixic acid and **Novobiocin**	Inhibit topoisomerases leading to inhibition of DNA synthesis in bacteria.
Puromycin	Mimics the aminoacyl region of aminoacyl tRNA and causes premature release of polypeptide chains during protein synthesis in bacterial and eukaryotic cells.
Tunicamycin	Inhibits N-linked glycosylation events in animal cells.

VII. REVIEW QUESTIONS ON PROTEIN BIOSYNTHESIS

DIRECTIONS: For each of the following multiple-choice questions (1 - 39), choose the ONE BEST answer.

1. If GGC is a codon in mRNA (5′- 3′ direction), which one of the following would be the anticodon (5′- 3′ direction) in tRNA?

A. GCC
B. CCG
C. CCC
D. CGC
E. GGC

Ans A: Codons and anticodons pair in an antiparallel manner forming a small double helix. The answer, 5′- 3′, is GCC.

2. Which one of the following antibiotics will inhibit protein synthesis in both prokaryotic and eukaryotic organisms?

A. cycloheximide
B. tetracycline
C. streptomycin
D. emetine
E. puromycin.

Ans E: Puromycin acts upon both eukaryotic and procaryotic peptidyl transferases releasing small peptides ending in puromycin at their C-terminal ends.

3. tRNA molecules have at their 3′ termini the sequence:

A. CCA
B. CAA
C. CCC
D. AAC
E. AAA

Ans A: All functional tRNA's end in CCA at the 3′ end. The CCA-amino acyl moieties are recognized by peptidyl transferase.

4. The large subunit of mammalian ribosomes has a sedimentation constant of:

A. 40 S
B. 70 S
C. 30 S
D. 80 S
E. 60 S

Ans E: Prokaryotic ribosomes are composed of 50S and 30S ribosomal subunits. Eukaryotic ribosomes are composed of 60S and 40S subunits.

5. A cluster of ribosomes translating the same mRNA is called a(n):

A. episome
B. monosome
C. polysome
D. spliceosome
E. genome.

Ans C: A cluster of ribosomes linked together by a mRNA molecule is referred to as a polyribosome or polysome.

6. According to Blobel's hypothesis, all proteins destined for secretion contain a sequence of amino acids:

A. at their C-terminal ends that are primarily hydrophilic
B. at their C-terminal ends that are primarily hydrophobic
C. at their N-terminal ends that are primarily hydrophilic
D. at their N-terminal ends that are primarily hydrophobic
E. in the middle of the polypeptide chain that are primarily hydrophilic.

Ans D: Blobel's hypothesis states that a signal sequence of 20 - 30 amino acids rich in hydrophobic residues is at the N-terminus of proteins destined for secretion.

7. Most mammalian genes are "split", i.e., they contain coding and non-coding segments. After transcription into pre-mRNA the non-coding parts are removed or "spliced out". These segments are called:

A. spliceosomes
B. introns
C. exons
D. promoters
E. signal pieces.

Ans B: Introns are non-coding sequences in pre-mRNA's that are removed and degraded. Exons are coding sequences that are conserved and spliced together to form functional mRNA.

8. When the ribosome reaches one or more of the chain termination codons on mRNA, release factors act to change the specificity of peptidyl transferase so that the growing peptide chain is transferred to:

A. water
B. EF-G
C. a membrane-bound polysome
D. the next incoming amino acid
E. the A-site.

Ans A: Release factors interact with peptidyl transferase allowing it to transfer the nacsent polypeptide chain to water instead of to the alpha amino group of an incoming aminoacyl tRNA.

9. Amino acid activating enzymes couple amino acids to the ribose moiety of the terminal adenylate residue of tRNA by forming an ester bond between the amino acid's alpha carboxyl and the ribose's:

A. 1' OH
B. 2' OH
C. 3' OH
D. 4' OH
E. 5' OH

Ans C: All amino acids are "activated" by forming an ester bond between the amino acid carboxyl and the 3'-OH of ribose at the terminal adenosine of tRNA.

10. In N-linked glycoproteins, the initial "mannose rich" oligosaccharide is assembled in the cytosol, transferred through the membranes of the endoplasmic reticulum and to the intra-luminal polypeptide chain with the aid of:

A. tocopherol phosphate
B. vitamin K phosphate
C. dolichol phosphate
D. isoprenol phosphate
E. vitamin A phosphate.

Ans C: Precursors of N-linked oligosaccharides of glycoproteins are assembled in cytosol on dolichol phosphate. Completed side chains are then transferred to specific asparagine residues on the polypeptide acceptor.

11. Chain termination codons in mRNA are recognized by which proteins?

A. release factors
B. restriction enzymes
C. elongation factors
D. cap binding protein
E. initiation factors.

Ans A: Chain termination codons recognize and bind release factors.

12. The codon UUC is specific for the amino acid:

A. leucine
B. lysine
C. proline
D. phenylalanine
E. hydroxyproline.

Ans D: UUC is one of the two codons specific for phenylalanine. The other is UUU.

13. The genetic code is:

A. non-degenerate
B. species specific
C. a triplet code
D. punctuated
E. translated by RNA polymerase.

Ans C: The genetic code is universal, non-punctuated, degenerate and a triplet code.

14. A mutation resulting from a deletion of one or more nucleotides results in a:

A. translocation
B. inversion
C. frameshift change
D. transversion
E. transition.

Ans C: The addition or deletion of one or more nucleotides causes changes in the reading frame and are referred to as frameshift mutations.

15. Which one of the following statements about eukaryotic mRNA is NOT correct?

A. contains an unusual structure at its 5′ end
B. has a long stretch of poly A at its 3′ end
C. is transcribed as a hnRNA precursor
D. is transcribed by RNA polymerase I
E. is transcribed in the nucleus.

Ans D: RNA Pol I transcribes the 45S RNA precursor leading to the major rRNAs. RNA Pol II transcribes mRNA and RNA Pol III transcribes tRNA.

16. tRNA molecules have structural elements which recognize:

A. mRNA codons
B. aminoacyl synthetases
C. elongation factor Tu
D. ribosomal subunits
E. all of the above.

Ans E: tRNAs are complex structures involved in many different interactions including all those mentioned.

17. RNA synthesized in the nucleolus will most likely end up:

A. in eukaryotic chromatin
B. in eukaryotic ribosomes
C. as mRNA
D. in prokaryotic ribosomes
E. degraded to tRNA.

Ans B: The nucleolus is the site of rRNA synthesis and ribosomal assembly in eukaryotic cells.

18. Which of the following has a "poly-A tail"?

A. bacterial mRNA
B. eukaryotic mRNA
C. rRNA
D. DNA
E. bacterial tRNA.

Ans B: Only eukaryotic mRNA molecules have poly (A) tails of 100 - 250 nucleotides long. Prokaryotic mRNA molecules lack poly(A) tails.

19. Which of the following compounds will be used to initiate eukaryotic protein synthesis?

A. Acetyl seryl tRNA.
B. Acetyl serine.
C. Formyl methionine
D. Formyl methionyl tRNA$_f$
E. Methionyl-tRNA$_i$.

Ans E: Chain initiation in eukaryotes requires Met-RNA$_i$, where 'i' stands for initiation. The methionine is not formylated.

20. A point mutation resulting in a single base change in a mRNA codon will most likely cause:

A. inactivation of ribosomes
B. no effect
C. inactivation of IF-3
D. mRNA hydrolysis
E. an amino acid change in a polypeptide chain.

Ans E: The most probable effect of a single base change is to change a codon sequence leading to the substitution of one amino acid for another.

21. In eukaryotic cells, those proteins destined to be secreted from the cell are formed:

A. in the nucleolus
B. in the mitochondria
C. on membrane-bound polysomes
D. on free ribosomes
E. on free polysomes.

Ans C: Proteins destined to be secreted from cells or to become part of cell membranes are synthesized on membrane-bound polysomes. Free polysomes synthesize "housekeeping" proteins.

22. Eukaryotic ribosomes sediment with a value of:

A. 30S
B. 40S
C. 70S
D. 80S
E. 50S

Ans D: Intact free eukaryotic ribosomal complexes sediment in a gradient with a value of about 80S.

23. When the signal recognition particle (SRP) is bound to the signal peptide on the growing polypeptide chain:

A. the elongation rate is accelerated
B. the elongation rate is decreased slightly
C. elongation is stopped completely
D. chain termination occurs
E. the signal peptide is cleaved off.

Ans C: When the signal piece extends from the 60S subunit a sufficient distance it is bound by the SRP and a further elongation is stopped until the SRP is removed.

24. An anticodon is:

A. the part of a DNA molecule which codes for chain termination
B. a 3-nucleotide sequence of a mRNA molecule
C. a specific part of a tRNA molecule
D. a nucleotide triplet of a rRNA molecule
E. the portion of a ribosomal subunit which interacts with the amino acid activating enzyme.

Ans C: An anticodon is that part of a tRNA molecule that will base pair with a specific codon on a mRNA molecule.

25. During elongation, for each amino acid added to the growing peptide chain, the energy required per each ribosome is furnished by:

A. 1 GTP
B. 2 GTP
C. 3 GTP
D. 4 GTP
E. 1 ATP

Ans B: One GTP molecule is hydrolyzed when an amino acyl tRNA is inserted into the A-site. A second is expended during the translocation process.

26. All tRNAs have the same:

A. base sequence
B. base composition
C. axial length
D. anticodon
E. amino acid activating enzymes.

Ans C: The axial distance between the amino acid and the anticodon regions is nearly the same for all aminoacyl tRNAs.

27. Diphtheria toxin inhibits:

A. peptide chain initiation
B. peptide chain elongation
C. peptide chain termination
D. DNA supercoiling
E. mRNA splicing.

Ans B: Diphtheria toxin catalyzes the covalent attachment of ADP to the catalytic unit of eEF-G, irreversibly inactivating it. Translocation, and therefore elongation, is blocked in eukaryotic cells.

28. Which of the following is a termination codon?

A. UAA
B. AUG
C. UUU
D. AAA
E. GCA

Ans A: The termination codons are UAA, UAG and UGA.

29. Which of the following factors are required for addition an amino acid to a growing polypeptide chain during protein biosynthesis in *E. coli?*

A. IF-1, IF-2 and IF-3
B. Poly(A) polymerase
C. R1
D. EF-Tu, EF-Ts and EF-G
E. ATP.

Ans D: EF-Tu, EF-Ts and EF-G are elongation factors in prokaryotes.

30. The genetic code is degenerate because:

A. multiple species of ribosomes exist
B. multiple species of tRNA exist for most amino acids
C. there is a great inaccuracy during the process of transcription
D. a common codon exists for at least two amino acids
E. the code is not universal.

Ans B: Except for methionine and tryptophan all amino acids have multiple codons assigned to them. Each codon is usually recognized by its own specific tRNA.

31. The 5'-"cap" on eukaryotic mRNA is essential for:

A. termination of translation
B. initiation of translation
C. transport of newly-formed mRNA from the nucleus
D. initiation of transcription
E. hydrolysis of GTP.

Ans B: All functional eukaryotic mRNA's have a 5' cap structure which is essential for the binding of the mRNA to the 40S ribosomal subunit initiation complex.

32. Each ribosome in a polysome is:

A. moving in the 3' to 5' direction on the mRNA
B. synthesizing many polypeptide chains
C. synthesizing only one polypeptide chain
D. dissociated
E. inhibited by actinomycin D.

Ans C: Each ribosome, as it passes 5' to 3' along a mRNA, synthesizes one polypeptide chain. But several ribosomes may utilize a single mRNA molecule, making several copies of the polypeptide.

33. The majority of the peripheral sugars of the oligosaccharide side chains of complex glycoproteins are attached in the:

A. rough endoplasmic reticulum
B. smooth endoplasmic reticulum
C. lysosomes
D. Golgi membranes
E. plasma membranes.

Ans D: The initial "mannose core" of N-linked oligosaccharide side chains is assembled in the rough ER. The peripheral N-acetylglucosamine, galactose, fucose, and N-acetylneuraminic acid are added in the Golgi membranes.

34. Proteins are synthesized:

A. from the N-terminal to the C-terminal direction
B. from the C-terminal to the N-terminal direction
C. from rRNA
D. from 19 amino acids
E. in the nucleus.

Ans A: All polypeptides start with methionine or formyl-methionine and add amino acids to the carboxyl end of the growing polypeptide chain. Growth is N to C.

35. An antibiotic that binds to the 50S subunit of a bacterial ribosome and blocks the A-site preventing the entrance of a new aminoacyl-tRNA complex is:

A. tetracycline
B. chloramphenicol
C. penicillin
D. dihydrostreptomycin
E. cycloheximide.

Ans A: The only antibiotic listed that acts in this way is tetracycline.

36. The 70S ribosome contains about 50 separate proteins and one clearly definable enzymatic activity involved in peptide bond formation. The name of that activity is:

A. 30S dependent ATPase
B. aminoacyl-tRNA synthetase
C. elongation factor G
D. peptidyl transferase
E. IF-3.

Ans D: The 50S subunit contains a latent catalytic activity, which when activated in the complete 70S ribosome/polysome, is known as peptidyl transferase. Current evidence links this activity with the 23S rRNA in prokaryotes.

37. Nascent glycoproteins having certain mannose residues phosphorylated are destined to be:

A. membrane proteins
B. secreted from the cell
C. transported to the nucleus
D. transported to lysosomes
E. degraded to amino acids.

Ans D: Phosphorylation of certain mannose residues of oligosaccharide side chains of glycoproteins at the 6-position in the membranes of the Golgi is a signal that these glycoproteins are destined to associate with pre-lysosomal vesicles to become lysosomal enzymes.

38. Small nuclear RNAs (snRNA) are involved in:

A. DNA synthesis
B. RNA synthesis
C. RNA splicing
D. aminoacylation
E. recombination.

Ans C: Several snRNPs combine to form a complex called a spliceosome which splices out introns from pre-mRNA molecules and ligates the exons together to form functional mRNAs.

39. Which of the following antibiotics binds to 70S ribosomal subunits, and if the concentration is high enough, will cause mis-reading of the genetic code?

A. cycloheximide
B. puromycin
C. chloramphenicol
D. dihydrostreptomycin
E. tetracycline.

Ans D: Streptomycin derivatives interfere with the relationships between ribosomes, mRNA and amino acyl tRNA, causing mis-reading of the genetic code. That is, the "proofreading" aspects of protein synthesis are disturbed.

DIRECTIONS: For each set of questions, choose the ONE BEST answer from the lettered list above it. An answer may be used one or more times, or not at all.

Questions 40 - 43:

A. transfer RNA (tRNA)
B. ribosomal RNA (rRNA)
C. Both
D. Neither

40. In bacteria, is found as 16S and 23S complexes.

Ans B: Bacterial rRNA consists of 5S, 16S and 23S species.

41. Forms ester bonds with activated amino acids.

Ans A: Activated amino acids form ester bonds with the 3'OH of the terminal adenosines of tRNA molecules.

42. Contains amino acid-specific codons.

Ans D: Amino acid specific codons are found only in mRNA molecules.

43. Synthesis is inhibited by actinomycin D.

Ans B: Ribosomal rRNA synthesis is quite sensitive to actinomycin D. tRNA synthesis is much more resistant.

Questions 44 - 47:

A. EF-G
B. peptidyl transferase
C. Both
D. Neither

44. Forms peptide bond.

Ans B: Peptidyl transferase is the activity that actually forms the peptide bond. EF-G is a translocase.

45. Inhibited by chloramphenicol in bacteria.

Ans B: Chloramphenicol interferes with peptidyl transferase activity in prokaryotes.

46. Requires GTP.

Ans A: EF-G acts as a translocase. It requires GTP as an energy donor.

47. Participates in translocation of peptidyl tRNA.

Ans A: EF-G catalyzes translocation of the peptidyl-tRNA/mRNA complex from the A site to the P site.

14. MEDICAL GENETICS

Sara L. Tobin

with figures by **Ann Boughton**, Thumbnail Graphics

Medical genetics involves the study of chromosomal DNA abnormalities that cause genetic diseases, such as mutations or misdistribution of chromosomes. DNA sequences vary among individuals, accounting for our individuality in appearance, personal biochemistry, resistance to disease, and response to medication. Deleterious mutations are more serious than these individual variations and result in the appearance of disease symptoms in affected family members. We will review the patterns of genetic inheritance, recombinant DNA techniques, and the use of molecular genetics to predict whether or not individuals or their children may inherit a genetic disease.

I. BASIC GENETICS

The cells of a human are **diploid**, containing 46 chromosomes. Twenty-three of these chromosomes are provided by the **haploid** germ cell (egg or sperm) contributed by each parent. For instance, one of your two copies of chromosome 15 is derived from your father and the other from your mother. You have 22 pairs of **autosomes**, numbered 1 - 22, and a pair of **sex chromosomes** such that females have two X chromosomes and males have one X and one Y chromosome. Each chromosome is composed of a long, continuous strand of DNA complexed with proteins. Genes are found at discrete locations along each chromosomal piece of DNA. A pair of autosomes represents two sets of genes in a particular order along the DNA strand.

Therefore we carry two (possibly different) versions of each gene governing a particular characteristic (except for genes on sex chromosomes). These different versions of a gene are called **alleles**. For instance, blue and brown are two alleles for the gene governing eye color. A possible distribution of alleles for four different genes on chromosome 7 is shown in Figure 14 - 1. An individual is **homozygous** if both alleles for a given gene are the same or **heterozygous** if the two alleles are different.

The inheritance pattern of a genetic disease depends on whether the mutation is **dominant** or **recessive** and on its chromosomal location. Recessive mutations may be masked if a normal version (allele) of the gene is present. Dominant mutations cause the disease to be evident even though the normal gene is present. From a biochemical point of view, recessive mutations result in the absence of a gene product or the production of an abnormal gene product, but the normal allele produces sufficient protein to allow the cell to perform its normal function. Therefore a recessive mutant allele is masked by the normal allele in the heterozygous state, but the recessive trait is exposed in homozygotes due to the altered gene products produced by both mutant genes. In contrast, dominant inheritance patterns can result from insufficient product from the normal allele, from interference with a normal cell structure or biochemical process by the mutant gene product (despite the presence of the normal protein), or from certain alterations in the ways genes are

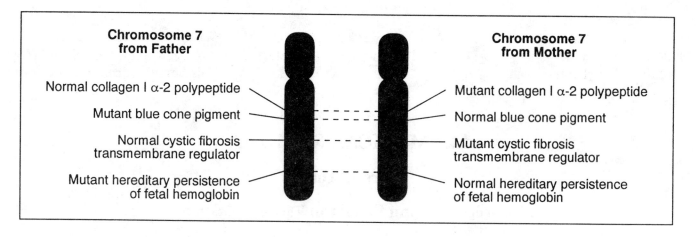

Figure 14 - 1: Allelism.

This diagram shows paternal and maternal copies of human chromosome 7 in an individual. The approximate positions of four autosomal recessive genes known to be on chromosome 7 are indicated. Note that each gene occupies the same position on each of the two homologous chromosomes. This person is heterozygous for each of these genes. However, because the homologous chromosome has a normal gene, he or she will be unaffected by any of these four genetic diseases. The normal and mutant alternatives of a particular gene are called alleles. Some human genes have several dozen different alleles. For instance, ΔF508 is the most common mutant allele encoding the cystic fibrosis transmembrane regulator (CFTR) gene. When both CFTR alleles are mutant, the individual will be affected by cystic fibrosis.

regulated. Mutations may be found on an autosomal chromosome, the X chromosome, or in the mitochondrial genome. In general, a genetic disease will be inherited in one of five different patterns, as summarized in the pedigrees drawn below, which use standard genetic symbols, as seen in Figure 14 - 2.

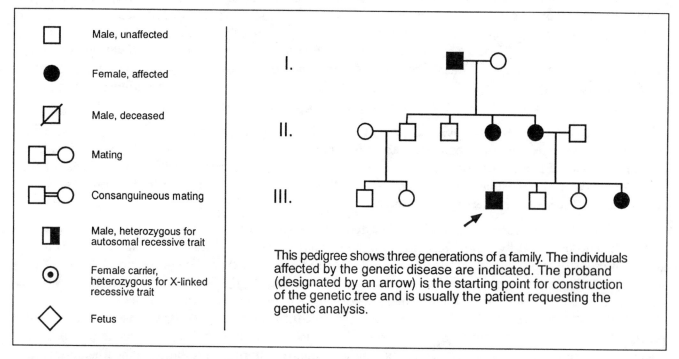

This pedigree shows three generations of a family. The individuals affected by the genetic disease are indicated. The proband (designated by an arrow) is the starting point for construction of the genetic tree and is usually the patient requesting the genetic analysis.

Figure 14 - 2: Genetic Pedigree Symbols.

The most common standard pedigree symbols are diagrammed on the left. The right panel shows a typical pedigree.

Autosomal recessive inheritance patterns (Figure 14 - 3) are seen when a mutant gene is located on an autosome and is recessive (not evident) in the presence of a normal allele. Examples of autosomal recessive diseases are cystic fibrosis and maple syrup urine disease (a defect in metabolism of branched-chain keto acids). Note that the standard genetic symbols (Figure 14 - 2) are utilized in this and subsequent pedigrees.

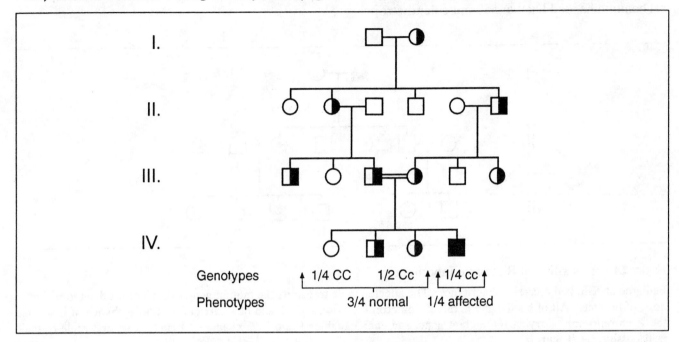

Figure 14 - 3: Autosomal Recessive Inheritance.

Pedigree of autosomal recessive inheritance, including a marriage between two first cousins. A recessive gene from generation I is transmitted through two offspring, resulting in a child affected with the disease in generation IV. Note the ratios of genotypes and phenotypes in generation IV resulting from a mating between two heterozygous individuals.

Autosomal dominant inheritance patterns (Figure 14 - 4) occur when a dominant mutation is located on an autosome. The mutation is evident even in the presence of a normal allele. Examples of genetic diseases that are inherited as autosomal dominants are Type I Alzheimer's disease (ADI) and Huntington disease.

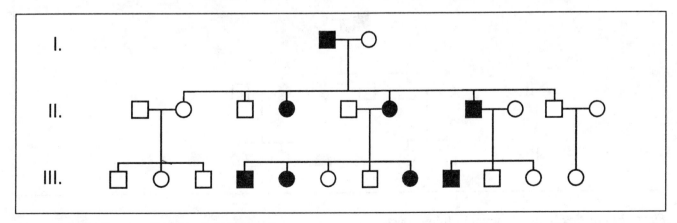

Figure 14 - 4: Autosomal Dominant Inheritance

Pedigree of autosomal dominant inheritance. On average, half of the offspring of affected persons are affected. The condition is transmitted only by affected persons and never by unaffected family members. An exception to transmission only by affected persons may occur in cases of reduced penetrance, in which a person may carry the mutant allele but appear unaffected. Equal numbers of males and females are affected, and male-to-male transmission occurs.

In the case of **sex-linked recessive** inheritance patterns (Figure 14 - 5), the gene is located on the X chromosome and is recessive in the presence of a normal allele. However, because males carry only a single X chromosome, a normal allele is not present. Therefore these diseases are evident predominantly in males. Examples of recessive sex-linked genetic diseases are male pattern baldness, red/green color blindness, classical hemophilia, and Duchenne's muscular dystrophy.

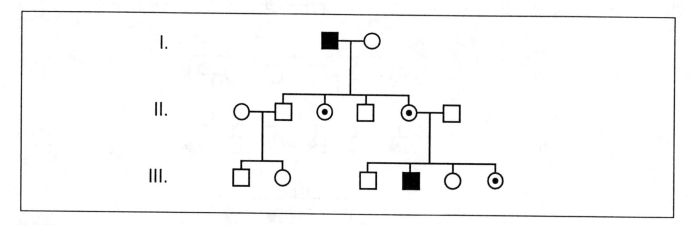

Figure 14 - 5: Sex-linked Recessive Inheritance

Pedigree of X-linked recessive inheritance. The affected male transmits the mutant gene to each of his daughters, but to none of his sons. All of his daughters therefore become carriers. Each son of a carrier has a 50% chance of inheriting the X chromosome carrying the mutant gene, and each daughter has a 50% chance of being a carrier. Male-to-male transmission is not seen.

Sex-linked dominant patterns of inheritance (Figure 14 - 6) occur when a dominant mutation is located on the X chromosome. Therefore the disease is evident in all who inherit the mutant allele. However, these diseases are often less severe in females due to the presence of the normal allele. Examples of sex-linked dominant genetic diseases are amelogenesis imperfecta (thin tooth enamel) and Charcot-Marie-Tooth neuropathy (at xq13), which causes degeneration of spinal nerve roots.

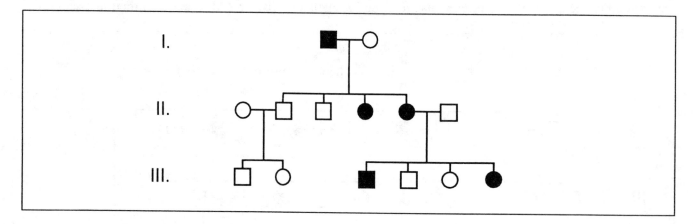

Figure 14 - 6: Sex-linked Dominant Inheritance

Pedigree of X-linked dominant inheritance. The daughters of affected males are always affected, and the sons of affected males are never affected. Half the offspring of affected females are affected. The inherited condition is often less severe in affected females.

Because mitochondria for the developing embryo are supplied by the mother during the process of oogenesis, mutations in the mitochondrial genome (see Figure 14 - 7) are inherited in a maternal pattern. This means that both male and female offspring of an affected mother will be affected, but males will not pass the trait on to their children. Two examples of genetic diseases that are due to mutations in the mitochondrial genome are mitochondrial myopathy (weakened muscles) and Leber's hereditary optic atrophy (a degenerative eye disease).

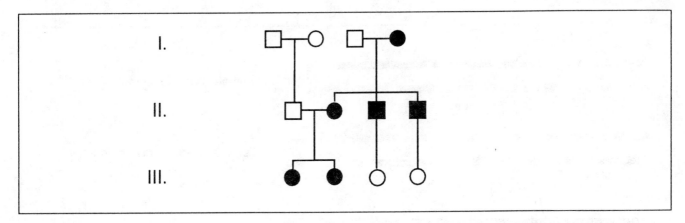

Figure 14 - 7: Mitochondrial Inheritance

Pedigree of a family with a disease resulting from a defective gene in the mitochondrial genome. Note that the mode of transmission is maternal because mitochondria pass from generation to generation through the egg. Therefore both male and female offspring of an affected mother are affected, but males cannot transmit the disease.

During the formation of germ cells, the 46 chromosomes are allocated in randomly distributed pairs. For example, during spermatogenesis in the potential father, half his sperm will receive the copy of chromosome 17 he inherited from his father, and half of his sperm will contain the chromosome 17 he inherited from his mother. The same random pattern of distribution occurs for each of the other 22 chromosome pairs. This random reallocation is known as the principle of **independent assortment**, in which each member of a given pair of chromosomes is distributed independently of the members of every other pair. The resulting germ cells carry a mixture of paternally- and maternally-derived chromosomes. Therefore in the inheritance of a mutation causing a disease, a heterozygous parent (carrying one normal gene and one disease gene) will generate germ cells, half of which carry the disease gene and half of which contain the normal gene.

In addition to the chromosomal redistribution mechanism represented by independent assortment, there is also a second scrambling process called **recombination** that is important for our ability to predict the inheritance of mutant alleles. Genes that are present on the same chromosome are said to be **linked** because they are located on the same continuous piece of DNA. If recombination did not exist and the genes for hair color and eye color were on the same chromosome, they would always be inherited together (brown hair with brown eyes, for example). However, recombination or "crossing over" permits the exchange of segments of chromosomal DNA between homologous chromosomes (for instance, the maternal and paternal copies of chromosome 18) during the process of generating germ cells. Therefore, as shown in Figure 14 - 8, particular alleles can be shifted from one chromosome to the other by breaking and rejoining of homologous chromosomes. This process does not, however, alter the order of genes on the chromosome.

Recombination is important in medical genetics because it influences our ability to predict the inheritance of a mutant gene. Recombination events are roughly proportional to the distance between two genes on a chromosome; the greater the distance between two genes, the higher the probability that recombination will occur between them, shifting one of the alleles to a homologous chromosome. In fact, **linkage** of two genes is generally expressed on the basis of the frequency of recombination. The percentage of recombination events between two genes provides a rough measurement of the physical distance between them: 20% recombination equals 20 centiMorgans, with one **centiMorgan** equal to approximately one million base pairs. Thus genes that are very close to each other will tend to be inherited together and are said to be genetically linked. In contrast, genes that are separated by 50 centiMorgans or more will be exchanged so often by recombination events that the alleles will be inherited randomly relative to each other, even though they are located on the same chromosome.

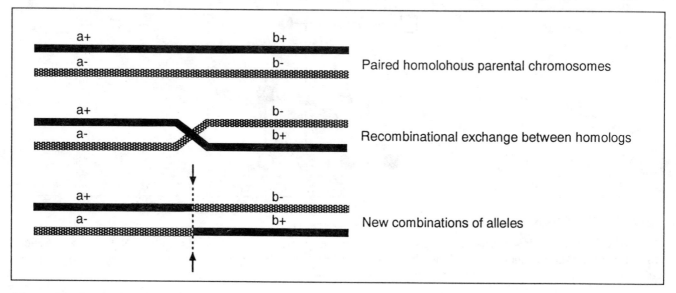

Figure 14 - 8: Recombination

Recombination events can move alleles from one parental chromosome to the homologous chromosome inherited from the other parent. These "crossing over" events can result in new combinations of linked alleles. The rate of recombination is roughly proportional to the distance along the chromosome separating two genes.

Important Terms and Concepts:

allele	autosomal inheritance
heterozygous	sex-linked inheritance
homozygous	mitochondrial inheritance
independent assortment	centiMorgan
recombination	diploid
genetic linkage	haploid
recessive	sex chromosome
dominant	autosome

II. MUTATIONS

A **mutation** is any change in DNA, whether or not it alters a gene. Mutations can be caused by a variety of mutagenic agents, such as X-rays, ultraviolet light, chemical **carcinogens**, or environmental carcinogens.

Mistakes in DNA replication or repair can also lead to mutational events (See Chapter 12, pp. 212 - 213 for details). If mutations occur in germ line tissue, the changes can be inherited by the offspring. Somatic mutations will not be inherited, but they can lead to the development of cancers. Cancer is fundamentally a genetic disease, resulting from the inactivation of tumor suppressor genes or the activation of oncogenes. Because of the linkage between mutagenicity and carcinogenicity, the identification of potential mutagens is important for human health.

Mutagenic events can be large in scale, such as rearrangements that involve large pieces of chromosomal DNA. Movement of a chromosome segment to a different position is called a **translocation**. **Insertion** events are characterized by addition of DNA to a particular site, while **deletions** remove a segment of DNA. Single-base changes occupy the most minute of DNA alterations. If **single-base changes** occur within the protein-coding region of a gene, the ability of the cell to produce a functional protein can be affected. As summarized in the Figure 14 - 9, **missense**, **nonsense**, or **frameshift** mutations fall into this latter category.

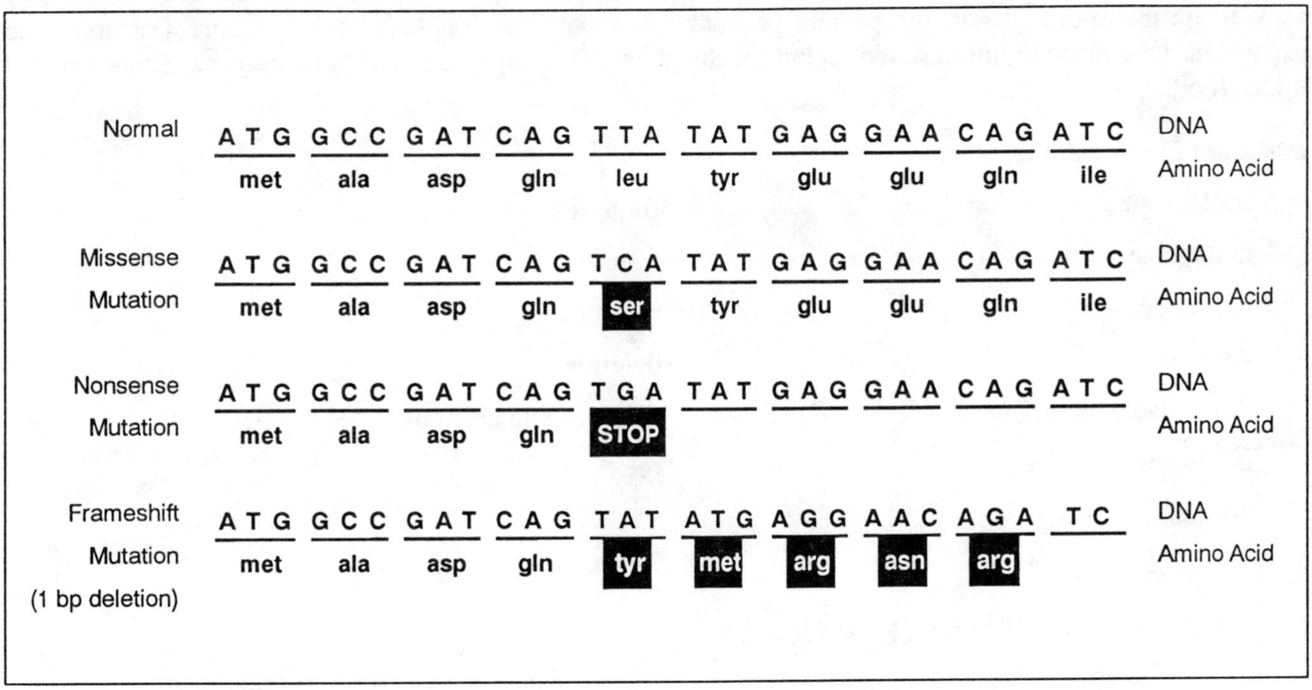

Figure 14 - 9: Single-Base Mutations

This figure illustrates the effects of mutational events involving a single base change on the ability of a gene to encode a functional protein. Note: the "sense" DNA strand specifying the amino acid sequence of an encoded protein is shown.

Mutagenicity of substances can be estimated by the **Ames test**, which detects DNA alterations in special bacterial strains. Suspect compounds can then be subjected to further testing to establish whether or not they are mutagenic in humans. This test has also been used to help identify compounds that serve as anti-carcinogens.

In addition to mutagenic chemicals, mutations can also arise from insertion of **transposable** elements or **retroviral** DNA sequences into chromosomal DNA. Both transposable elements (jumping genes) and retroviruses must become integrated into the chromosome in order to be replicated. In addition to their insertional ability, transposable elements have the capacity to excise from the chromosomal DNA. Often, these excision events are inexact and do not restore the original chromosomal DNA sequence. This can result in an insertion composed of remnants of the original transposable element. Alternatively, inexact excision can also result in the deletion of host chromosomal DNA sequences.

A third category of mutational event is the newly discovered **"triplet repeat expansion."** This class of mutations was first discovered in humans as a result of the Human Genome Project. These expansion events have currently been found in four different genes and are characterized by expansion of a trinucleotide repeat. For each of these four mutations characterized to date, a normal number of triplet repeats, usually in the range of 5 - 50, is associated with the normal allele of these genes. However, expansion of the number of triplet repeats into the range of several hundred causes the mutant allele. The four mutations that have been associated with the expansion of triplet repeats are listed below:

> Fragile-X Syndrome (X-linked recessive; mental retardation)
> Spinobulbar Muscular Atrophy (X-linked recessive; androgen receptor)
> Myotonic Dystrophy (autosomal dominant; myotonin kinase)
> Huntington Disease (autosomal dominant; uncontrollable shaking, dementia)

Expansions in the androgen receptor gene and in the Huntington disease gene cause additional amino acids to be incorporated into the protein product. However, for fragile-X and myotonic dystrophy, the expansions take place in untranslated regions of the gene. This important mutational process is not yet well understood.

Important Terms and Concepts:

triplet repeat expansions	frameshift
mutagenic	translocation
carcinogenic	insertion
Ames test	deletion
single-base changes	transposable element
missense	retrovirus
nonsense	

III.   RECOMBINANT DNA TECHNIQUES

Recombinant DNA techniques permit the amplification of specific segments of DNA. The resulting multiple identical DNA copies (clones) can then be isolated, studied, and used for production of medicines or for gene therapy. A recombinant DNA "flow chart" is shown in Figure 14 - 10. This figure is designed to help you understand how these techniques fit together. It may be useful for you to refer back to the figure as you review the recombinant DNA techniques presented in this section.

Recombinant DNA techniques grew out of small, seemingly unrelated discoveries that have led to a revolution in both biology and medicine. First, bacteria were found to house small, circular molecules of DNA called **plasmids** that carry genes conferring **antibiotic resistance**. The plasmids replicate independently of the bacterial chromosome so that multiple copies of a plasmid can be found in a single bacterium. These plasmids can be exchanged between bacteria, and this accounts for the spread of strains that are resistant to antibiotics. Second, **restriction endonucleases** (also called restriction enzymes) were found to serve as the basis of a bacterial defense against foreign DNA. These enzymes cleave foreign DNA at sequence-specific recognition sites. The host bacterium protects its own DNA from cleavage by adding methyl groups to the recognition sites in its genome. As will be summarized below, these two discoveries were utilized in

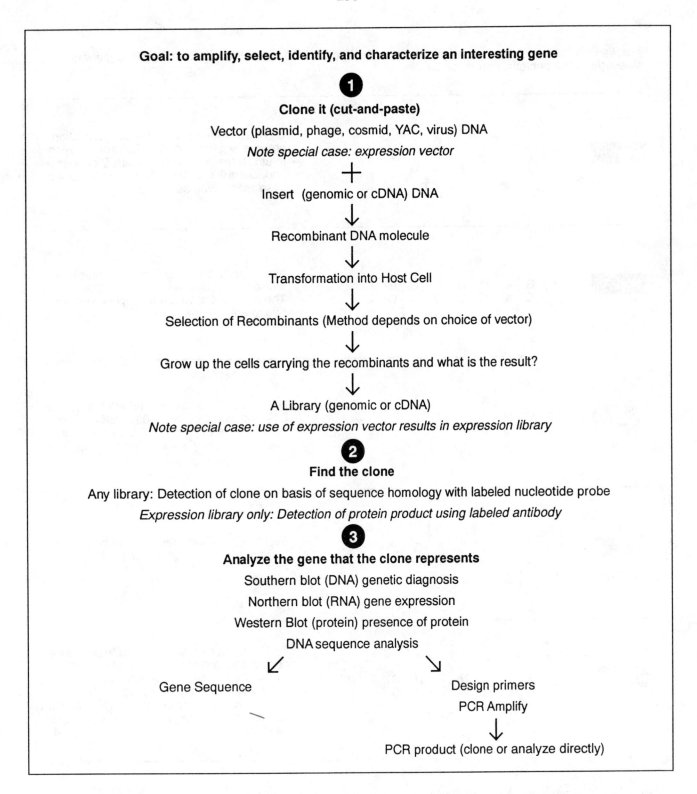

Figure 14 - 10: Recombinant DNA Flow Chart

This diagram illustrates the relationships between the basic steps in recombinant DNA cloning techniques, including creation of a library, isolation of the desired clone, and analysis of the cloned DNA.

Arrows indicate the sites of cleavage, and asterisks indicate the bases that are methylated to prevent cleavage of the bacterial chromosome.

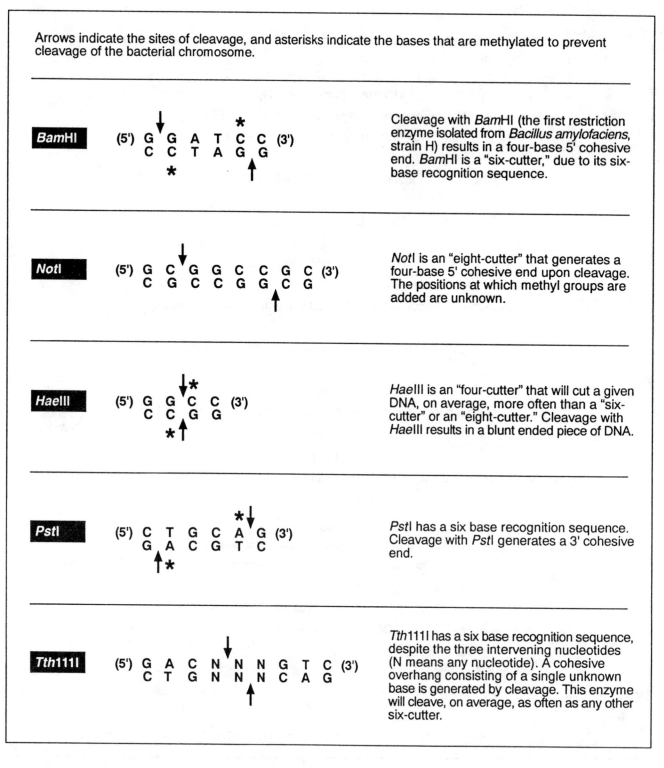

BamHI

(5') G G A T C C (3')
C C T A G G

Cleavage with *Bam*HI (the first restriction enzyme isolated from *Bacillus amylofaciens*, strain H) results in a four-base 5' cohesive end. *Bam*HI is a "six-cutter," due to its six-base recognition sequence.

NotI

(5') G C G G C C G C (3')
C G C C G G C G

*Not*I is an "eight-cutter" that generates a four-base 5' cohesive end upon cleavage. The positions at which methyl groups are added are unknown.

HaeIII

(5') G G C C (3')
C C G G

*Hae*III is an "four-cutter" that will cut a given DNA, on average, more often than a "six-cutter" or an "eight-cutter." Cleavage with *Hae*III results in a blunt ended piece of DNA.

PstI

(5') C T G C A G (3')
G A C G T C

*Pst*I has a six base recognition sequence. Cleavage with *Pst*I generates a 3' cohesive end.

Tth111I

(5') G A C N N N G T C (3')
C T G N N N C A G

*Tth*111I has a six base recognition sequence, despite the three intervening nucleotides (N means any nucleotide). A cohesive overhang consisting of a single unknown base is generated by cleavage. This enzyme will cleave, on average, as often as any other six-cutter.

Figure 14 - 11: Restriction Endonuclease Recognition Sequences

Restriction endonucleases are bacterial enzymes that recognize short "palindromic" sequences in double-stranded DNA and cleave the DNA, leaving "sticky ends" or "blunt ends," depending on the enzyme. A few examples are illustrated in the figure to demonstrate protection by methylase modification, the palindromic nature of recognition sites, the modes of cleavage resulting in different cut ends, and the relationship between frequency of cleavage and length of recognition site. Note: Do NOT memorize specific cleavage sites.

the formulation of the recombinant DNA techniques that are responsible for many recent advances in biology and medicine, as well as the emergence of the biotechnology industry.

Restriction enzymes serve as the "scissors" for the "cut-and-paste" recombinant DNA techniques. They recognize and cleave at sites that are described as **palindromic**; that is, at the recognition site, the nucleotide sequence reads the same from 5′ to 3′ as from 5′ to 3′ on the complementary strand (see examples in Figure 14 - 11). There are several hundred different restriction enzymes isolated from different bacterial strains. The lengths of their recognition sequences vary from 4 to 10 or more bases.

A restriction enzyme that recognizes a 4-base site (a four-cutter) will cleave more often, on average, than a "six-cutter." As these enzymes cleave, they can leave ends that are staggered (with protruding single-stranded DNA) or blunt. As shown in Figure 14 - 12, DNAs from differing sources, cut with the same restriction enzyme, generally have compatible cohesive ends and may be joined (ligated) together by DNA **ligase** (the "paste" of "cut-and-paste").

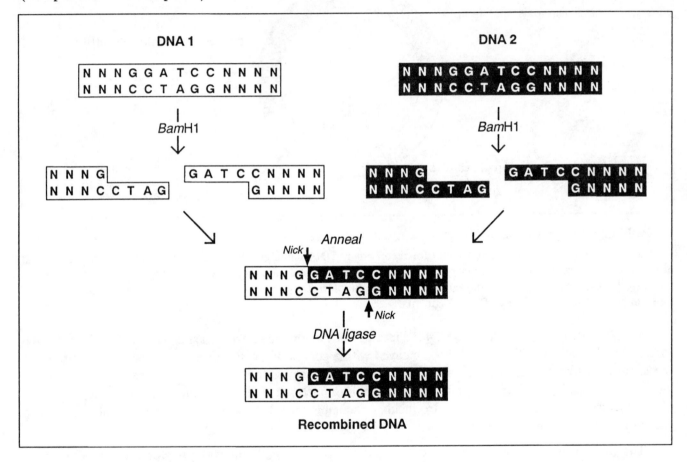

Figure 14 - 12: Recombination of DNA Fragments

DNAs from two different sources, such as vector DNA and prospective insert DNA, can be cleaved with restriction endonucleases that generate compatible cut ends. In this example, both DNAs have been cleaved with *Bam*HI, which leaves a single-stranded protrusion called a "sticky end" or cohesive end. The DNAs are mixed in a test tube, and base pairing stabilizes annealing of the cohesive ends. The enzyme DNA ligase seals the nicks in the DNA backbone, creating a strand of double-stranded DNA that is a new combination (recombination) of DNAs.

Therefore DNA from a circular plasmid molecule can be cleaved with a restriction enzyme and joined with a piece of DNA from another source that was cleaved with a restriction enzyme that generates a compatible end. When the DNAs from these two sources are combined, "recombinant DNA" has been formed.

The next goal is to amplify this DNA with the use of a vector. Vectors must have the ability to use host cell machinery to replicate multiple copies, convenient restriction sites for ligation of foreign DNA, and a mechanism for selecting recombinant molecules. A plasmid, in this example, serves as a **vector** for the amplification of a foreign DNA **insert**. Therefore plasmids are engineered to contain a very active origin of replication so that multiple plasmids are formed in each bacterial host cell. In addition, plasmid vectors are designed to contain unique restriction endonuclease sites that are conveniently located in the vector DNA. Antibiotic resistance genes are one mechanism for selection of recombinant molecules, and are included in most plasmid vectors. A map of a common plasmid vector is shown in Figure 14 - 13.

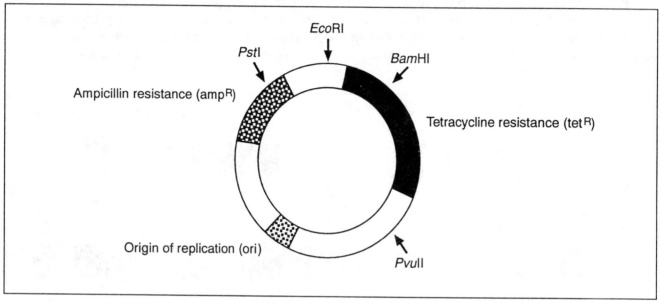

Figure 14 - 13: A Typical Plasmid Vector

pBR322 is a common circular plasmid vector for recombinant DNA cloning. This plasmid has some of the features that are important for such a vector: an origin of replication capable of generating multiple copies; genes providing resistance to the antibiotics ampicillin and tetracycline; and unique restriction sites such as *Eco*RI, *Bam*HI, *Pvu*II, and *Pst*I that can be used to accept foreign DNA fragments.

Segments of foreign DNA can be ligated into any of the unique restriction sites of a vector. The cleaved foreign DNA is mixed with plasmid DNA cleaved with a compatible restriction endonuclease that produces the same cohesive end to permit annealing, and the two are ligated together. The resulting mixture of ligated molecules is then introduced into a culture of host bacteria. The ligated molecules enter a proportion of the bacterial cells. The process of introducing a recombinant DNA molecule into a bacterial cell is called **transformation** (Figure 14 - 14).

The host bacteria are not resistant to antibiotics until a plasmid enters the bacterial cell and begins to express an antibiotic resistance gene. In the case of the intact plasmid vector pBR322, there are two antibiotic resistance genes that confer resistance to the antibiotics ampicillin and tetracycline. A fragment of foreign DNA inserted, for example, into the tet resistance gene will inactivate that gene, rendering the bacterium sensitive to tetracycline, but leaves the ampicillin resistance gene intact. Therefore, as shown in Figure 14 - 15, the antibiotic resistance characteristics of the transformed bacteria can be used to select the

Legend to Figure 14 - 14

After ligation of vector and insert DNAs, the resulting mixture is introduced into bacterial cells. Cells receiving a plasmid then take on antibiotic resistance characteristics of the plasmid construct. Transformed cells can be used to grow many identical copies of the recombinant molecule they carry. Recombinant DNAs are isolated and studied in detail.

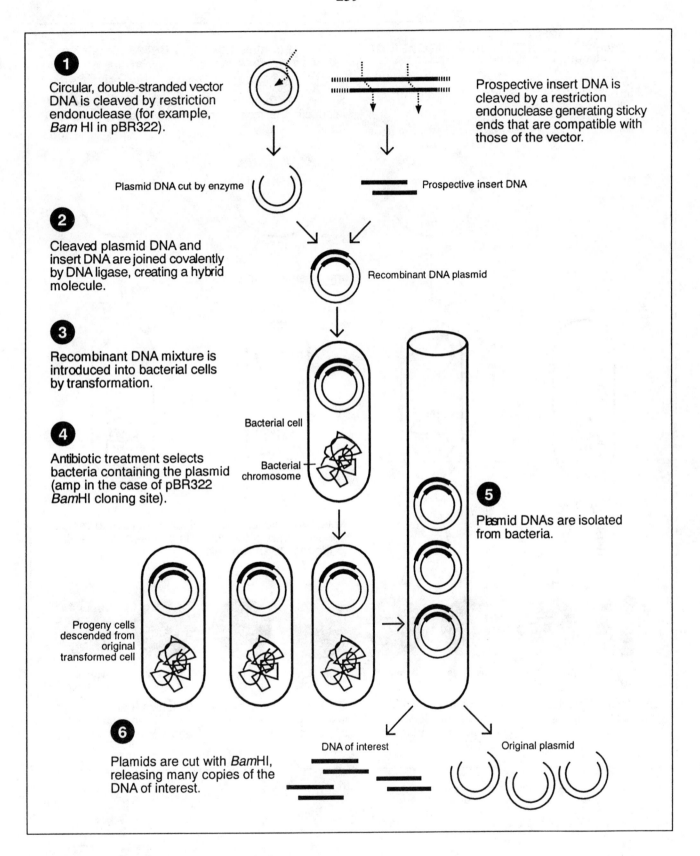

1 Circular, double-stranded vector DNA is cleaved by restriction endonuclease (for example, *Bam* HI in pBR322).

Prospective insert DNA is cleaved by a restriction endonuclease generating sticky ends that are compatible with those of the vector.

Plasmid DNA cut by enzyme

Prospective insert DNA

2 Cleaved plasmid DNA and insert DNA are joined covalently by DNA ligase, creating a hybrid molecule.

Recombinant DNA plasmid

3 Recombinant DNA mixture is introduced into bacterial cells by transformation.

Bacterial cell

Bacterial chromosome

4 Antibiotic treatment selects bacteria containing the plasmid (amp in the case of pBR322 *Bam*HI cloning site).

5 Plasmid DNAs are isolated from bacteria.

Progeny cells descended from original transformed cell

6 Plamids are cut with *Bam*HI, releasing many copies of the DNA of interest.

DNA of interest

Original plasmid

Figure 14 - 14: Transformation

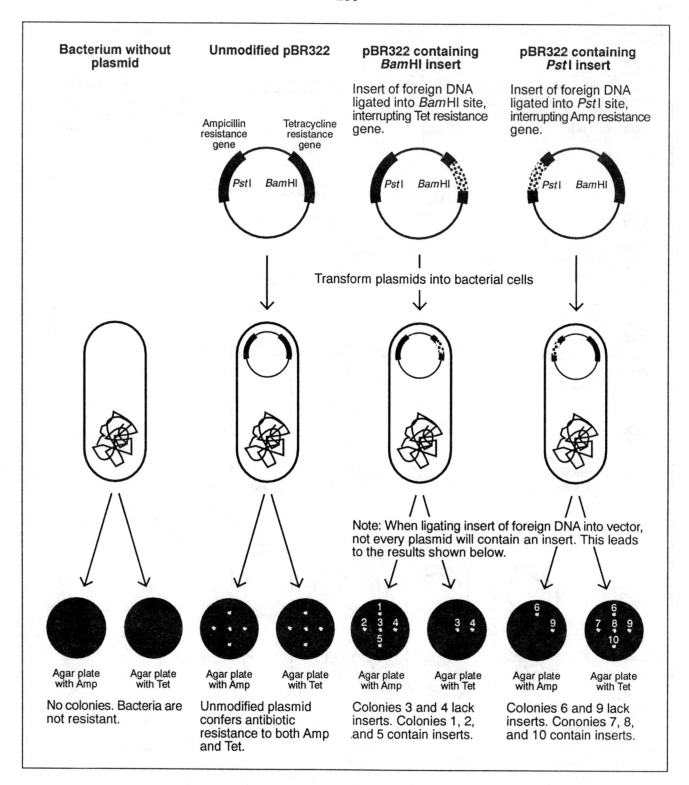

Figure 14 - 15: Antibiotic Selection

Legend to Figure 14 - 15

This figure illustrates the effects of antibiotic selection on bacterial cells that have been subjected to transformation. The bacterial cells are not antibiotic resistant until they are transformed with the plasmid. The mixture of ligated plasmid and insert DNAs is transformed into a population of bacteria which is then treated with the antibiotic corresponding to the unmodified resistance gene. Any non-transformed cells do not grow in the presence of the antibiotic and are eliminated. Note that cloning into one of the two antibiotic resistance genes inactivates that gene, but leaves the other intact. Therefore all transformed cells can form colonies in the presence of the first antibiotic. Subsequently, individual colonies can be tested for lack of resistance to the antibiotic corresponding to the interrupted gene. Those cells that are able to form colonies in the presence of both antibiotics are presumed to contain the intact plasmid and are discarded.

bacterial cells that have been transformed by their ability to form colonies in the presence of ampicillin. Subsequently the bacteria transformed by a vector containing an insert can be identified by their loss of resistance to tetracycline. Of course, cloning into a restriction site within the ampicillin resistance gene will cause sensitivity to ampicillin for transformants that are carrying an insert of foreign DNA.

Thus the bacteria containing foreign DNA can be identified by their sensitivity to the antibiotic corresponding to the interrupted resistance gene. Vectors other than pBR322 utilize different selection strategies. For instance, **phage** vectors are designed such that only the phage DNAs that acquire an insert are able to form infectious particles. **Cosmid** vectors can be grown either as plasmids or as phage and are able to carry large inserts. **Yeast artificial chromosomes (YACs)** are recombinant eukaryotic chromosomes that consist of cloned centromere and telomere sequences, an origin of replication, and a selectable marker gene. They are capable of carrying extremely large inserts of foreign DNA and are usually used for the initial cloning of chromosomal DNAs from organisms that have large genomes, such as humans.

Chromosomal DNA from an organism (in this case, human DNA) can thus be cleaved into fragments of a manageable size that are subsequently ligated into vector DNA. Transformation of this population of cloned fragments into a culture of host bacteria or yeast permits the amplification of all the cloned DNA. Provided that the chromosomal DNA is evenly represented and that a sufficient number of transformants are obtained, this set of cloned DNA fragments can represent the entire human genome. This population of clones containing chromosomal DNA is now called a **genomic library**. A second kind of library is a **cDNA library**. (Figure 14 - 16).

In this case, the insert DNA fragments are copies of messenger RNA transcripts, rather than segments of chromosomal DNA. Single-stranded DNA copies of the mRNA templates are synthesized using reverse transcriptase, the enzyme used by retroviruses to copy their genome into DNA after infection. The designation cDNA thus stands for "copy DNA" derived from mRNA. After removal of the RNA template, the single-stranded DNAs are then made double-stranded by the action of DNA polymerase, and these double-stranded DNAs are inserted into an appropriate vector. Genomic and cDNA libraries thus differ in that introns are present in genomic libraries and are generally absent from cDNA libraries due to excision of introns during mRNA processing. In addition, regulatory sequences upstream from the mRNA transcriptional start point are present in a genomic library, but are absent in a cDNA library. A further consideration is that the representation of a gene in a genomic library will correspond to its frequency in the genome. However, in a cDNA library, the proportion of clones representing a particular gene will depend on the abundance of that RNA transcript in the tissue from which the RNA was isolated.

In order to locate a clone representing the gene of interest, the desired clone must be identified specifically among the thousands of other recombinant DNA clones containing inserts that are not relevant. Usually this is carried out by separating the individual bacterial transformants or the individual phages on an agar plate. The individual clones represented by the host colonies or phage plaques are transferred to a membrane of the same shape and size as the agar plate so that the cloned DNAs are bound to the membrane at the same position as the bacterial colony or phage plaque on the plate. The DNAs from the individual

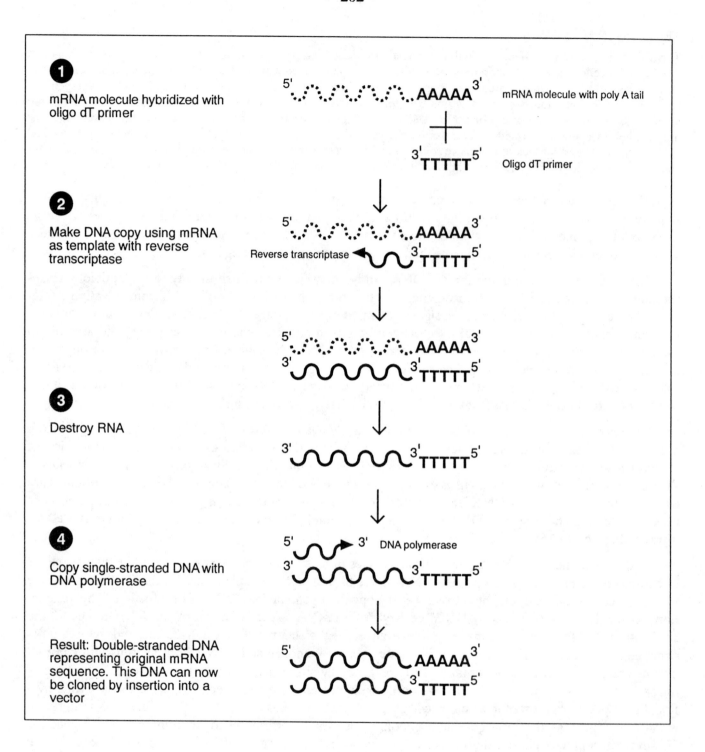

Figure 14 - 16: cDNA Cloning

In addition to the cloning of chromosomal DNA fragments, sequences transcribed into RNA can also be cloned. This figure illustrates the basic operations of cDNA cloning, in which messenger RNA (mRNA) transcripts are copied into a complementary DNA strand (cDNA stands for "copy DNA"). After removal of the RNA by RNAse treatment, a second DNA strand complementary to the first is synthesized, and this double-stranded DNA may be cloned into an appropriate vector.

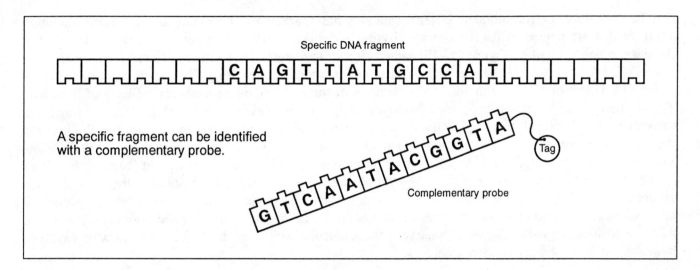

Specific DNA fragment

| C | A | G | T | T | A | T | G | C | C | A | T |

A specific fragment can be identified
with a complementary probe.

| G | T | C | A | A | T | A | C | G | G | T | A | Tag |

Complementary probe

Figure 14 - 17: Probe

A probe is a labeled RNA or DNA sequence that is single-stranded and therefore able to hybridize with a complementary nucleotide sequence. Under appropriate hybridization conditions, when the nucleotides of the probe find their complements in the target DNA or RNA, a stable hybrid is formed. The location of the probe can be detected by its label or "tag," which is usually radioactive.

clones are subjected to **denaturation**, a procedure that separates the complementary base pairs of double-stranded DNA, resulting in single-stranded DNA strands. The individual clone containing the gene of interest is then identified, usually by a **hybridization** procedure based on nucleic acid sequence homology, as described below.

First, a nucleotide sequence **probe** that is usually labeled with a radioactive tag or occasionally with a nonradioactive detection reagent is used as a tool to identify the desired clone (see Figure 14 - 17). A probe can be a cloned DNA fragment (usually several hundred nucleotides long), a synthetic nucleotide sequence, or a population of labeled RNAs. Detection is based on the fact that G always pairs with C and A always pairs with T in double-stranded nucleic acids. In single-stranded nucleic acids, the sequence of bases is exposed. Therefore single-stranded nucleic acids can be incubated together under appropriate conditions, and when complementary sequences come together, a stable **double-stranded hybrid** is formed. The cloned DNAs placed on the membrane filter for identification have already been subjected to denaturation and are available for annealing or hybridization. The probe sequences are also single-stranded. Under proper conditions of heat and salt, complementary sequences can anneal by base pairing and form a stable, double-stranded hybrid. The label on the probe permits the precise location of the desired clone to be determined. This process is illustrated in the figure below, using X-ray film to detect the location at which the radioactively labeled probe has hybridized. When the film is aligned with the colonies or plaques on the original plate, the colony containing the recombinant DNA clone with the sequence of interest is identified (Figure 14 - 18). The bacterium containing the clone can then be grown up in large quantities, and the cloned DNA can be subjected to analysis using the techniques presented later in this chapter.

In addition to the above vectors and nucleic acid hybridization detection methods, antibody detection procedures are sometimes used. In these cases, a special class of vector, the **expression vector**, is engineered to produce a protein using cloned DNA as template. If the inserted DNA is ligated in the appropriate reading frame for synthesis of the protein, the host cell will translate a "recombinant" protein. Then the desired clone can be detected using a labeled antibody that can recognize the protein of interest. This procedure normally involves filter techniques to detect colonies expressing the antigen that reacts with the antibody.

The invention of recombinant DNA techniques has sparked the development of several analytical advances that are important for the characterization and diagnosis of human genetic diseases. The first of these is the **Southern blot**, (Figure 14 - 19) named after its inventor, E. S. Southern. In this technique, either genomic DNA or cloned DNA is cut with a restriction enzyme. The resulting DNA fragments are then subjected to gel electrophoresis, a method for separation of the restriction fragments on the basis of their size. The negatively charged DNAs are pulled through a gel matrix by an electric current. The smaller fragments can move through the gel more rapidly than the larger molecules, because they can pass through the pores in the gel with less resistance. The separated fragments are then denatured, usually by soaking the gel in alkali. The resulting single stranded DNAs, still held in position in the gel, are then transferred (blotted) to a filter membrane. The DNAs on the filter can be hybridized with a labeled single-stranded probe. If the probe sequence is homologous to the sequence of a denatured DNA fragment, the molecules of labeled probe will accumulate at the position of those DNA fragments on the filter. By comparing the position of the signal with the positions of molecular weight markers, the molecular weight of the DNA fragment with sequence homology to the probe can be identified.

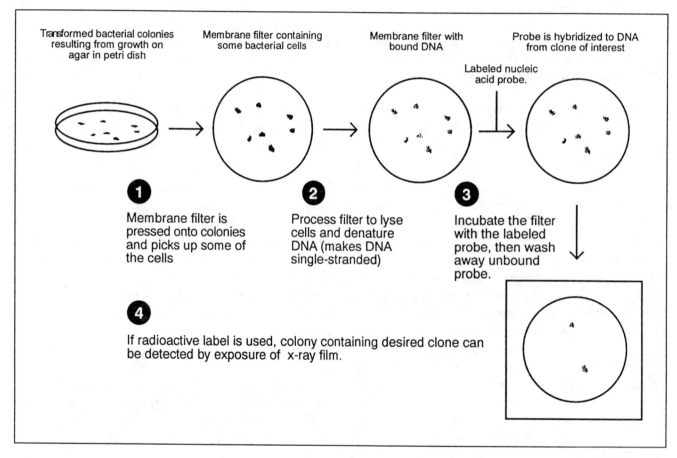

Figure 14 - 18: Detection of the Specific Recombinant DNA Clone of Interest

Identification of a specific recombinant DNA clone usually involves nucleic acid hybridization of a labeled probe to the DNA in the bacterial cells of a recombinant DNA library. Each bacterium usually contains only a single plasmid, so identification can be carried out by separating the mixture of clones into single colonies on an agar plate, transferring DNA from the colonies to a filter, denaturing the DNA into single stranded molecules, and hybridizing the filter to the labeled probe. DNA from bacterial colonies carrying a recombinant DNA clone homologous to the probe are then detected and recovered from the original agar plate.

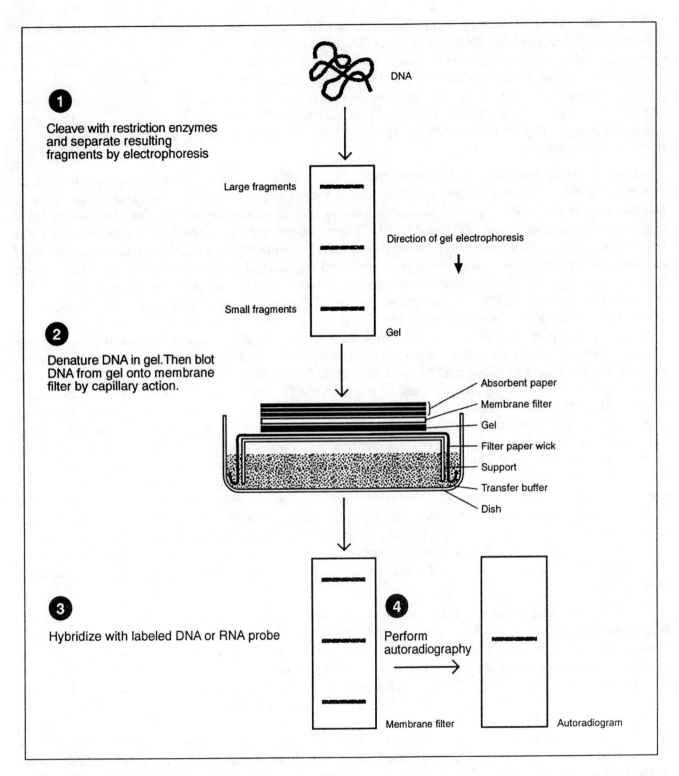

Figure 14 - 19: Southern (DNA) Blotting Technique

Southern blotting involves cleavage of DNA with a restriction endonuclease, separation of the resulting fragments by electrophoresis through a gel, and denaturation of the DNA in the gel. The DNA is then transferred to a filter, and detected with a labeled probe. If the probe is radioactively labeled, the signal is detected by autoradiography using X-ray film. Comparison of the location of the signal with DNA markers of known size provides an estimate of the molecular weight of the DNA fragment carrying a sequence homologous to the probe.

Two other kinds of blotting techniques are useful for molecular analysis. The **Northern blot** (as opposed to a Southern) is an **RNA blot** in which RNAs extracted from a particular tissue or stage of development are subjected to gel electrophoresis, separated by size, blotted onto a filter, and hybridized with a labeled probe. In this case, the position of the signal gives the size of the RNA transcript, and the strength of signal is proportional to the number of transcripts present in the RNA sample. In contrast, **immunoblotting (Western blotting)** involves separation of proteins by gel electrophoresis, transfer of the proteins to a membrane, and detection by binding of a labeled antibody. This procedure shows the mobility (and therefore the size) of the detected protein.

DNA sequence analysis is another powerful technique used to establish the nucleotide sequence of a segment of DNA. The most common method, called Sanger sequencing or **dideoxy sequencing**, involves amplification of the DNA fragment, usually by cloning into the bacteriophage vector M13, which has a single-stranded circular genome (Figure 14 - 20). Therefore the starting material is a single-stranded DNA insert within the circular genome of the phage vector. As shown in Figure 14 - 20, because the sequence of the vector is known, a primer (a synthetic single-stranded sequence of deoxynucleotides) has been designed to bind to the vector DNA immediately upstream of the insert. This primer provides a start point for DNA polymerase to synthesize a DNA strand complementary to the insert by incorporating nucleotides in a $5'$ to $3'$ direction. The polymerase is supplied with nucleotide triphosphates (one of which is labeled, usually radioactively), template, and primer. So how does this reaction reveal the sequence of the template? Four identical reactions are set up in tubes labeled A, C, G, and T. The A tube contains, in addition to the components listed above, a modified dATP called dideoxyATP (ddATP). Whenever a molecule of ddATP is incorporated into the growing strand instead of the usual dATP, chain termination occurs because addition of the next nucleotide is blocked. Therefore a staggered set of randomly terminated chains is generated, each ending at a dideoxyadenosine residue. In a similar fashion, the C tube contains the reaction components and a ddCTP that terminates the growing chains at a C residue, etc. The products of the four reactions are loaded into four lanes of a high resolution polyacrylamide gel and are separated by electrophoresis. Each lane then results in a "ladder" of fragments terminated with the particular dideoxynucleotide in that reaction mixture. The shorter fragments migrate through the gel more easily and are therefore closest to the bottom of the gel, so by comparing the relative positions of the bands, the sequence can be determined. Because DNA polymerase copies from $5'$ to $3'$, it follows that as you read the sequence from the bottom of the gel, the $5'$ to $3'$ DNA sequence complementary to that of the insert is determined. Depending on the sophistication of the sequencing technology utilized, from 200 to several hundred bases can be read from a single set of reactions.

PCR (polymerase chain reaction) is an extremely sensitive analytical technique that can be utilized to amplify specific regions of DNA without cloning into a vector. The PCR reaction is versatile; for example, dinosaur DNA and minute, semi-degraded forensic samples have been amplified successfully using PCR. PCR is especially important for genetic diagnosis, but certain conditions must be met for a PCR reaction to be utilized. First, sequence information must be available for the DNA segment to be amplified. Therefore a version of the gene to be amplified must have been cloned and sequenced previously. This sequence information is necessary in order to design the primers that are used to initiate the action of polymerase. The primers must bind to opposite strands of the segment of DNA to be amplified such that their $3'$ ends are directed toward each other. The polymerase will then begin at a bound primer and extend a DNA strand toward the complement of the binding site of the other primer, as shown in Figure 14 - 21 (page 269).

The PCR reaction is initiated by heat denaturation of DNA to separate the strands. The reaction is cooled, permitting the primers to bind to their complementary sequences in the denatured DNA. A heat-stable polymerase that can survive the denaturation cycles then extends a growing strand of DNA in a $5'$ to $3'$ direction from each primer. The reaction proceeds for a time sufficient to copy the DNA strand beyond the binding position of the other primer on the opposite strand. Therefore, the distance between the primer

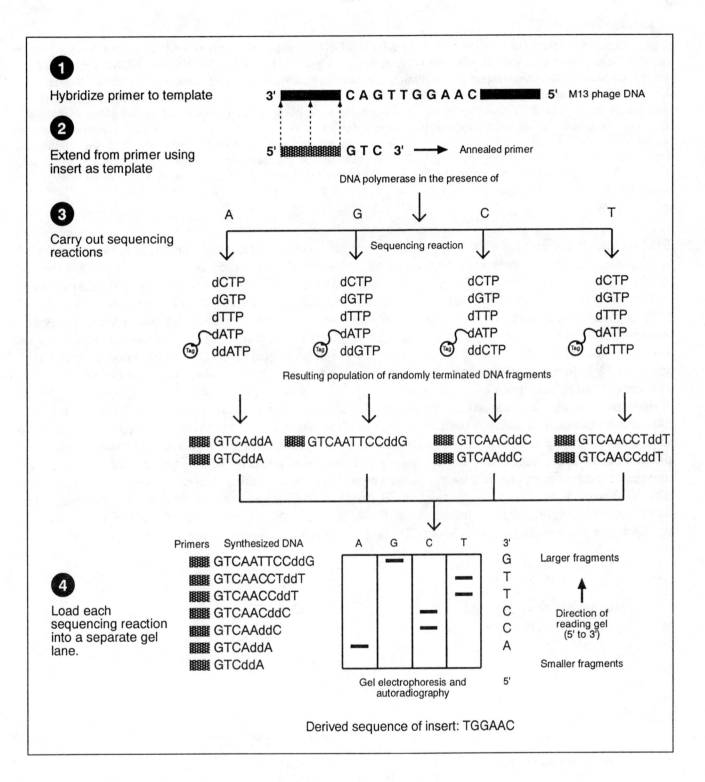

Figure 14 - 20: DNA Sequence Analysis

Legend to Figure 14 - 20

Sanger sequencing permits the sequence of nucleotides in a DNA fragment to be determined. The DNA is cloned into a bacteriophage (M13) with a single-stranded DNA genome. A primer is hybridized to the phage DNA adjacent to the insert, and DNA polymerase begins at the primer and extends a newly synthesized molecule in a 5′ to 3′ direction. There are four separate reactions, each containing a mixture of four deoxynucleotides and a single dideoxynucleotide. Random incorporation of a dideoxynucleotide into a growing strand causes chain termination at that dideoxynucleotide. The four reactions result in the creation of four populations of DNA fragments, each composed of DNA strands randomly terminated at a different nucleotide. The fragments are then separated on the basis of size by electrophoresis through a high resolution polyacrylamide gel. Because the shortest fragments move most quickly through the gel, and because these fragments are closest to the beginning of the growing strand, the sequence of a DNA fragment can be read from the bottom of the gel toward the top, yielding the sequence in a 5′ to 3′ direction.

binding sites is very important for successful PCR amplification because short sequences are amplified more readily than sequences longer than 2000 base pairs. A second heating cycle separates the strands, followed by cooling to allow annealing of the primers to the original strands as well as the newly synthesized strands, and another round of DNA synthesis. Thus each heating/cooling/synthesis cycle doubles the number of DNA strands. It is important to note that the binding sites for the primers delineate the region of DNA that is amplified. Often 30 to 50 cycles of amplification are carried out, resulting in the production of microgram quantities of the DNA fragment of interest. This DNA can be subjected to various restriction enzyme analyses, to DNA sequencing, or the amplified fragment can be cloned for other purposes. It is logical to ask why one would want to carry out PCR if the gene already had to be sequenced to design the primers. The answer is that once the normal gene has been sequenced, PCR amplification can be used to determine the exact nature of a mutation carried by an individual patient. This is possible because primers based on the sequence of the normal gene can be utilized for generation of a PCR product using DNA from the patient as the template. Prior to the invention of PCR techniques, preparation of an entire genomic library from each individual patient would have been necessary, followed by isolation of the desired clone. This would have required months of work, in contrast to an overnight PCR reaction. PCR is also used for forensic analysis because the technique can amplify large quantities from DNA in a minute sample. This incredible sensitivity also requires great care in the preparation of PCR reactions to avoid the amplification of contaminants.

Important Terms and Concepts:

dideoxynucleotide sequencing	cDNA library
PCR	denature
probe	hybridize
vector	Southern blot
insert	RNA blot (Northern)
ligase	immunoblot (Western)
restriction endonuclease	palindrome
plasmid	transformation
bacteriophage (phage)	YAC (yeast artificial chromosome)
antibiotic resistance gene	cosmid
genomic library	expression vector

- 269 -

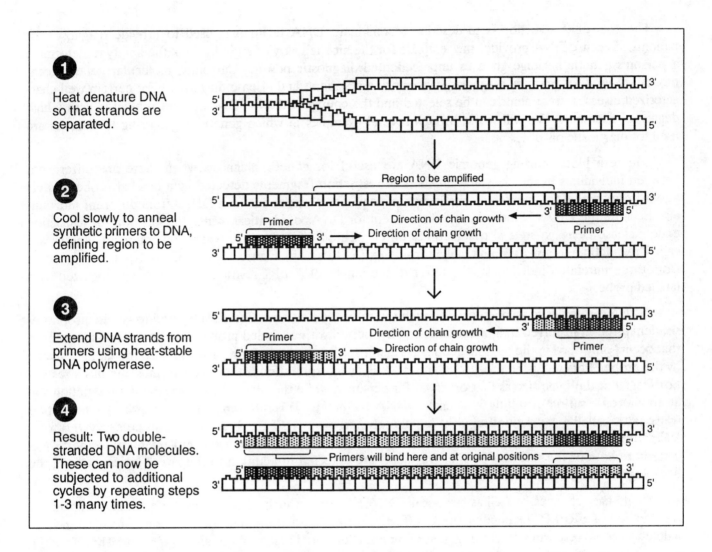

Figure 14 - 21: Polymerase Chain Reaction (PCR) Technique

Beginning with extremely small amounts of DNA or RNA template, PCR permits the amplification of a specific segment to levels that can be visualized easily by gel electrophoresis. To initiate the procedure, two synthetic primers flanking the DNA sequence to be amplified are placed in a reaction containing the DNA, a heat stable polymerase, and deoxynucleotide triphosphates. The DNA chains are separated by rapid heating and cooled slowly to permit the primers to bind to the complementary chains so that their 3' ends "face" each other. A synthesis phase follows, during which the polymerase copies the complementary strands by extending from the annealed primers. After a cycle of synthesis, the heat denaturation, annealing, and synthesis steps are repeated for up to forty cycles. The resulting PCR product, which represents the specific nucleic acid fragment located between the two oligonucleotide primers, can be subjected to gel electrophoresis, analyzed by Southern blotting, cleaved with restriction endonucleases, subjected to DNA sequence analysis and/or cloned directly. Thus PCR is an exceedingly powerful technique that requires only minute amounts of starting material and results in several million-fold amplification.

IV. GENETIC DIAGNOSTIC TECHNIQUES

The goal of this section is to describe recombinant DNA techniques used to provide diagnoses for patients. Because DNA provides the template for the biochemistry of an individual, the ability to determine a person's genetic heritage gives us unprecedented diagnostic power. Currently, molecular genetic techniques are used for prenatal diagnosis, cancer risk assessment, and disease diagnosis. For certain well-characterized diseases, treatments can be selected and the course of disease development predicted with a high degree of accuracy. We will work through several examples in which genetic engineering techniques are used for diagnostic purposes.

Southern blots utilizing genomic DNA are useful for genetic diagnosis when there are differences between individuals in the sizes and numbers of restriction fragments detected by a labeled probe. Alterations in restriction fragments are a reflection of sequence changes between DNAs from different individuals. Because the DNAs of individuals differ (except for DNAs of identical twins), the locations of restriction endonuclease cleavage sites also sometimes differ. These differences in restriction fragment lengths are called **restriction fragment length polymorphisms (RFLPs)**. In Figure 14 - 22, DNA has been extracted from three unrelated individuals who differ in the length of *Eco*RI restriction fragments recognized by a labeled probe.

All DNA samples have been cleaved with the restriction enzyme *Eco*RI, subjected to gel electrophoresis, denatured, transferred to a membrane, and hybridized with a labeled probe known to detect a sequence that occurs only once in the haploid genome. Person 1 is homozygous for a particular size fragment detected by the probe. Person 2 is heterozygous; the most likely interpretation is that one chromosome has a restriction fragment the same size as that observed for person 1, while the other chromosome has a restriction site at an altered location, resulting in a smaller DNA fragment. This difference in the length of a restriction fragment is called a restriction fragment length polymorphism (RFLP). Person 3 is also heterozygous for a RFLP. In this case, a restriction site apparently cleaves within the sequence complementary to the probe. Therefore both of the resulting smaller fragments are detected in addition to the larger fragment from the homologous chromosome.

Single-base changes, as well as large-scale DNA alterations, such as insertions or deletions, can result in the creation of a RFLP. Depending on the effect of the mutational event on the location of restriction sites, a deletion could result in a larger fragment or an insertion could result in a smaller fragment. Therefore the nature of a mutational event cannot be deduced from an increase or a decrease in the sizes or numbers of the restriction enzyme cleavage fragments detected by a probe. Single-base changes can also cause a RFLP if the mutation creates or deletes a restriction site.

These polymorphic sites can be observed when DNAs from different individuals are compared (as in Figure 14 - 22) and thus can serve as landmarks in the chromosomal DNA. When a RFLP occurs close to a gene (linked), it can serve as a marker for the inheritance of an allele of that gene. Therefore RFLP analysis

Legend to Figure 14 - 22 (next page)

RFLPs are observed when DNAs are compared by restriction endonuclease cleavage, followed by Southern blotting. The probe, usually several hundred nucleotides in length, recognizes a sequence occurring only once in the haploid human genome. A RFLP occurs when the probe detects DNA fragments of differing lengths in the same individual or in DNA samples from a population. In this illustration, person 1 inherited *Eco*RI restriction fragments of identical length from each parent and is therefore homozygous for the RFLP. Person 2 inherited a different length restriction fragment from each parent and is heterozygous. Person 3 is also heterozygous, having inherited one *Eco*RI fragment similar in size to that of the large RFLP from persons 1 and 2, along with two *Eco*RI fragments created by cleavage within the region recognized by the probe. Consequently three fragments are detected on the Southern blot. Such differing patterns of cleavage reflect sequence variation. RFLPs can therefore serve as landmarks on a chromosome.

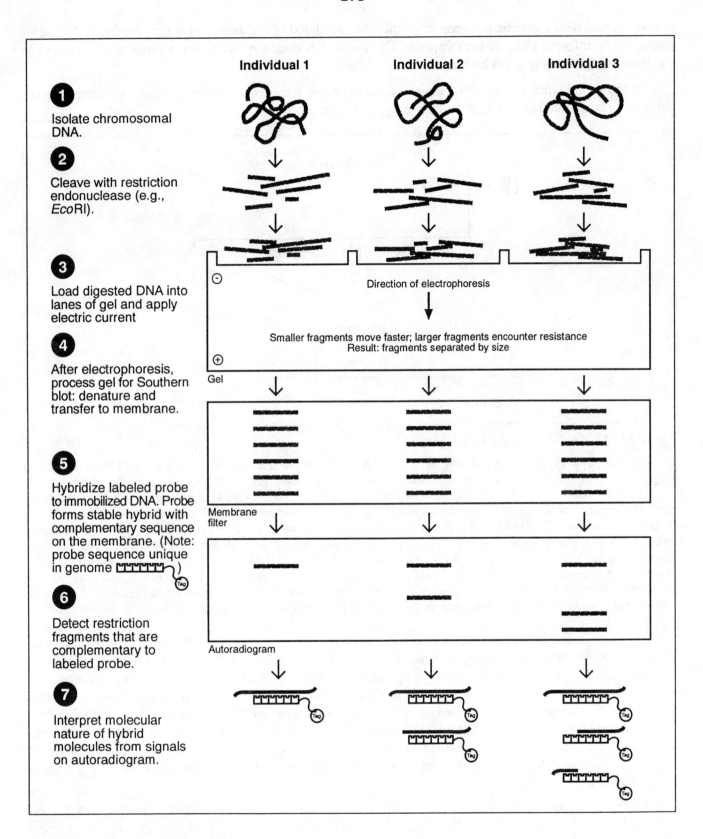

Figure 14 - 22: Restriction Fragment Length Polymorphisms (RFLPs).

can be coupled with a genetic pedigree to predict the likelihood of a patient developing a genetic disease or passing a mutant gene to his or her offspring. Of course, it is usually better to use a probe corresponding to the gene itself, but many genes have not yet been cloned.

As shown in Figure 14 - 23, DNA samples have been taken from members of a family. By examining the pattern of restriction fragments, probable genotypes can be predicted.

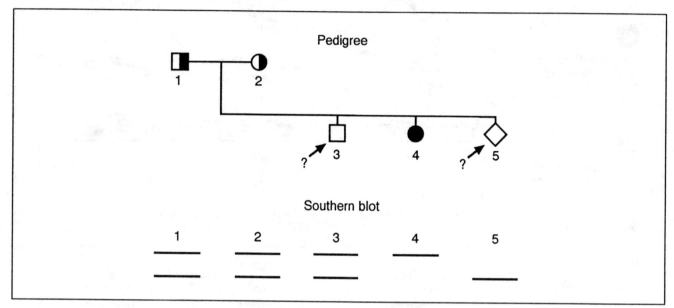

Figure 14 - 23: Molecular Genetic Diagnosis.

If a RFLP is physically linked to a disease gene (close together on the same chromosome), the RFLP can serve as a means of tracking the inheritance of the chromosome carrying the mutant gene. This figure shows DNA from a family in which DNA samples from each family member have been subjected to cleavage by a restriction enzyme, followed by Southern blotting and detection of the labeled fragments. The parents have one child affected by a serious autosomal recessive genetic disease. The mother is currently pregnant, and they wish to know whether their older child is a carrier and the probability that the fetus will be affected by the disease. We will assume that the probe hybridizes within the gene causing the disease. The parents must be heterozygous because they have an affected child. Each parent shows two different fragment sizes. The affected child has inherited the larger fragment from each parent, implying that this fragment is associated with the mutant gene. It then follows that the older child is presumably a carrier, while the fetus will be unaffected because it is homozygous for the smaller RFLP associated with the normal gene.

When sufficient DNA is available to carry out a Southern blot, complex RFLP patterns (**DNA finger-printing**) are also used for forensic analysis. These complex patterns may be generated with the simultaneous use of several probes or with a probe that detects genomic sequences that are repeated many times. Certain of these repetitive sequences are called **VNTR sequences** (VNTR = variable number of tandem repeats). The number and arrangement of these repeats change rapidly, causing the signal resulting from hybridization of a VNTR probe to be highly individualistic. These complex patterns are not generally used for genetic diagnosis because they would be difficult to interpret. However, they can be very useful for establishing a probability that a DNA sample (at a crime scene, for instance) might be derived from a particular individual.

In addition to use of the Southern blot for DNA diagnosis, the polymerase chain reaction (PCR) is also very useful. As discussed above, PCR permits the amplification of a nucleotide sequence between the positions of two synthetic single-stranded DNA primers. Therefore, the gene or segment of DNA to be amplified must already have been cloned and sequenced in order for the primers to be designed. For example, the most common mutation causing cystic fibrosis is a deletion of three base pairs at codon 508. Therefore genetic diagnosis of a fetus or of individuals at risk can utilize PCR amplification of a small DNA interval that includes codon 508. When the PCR products are subjected to high resolution gel electrophoresis, those amplified from DNA carrying the deletion will be three base pairs shorter than the product copied from a normal sequence. This detailed diagnosis is possible because the cystic fibrosis gene has already been cloned and sequenced, so appropriate primers could be designed and synthesized. Note, however, that the ΔF508 allele accounts for only 70 - 80% of all mutations causing cystic fibrosis. Most of the remaining cystic fibrosis alleles would generate a PCR product of normal length following amplification with these primers.

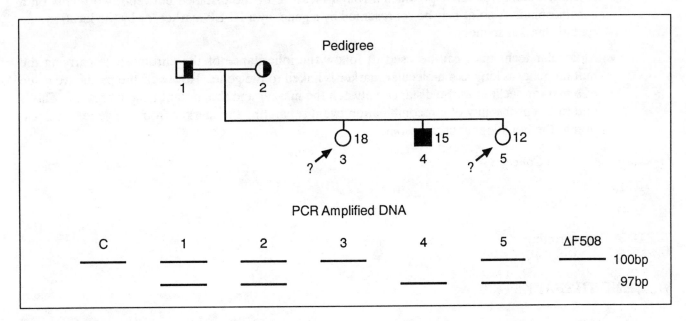

Figure 14 - 24: Genetic Diagnosis by PCR Analysis.

This figure illustrates the use of PCR to detect the most common allele for cystic fibrosis, the ΔF508 allele. This mutation involves the deletion of the three base pairs that normally encode amino acid 508 of the cystic fibrosis transmembrane regulator protein. These parents have a 15 year old son affected by cystic fibrosis. Therefore both parents are presumably heterozygous for a cystic fibrosis mutation. They wish to know the genetic status of their daughters, ages 18 and 12. DNAs from each family member were subjected to PCR amplification using primers that generate a 100 bp fragment using the normal sequence as template, but a 97 bp fragment using the ΔF508 allele. The results shown above indicate that daughter 3 is not a carrier and daughter 5 is a carrier. C = normal control; ΔF508 = homozygous ΔF508 allele.

It is important to distinguish between the two major methods of molecular diagnosis, the Southern blot of genomic DNA and gel analysis of PCR generated DNA. A Southern blot gives the molecular weight of a restriction fragment generated by the cleavage by a particular enzyme. The absence of a signal on a Southern blot (assuming that experimental techniques are working) may indicate that a deletion of chromosomal DNA has removed the target DNA fragment. In contrast, the absence of successful amplification by PCR may mean that a DNA rearrangement has altered the relative positions of the binding sites for the two primers. A second possibility is that one or both primer sequences may have been deleted. A third possibility is that damage to the DNA sample may not have provided a continuous strand of DNA to be copied.

A brief recap of important concepts:

- ♦ Because we are diploid, we carry a copy of each chromosome inherited from our father and a copy from our mother. This includes each of the 22 autosomes and the sex chromosomes. The two copies of chromosome 17, for instance, are termed homologous chromosomes.

- ♦ The principle of independent assortment means that for any pair of homologous chromosomes, our germ cells have a 50 : 50 chance of inheriting our paternally-derived chromosome versus our maternally-derived chromosome.

- ♦ Recombination frequency provides a rough measure of the distance between two points on a chromosome. A point may be represented by a gene, by a RFLP marker, or by the location of a specific marker sequence.

- ♦ Molecular techniques can be used to follow the inheritance of the chromosome carrying the mutant gene, as long as a molecular marker is linked to the gene. However, the predictive value of a marker declines as the distance between the marker and the mutant gene increases. This is due to the probability of a recombination event separating the marker from the gene by relocation to the homologous chromosome.

Important Terms and Concepts

RFLP

VNTR

DNA fingerprinting

V. GENE THERAPY

Gene therapy refers to protocols designed to ameliorate the effects of genetic disease. There are two basic kinds of gene therapy: **somatic gene therapy** and **germ line gene therapy**. In somatic gene therapy, a recombinant DNA construct containing the normal version of a gene is delivered to target tissue, permitting normal or improved function. There are several current examples of successful somatic gene therapy, such as the establishment of immune function in children born with adenosine deaminase deficiency or use of a respiratory virus (adenovirus) as a vector to carry the normal cystic fibrosis transmembrane regulatory protein into the lungs of cystic fibrosis patients. These treatments alter only somatic cells and do not change the genes that will be passed on to offspring. As long as the treatments are safe and effective, there are few ethical concerns about somatic treatments for genetic diseases. However, each somatic gene therapy protocol must solve the problem of safe and effective delivery of the normal gene product to the affected tissue. This means that the vector carrying the normal gene must be able to enter the targeted tissue(s) and produce sufficient normal product to alleviate the symptoms of the genetic disease.

In contrast, germ line gene therapy involves the correction of a genetic defect in germ line tissue to ensure that a normal gene, rather than the defective gene, will be transmitted to offspring. Germ line gene therapy is not yet being carried out on humans because of both technical and ethical concerns. Technical concerns include both safety and efficacy of treatment. For instance, if applied to humans, the protocol must not increase chances of birth abnormalities among offspring. Second, because the same techniques can be utilized for genetic improvement (in contrast to correction of a genetic disease), an ethical consensus must be reached. For some dominant mutations, the presence of the defective product disrupts the normal process. In this case, the defective gene would need to be inactivated or replaced. In contrast, for a recessive genetic disease, it would presumably be sufficient to add a copy of the normal gene.

Currently, there are two mammalian protocols that target germ line selection. In the first protocol, gene replacement events are selected. These methods have permitted the creation of "**knockout mice**" in which a normal gene is disrupted. The resulting **transgenic mice** are now providing animal models for human genetic diseases such as cystic fibrosis. Experimental therapies can be evaluated using these animals, and the effects of the disease can be studied in detail. The procedure begins with culture of cells from brown mice that are transformed with a gene altered by recombinant DNA techniques. In the example shown in Figure 14 - 25, the alteration is caused by insertion of the neomycin (antibiotic) resistance gene into an exon of the gene targeted for disruption. This insertion simultaneously disrupts the target gene and provides a selective tool. In addition, a thymidine kinase (TK) gene is attached to the construct at one end to confer sensitivity to the thymidine analogue gancyclovir if gene replacement does not occur.

Following the construction of the disrupted gene by researchers, the DNA is transformed into culture cells from black mice. Three possible events may occur. The most common outcome is that the transforming DNA has no apparent effect. In this case, the cells will be sensitive to the antibiotic and resistant to gancy-clovir. A second outcome is that the cells are transformed by random incorporation of the construct into the genome. If this occurs, the inactivated target gene (with inactivating insert of antibiotic resistance gene) and the thymidine kinase gene are both incorporated into chromosomal DNA. Therefore these cells are resistant to the antibiotic because of the introduction of the neomycin resistance gene. However, they are sensitive to gancyclovir because of the presence of the TK gene. The desired event (and the rarest) is the targeted gene replacement in which the disrupted gene replaces the normal gene in the chromosomal DNA. This is thought to occur via a recombination in which the homologous sequences of the construct and the endogenous gene line up with each other and an exchange event (probably a double recombination) takes place within the homologous regions. If this occurs, the disrupted gene, including the gene conferring resistance to neomycin, is incorporated into the genome. However, because the TK gene is outside the region of homology, it is excluded. The cells resulting from the targeted gene replacement therefore are resistant to both neomycin (or its analogue, G418) and to gancyclovir. Treatment of the culture of mixed transformed cells with these two selective agents therefore eliminates all cells except those in which the rare targeted gene replacement event has occurred.

The desired cells are then used to make transgenic mice by incorporating them into the embryos of mice with a different coat color marker. The embryos are then implanted into surrogate mothers. Mice showing brown colored areas have incorporated some of the altered cells, and they are bred to black mice to see whether any cells marked with brown coat color have been incorporated into germ cells (Figure 14 - 26). The production of a brown mouse (brown is dominant to black) indicates that the transformed cells carrying the disrupted gene have been incorporated into germ cells. Therefore successful production of a mouse in which the targeted gene has been inactivated has been accomplished. These mice are proving to have major

Legend to Figure 14 - 25 (next page)

This figure illustrates the basic steps involved in targeted gene disruption. In this example, the procedure utilizes recombinant DNA techniques to insert an antibiotic resistance gene into an exon of the gene to be disrupted. The thymidine kinase gene is attached to one end of the construct, flanking the DNA derived from the targeted gene. This construct is introduced into cultured embryonic stem (ES) cells from mice carrying a brown coat color marker. Three possible outcomes occur. The most common outcome is that the cultured ES cells do not incorporate the construct. The second most common event is that the construct, including the TK gene, inserts into the genome randomly. The third event is the rare targeted gene replacement. This is believed to result from crossing over within the homologous DNA regions of the cloned gene and the normal cellular gene. Therefore one of the two normal genes in the mouse cell is replaced with the disrupted gene. In this case, the TK gene is outside the region of homology and is not incorporated into the chromosome. As indicated in the figure, treatment of the transformed cells with antibiotic and gancyclovir results in the elimination of cells without an insertion as well as cells with a random insertion. Therefore only the cells that have incorporated the rare targeted insertion will survive. These cells are permitted to multiply in culture.

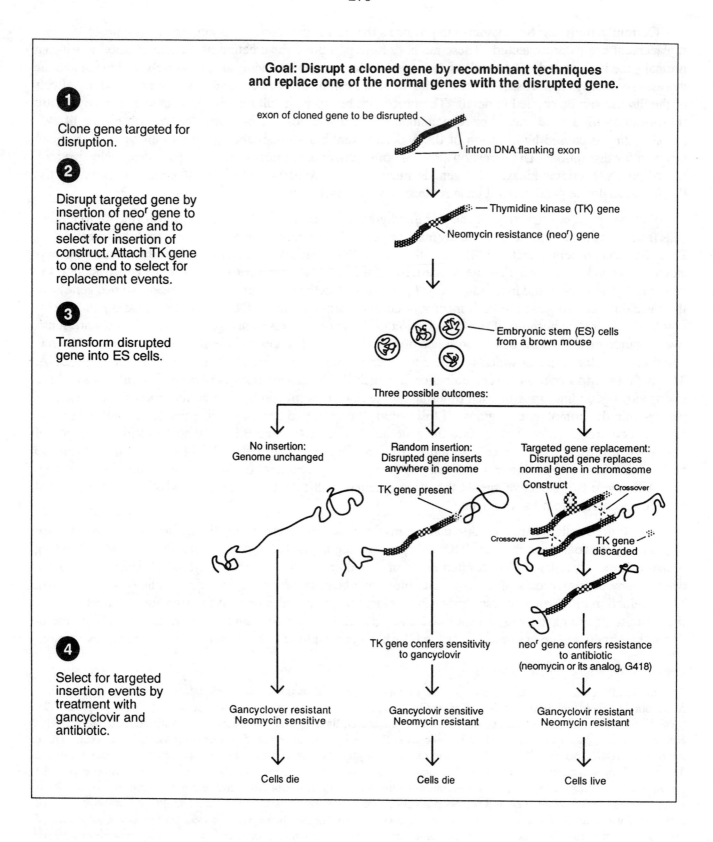

Figure 14 - 25: Targeted Gene Replacement in Embryonic Stem Cells.

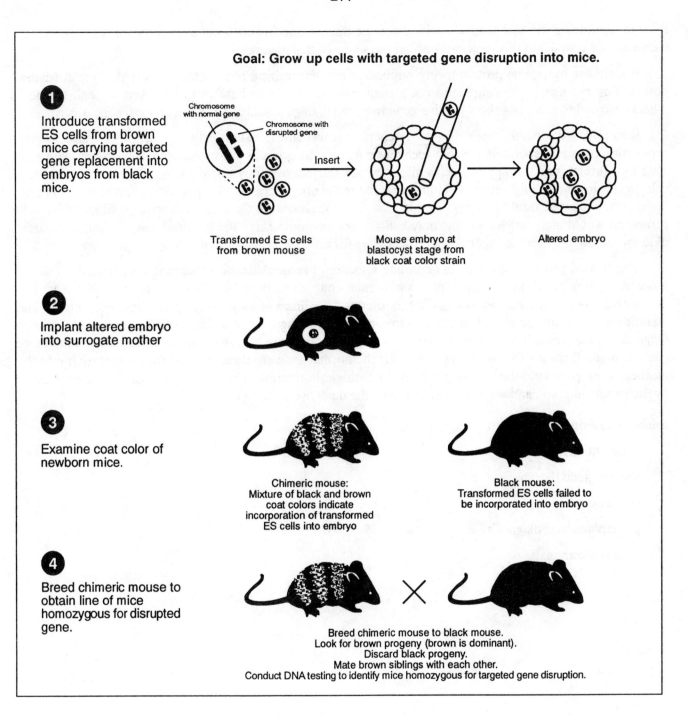

Goal: Grow up cells with targeted gene disruption into mice.

1 Introduce transformed ES cells from brown mice carrying targeted gene replacement into embryos from black mice.

Chromosome with normal gene

Chromosome with disrupted gene

Insert

Transformed ES cells from brown mouse

Mouse embryo at blastocyst stage from black coat color strain

Altered embryo

2 Implant altered embryo into surrogate mother

3 Examine coat color of newborn mice.

Chimeric mouse:
Mixture of black and brown coat colors indicate incorporation of transformed ES cells into embryo

Black mouse:
Transformed ES cells failed to be incorporated into embryo

4 Breed chimeric mouse to obtain line of mice homozygous for disrupted gene.

X

Breed chimeric mouse to black mouse.
Look for brown progeny (brown is dominant).
Discard black progeny.
Mate brown siblings with each other.
Conduct DNA testing to identify mice homozygous for targeted gene disruption.

Figure 14 - 26: Creation of a Transgenic Mouse.

Following the selection of ES cells from a brown mouse in which a targeted gene replacement event has occurred, the cells are introduced into a mouse embryo with a black coat color marker. The altered embryo is then implanted into a surrogate mother. Chimeric mice carrying the disrupted gene can be recognized because they display a mixture of both coat colors. The chimeric mice are subjected to a breeding protocol designed to recover mice homozygous for the gene replacement event. They are first bred to black mice, and brown progeny are selected because brown coat color is dominant to black. The brown siblings are mated to each other, and DNA testing is conducted to identify the mice that are homozygous for the disrupted gene.

roles in identifying the biochemical basis of genetic disease and in providing animal models in which treatments can be tested, but this subject is beyond the scope of this review.

If a similar transgenic protocol were applied to humans, ethical considerations would apply at several points. For instance, both manipulation of human embryos by adding heterologous cells or breeding experiments designed to eliminate the defective gene from the lineage would raise serious ethical concerns.

Recently, it has been possible for researchers to transplant the stem cells that are responsible for the production of mouse sperm into sterile recipient mice, resulting in the production of viable sperm. It is possible that future gene therapy protocols might exploit this discovery by transforming sperm-producing stem cells in culture and re-implanting them into a sterile recipient. Sperm resulting from the transformation procedure could then be used to impregnate females. However, because the stem cells are diploid and the transformation would alter only one copy, only half the sperm would carry the modification. Therefore a technique to isolate the modified sperm will be necessary for reliable production of transgenic progeny.

The second protocol resulting in germ line selection, **pre-implantation diagnosis**, is currently in use in humans. This is based on genetic diagnosis of human embryos fertilized *in vitro*. For couples who elect this alternative, oocytes are harvested from the mother and fertilized *in vitro* using sperm from the father. The resulting embryos are permitted to develop to the eight cell stage, when a single cell is removed for genetic diagnosis. PCR amplification and analysis of the PCR product are then used to determine the genotype of each embryo. Embryos that are homozygous for the normal gene are then selected for implantation into the mother. If all goes well, the resulting child is the biological offspring of his or her parents, will be unaffected by the genetic disease, and will not be able to pass the defective gene to offspring.

Important Terms and Concepts

germ line gene therapy

somatic gene therapy

knockout mice

preimplantation diagnosis

transgenic animals

VI. WELL-KNOWN GENETIC DISEASES

These diseases share a common history of appearance on board examinations.

Disease	Mode of Inheritance	Major Symptoms
Cystic fibrosis	autosomal recessive	abnormal salt balance, blockages of pancreatic duct and small intestine, and sticky mucus in lungs due to defective cystic fibrosis transmembrane regulator gene
Duchenne's muscular dystrophy	X-linked recessive	muscle weakness in young boys due to defect in dystrophin gene.
Hemophilia (classical or hemophilia A)	X-linked recessive	clotting disorder due to mutant factor VIII gene
Huntington disease	autosomal dominant	tremors (chorea) and dementia, onset in middle age due to triplet repeat expansion in Huntington gene
Hypercholesterolemia	autosomal dominant	hyperlipidemia due to defect in LDL cell membrane receptor gene
I-Cell disease (mucolipidosis II)	autosomal recessive	birth deformities, retarded development due to abnormal lysosomal enzyme targeting
Lesch-Nyhan syndrome	X-linked recessive	mental retardation, gout, spastic cerebral palsy, self-destructive biting of fingers and lips due to uric acid accumulation caused by defect in purine metabolism
Sickle cell anemia	autosomal dominant	anemia, malaria resistance in heterozygote due to hemoglobin A mutation (See Chapter 1).
Tay-Sachs disease	autosomal recessive	developmental retardation, paralysis, dementia, death at 2 - 3 yrs, characterized by "cherry-red" spot in center of retina due to accumulation of G_{M2}-ganglioside.
Thalassemias	usually autosomal dominant	varied mutations resulting in production of abnormal amounts of certain globin chains (See Chapter 1).

VII. REVIEW QUESTIONS ON MEDICAL GENETICS

> **DIRECTIONS:** For each of the following multiple-choice questions (1 - 10), choose the
> ONE BEST answer.

1. You have taken the medical history of a 45 year old male patient affected by the recent onset of uncontrollable shaking and have derived the genetic pedigree shown below. If the disease has a genetic basis, what is the probable mode of inheritance?

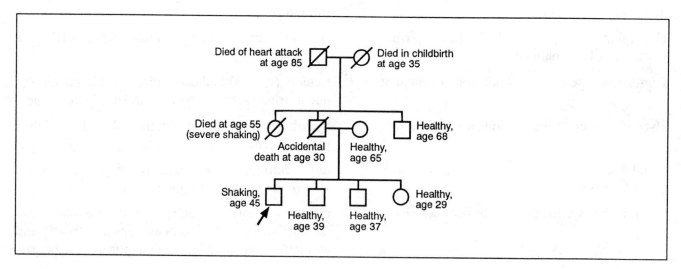

A. Autosomal recessive
B. Autosomal dominant
C. X-linked recessive
D. X-linked dominant
E. Mitochondrial

Ans B: Autosomal dominant because the disease is appearing in successive generations (paternal aunt), and the patient's X-chromosome is from unaffected parent. We assume that the father of the patient would have been affected, along with the paternal grandmother, had either survived. Some of the patient's siblings may develop the disease in a few years.

2. A RFLP marker is located near the pencillosis gene such that there is 5% recombination between the RFLP and the gene. Pencillosis is a serious autosomal recessive genetic disease having to do with compulsive theft of writing tools. What is the approximate physical distance between the marker and the gene, and what is the probability that an individual heterozygous for the marker is a carrier for the disease?

A. 10,000 kb and 90%
B. 5,000 kb and 90%
C. 5,000 kb and 95%
D. 1,000 kb and 95%

Ans C: A five percent recombination rate is equal to approximately five million base pairs, or 5000 kb, and the marker will be inherited with the gene 95% of the time.

3. The members of the family whose pedigree is shown below have requested genetic testing. They have had a child affected with cystic fibrosis and the mother is now pregnant. They wish to know the genetic status of the fetus. You administer a PCR amplification test for the ΔF508 mutation responsible for 70% of all cases of cystic fibrosis and obtain the data diagrammed below the pedigree. What do you tell the family?

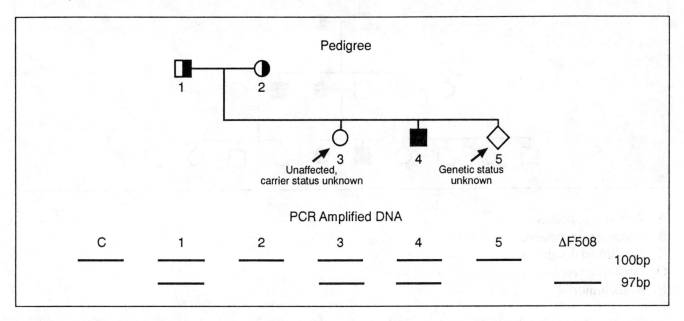

A. The fetus does not carry any mutant allele for the CF gene, will not be a carrier, and will be unaffected by the disease.

B. The fetus does not carry the ΔF508 mutation, has a 50% chance of being a carrier, and will be unaffected by the disease.

C. The fetus is heterozygous for cystic fibrosis, is a carrier for the disease, and will be unaffected by the disease.

D. The fetus is homozygous for two different CF alleles and will be affected by the disease.

Ans B: The mother carries a CF allele that is different from ΔF508. This test cannot differentiate between her CF mutation and her normal allele. However, the fetus did not inherit ΔF508 from its father, so it has a 50% chance of getting the mutant allele from its mother and in any case will be unaffected by the disease.

4. You must evaluate the genetic status of a family with a history of Huntington disease and wish to determine the number of triplet repeats in each individual. What would be the best way of those given below to determine the number of triplet repeats?

A. Carry out PCR amplification using triplet repeat primers and measure the size of the resulting fragment

B. Cut genomic DNA with a restriction enzyme and carry out Southern blotting using a VNTR probe

C. Carry out PCR amplification using primers flanking the triplet repeats and measure the size of the resulting fragment

D. Make a genomic library from the DNA of each individual, find the appropriate clones and carry out DNA sequence analysis.

Ans C: PCR primers on either side of the DNA segment containing the repeat would measure the repeat length. In alternative A, primers containing triplet repeat sequences would bind at various locations within the repeat and generate PCR products of heterogeneous lengths. Alternative D would work, but would be extremely expensive and time-consuming.

5. You are investigating inheritance of progressive muscle weakness in a family and have derived the genetic pedigree presented below. What is the probable mode of inheritance of this genetic disease?

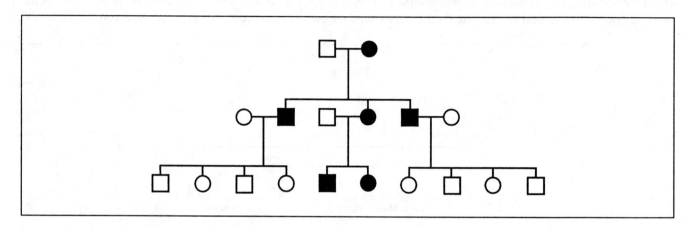

A. Autosomal dominant
B. Autosomal recessive
C. Sex-linked dominant
D. Sex-linked recessive
E. Mitochondrial

Ans E: The disease is inherited only through the maternal line, suggesting that the mutation is carried by the mitochondrial genome.

6. Genetic pedigrees describe the inheritance of genes among related individuals. Please look at the pedigree below. What is the genetic status of the individual indicated by the asterisk (*)?

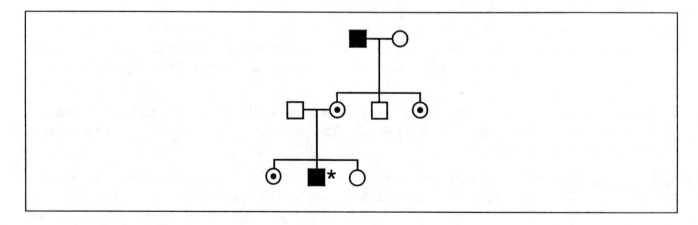

A. He is a male affected by a sex-linked mutation.
B. She is a female carrier of a sex-linked mutation.
C. He is a male homozygous for a dominant mutation.
D. She is a female heterozygous for an autosomal recessive mutation.

Ans A: The pedigree shows inheritance of a sex-linked recessive trait. All daughters of an affected father are carriers, and their sons have a 50% chance of being affected by the disease.

7. You are carrying out PCR analysis of the genomic DNA from three patients who have symptoms of Duchenne's muscular dystrophy. You obtain PCR amplified fragments (as shown below) for control DNA (from a normal boy) and for patient number 3.

     C   1   2   3

                                            —

       —

What is the most likely explanation for the lack of PCR amplified fragments for patients 1 and 2?

A. A deletion has taken place
B. One or both PCR primers cannot bind
C. The distance between the primers does not permit amplification
D. Unable to distinguish among the above alternatives

Ans D: The lack of PCR amplification is due to a DNA alteration, but any of the three given alternatives could have prevented successful PCR amplification.

8. The frequency of restriction endonuclease cleavage depends on characteristics of the recognition site. Please put the following restriction enzymes in order, from the most frequent to least frequent probability of cleavage, where N is any nucleotide, Pu is either purine, and Py is either pyrimidine.

1. Tth111I   GACNNNGTC
2. NotI   GCGGCCGC
3. HaeIII   GGCC
4. AvaI   CPyCGPuG

A. 3, 4, 1, 2
B. 4, 3, 2, 1
C. 3, 4, 2, 1
D. 4, 3, 1, 2

Ans A: AvaI will cut less often than HaeIII because of the additional specifications for purines and pyrimidines. Tth111I will cut more often than NotI because the presence of additional unspecified nucleotides in the recognition site do not alter its characteristic six-cutter cleavage pattern.

9. In the protocol involving the targeted replacement of genes in the mouse, a mouse developing from the injected blastocyst consists of cells originating from:

A. cultured mouse embryonic stem cells
B. mouse erythroleukemia cells
C. cells of the host mouse embryo
D. A and B only
E. A and C only

Ans E: The chimeric mouse is composed of cells derived from the transformed embryonic stem cells mixed with cells from the host embryo.

10. Please choose the correct reading of the sequencing gel diagrammed below:

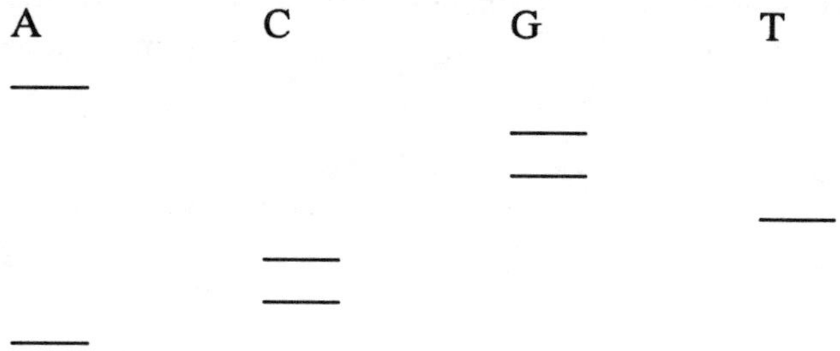

A	C	G	T
——			
		——	
		——	
			——
	——		
	——		
——			

A. 5′ AGGTCCA 3′
B. 5′ TCCAGGT 3′
C. 5′ ACCTGGA 3′
D. 3′ ACCTGGA 5′

Ans C: When a sequencing gel is read from bottom to top, you are reading the sequence from 5′ to 3′.

11. The Y chromosome contains:

A. thousands of genes, just like an autosome.
B. genes responsible for the formation of a male.
C. no genes at all.
D. genes that cause female development.

Ans B: The Y chromosome contains few genes, but is necessary for expression of male characteristics. The crucial gene appears to be a transcription factor that initiates the developmental program resulting in maleness.

12. You are treating a child with a genetic disease. His family has no previous history of the disease. When you carry out Southern blotting of DNAs from members of his family, you see the following pattern:

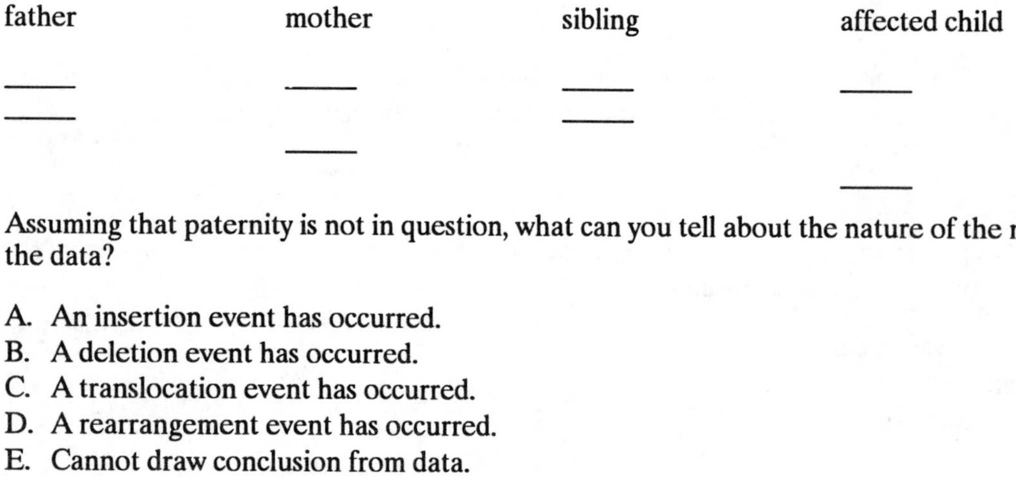

father	mother	sibling	affected child
——	——	——	——
——		——	
——			
			——

Assuming that paternity is not in question, what can you tell about the nature of the mutational event from the data?

A. An insertion event has occurred.
B. A deletion event has occurred.
C. A translocation event has occurred.
D. A rearrangement event has occurred.
E. Cannot draw conclusion from data.

Ans E: The mutation has resulted in a smaller RFLP in the affected child. However, this could be due to introduction of a restriction site by an insertion, creation of a smaller RFLP by a deletion, or relocation of a restriction site by a translocation or other rearrangement.

13. You are cloning *Bam*HI-cut genomic DNA into the plasmid vector pBR322, which has both an ampicillin resistance gene and a tetracycline resistance gene. If the inserts are ligated into the *Bam*HI site in the tetracycline resistance gene of the vector, the entire population of transformed bacteria should first be grown on:

A. media containing no antibiotic
B. media containing ampicillin
C. media containing tetracycline
D. media containing both ampicillin and tetracycline

Ans B: Because a genomic DNA insert will inactivate the tetracycline resistance gene, the transformants must be grown on ampicillin. This will select against any bacterial cells that were not transformed by a plasmid.

14. Sometimes the gene responsible for a particular genetic disease has not yet been cloned, so researchers must use a probe that is genetically linked to the gene they wish to follow. If a new RFLP is detected using a probe linked to the gene, this result may be interpreted to mean:

A. that a mutation has occurred in the gene
B. that a mutation has occurred in the region flanking the gene
C. that a mutation has not occurred
D. that either A or B may have occurred

Ans D: Detection of a new RFLP means that the length of the DNA fragment that hybridizes to the probe is different; therefore a mutation has occurred. However, without additional information, it is not possible to determine the location of the mutation itself.

15. You wish to clone the gene encoding a peptide hormone that will permit medical students to go without sleep while they are studying for their medical board examinations. You plan to express this gene in bacteria so that the hormone can be purified in commercial quantities. You have an antibody that recognizes the peptide hormone. What kind of recombinant DNA library will you screen?

A. Genomic library cloned into a cosmid vector
B. cDNA library cloned into an expression vector
C. cDNA library cloned into the plasmid pBR322
D. Genomic library cloned into a YAC vector

Ans B: If you plan to identify a clone using an antibody, you must use an expression library that will produce the protein.

16. If you are screening a library by nucleic acid sequence homology with a radiolabeled DNA probe, how are the colonies of interest identified?

A. microscope
B. X-ray film
C. scintillation counter
D. culture techniques

Ans B: The location of a radioactive label on a filter is detected with X-ray film (autoradiography).

17. You are carrying out Sanger DNA sequence analysis on a very large insert, so you would like to sequence as far as possible. How might you alter the ratio of dideoxynucleotides to deoxynucleotides in your DNA sequencing reactions to favor accomplishment of your goal?

A. Increase the concentration of dideoxynucleotides relative to that of deoxynucleotides
B. Decrease the concentration of dideoxynucleotides relative to that of deoxynucleotides
C. Increase the concentration of dideoxynucleotides relative to that of labelled dATP
D. Decrease the concentration of dideoxynucleotides relative to that of labelled dATP

Ans B: A decrease in the concentration of dideoxynucleotides will cause the growing chains to terminate less often. Therefore the average length of the terminated chains will be longer, permitting the sequence to be read further into the insert.

18. You carry out a Northern blot of human brain RNA using a probe representing a sequence that is present only once in the haploid genome. When you develop the autoradiogram, you see a signal in three different molecular weight size classes. What is the most likely interpretation of this result?

A. Transcripts of three different, unrelated genes
B. Transcripts of differing sizes from the same gene
C. Differential translation of mRNA size classes
D. A eukaryotic, polycistronic transcript

Ans B: When a probe is represented only once in the genome, then all transcripts that hybridize to the probe must be from the same gene. In this case, differential splicing is probably creating the different length transcripts.

19. In order for somatic gene therapy to be carried out successfully, the following factor is NOT required:

A. safety
B. delivery of product
C. appropriate levels of production
D. germ cell transformation

Ans D: Somatic gene therapy involves engineering of somatic cells for delivery of a functional product to ameliorate the effects of the genetic disease. Transformation of germ cells would only be required for germ line gene therapy.

20. The father of a son is affected by sex-linked red/green color blindness. What is the probability that his son will inherit the disease from his father?

A. 100%
B. 50%
C. 25%
D. 0%

Ans D: Sons must inherit their Y chromosome from their fathers. Because sex-linked genetic diseases are carried on the X chromosome, sons can inherit them only from their mothers.

21. Targeted gene replacement in mice is an example of germ line gene therapy. Currently, targeted gene disruption is being used to create animal models for human genetic diseases. If the embryonic stem cells become incorporated into a host embryo, how can this event be recognized?

A. The coat color of the baby mouse is black.
B. The coat color of the baby mouse is brown.
C. The coat color of the baby mouse is white.
D. The coat color of the baby mouse is mixed black and brown.

Ans A: The embryonic stem cells are from a brown mouse and are introduced into an embryo from a line of black mice. Therefore if the embryonic stem cells are incorporated, the coat color of the baby mouse is a mixture of brown and black.